United States Navy and Marine Corps Fighters 1918-1962' is published with the co-operation of the Department of the Navy, Washington, D.C., U.S.A.

Wings over the Waves

A fighter used by the U.S.N. and U.S.M.C. at the very middle of the period covered by United States Navy and Marine Corps Fighters 1918-1962. *By depicting the Squadron Commander's Grumman F3F-2 of VF-6 from the U.S.S.* Saratoga *(CV-3), J. D. Carrick has captured the most colourful era of U.S. naval fighters—the late 'thirties, to parallel the most colourful period of U.S.A.A.F. fighters—1944-45, painted for the companion volume* United States Army and Air Force Fighters 1916-1961.

Famous Fighter—Early 'thirties

Perhaps the most famous of the pre-war naval fighters were the 'Boeing Bipes' of the F4B series. This F4B-3 of the U.S. Marine Corps was used with special underslung dive-bombing racks by VB-4M, a squadron reorganised from VF-10M with F4B-4 fighters to meet a changing requirement. This dual role has been expressed in various ways; then it was a fighter-bomber, during the war fighters became strike aircraft and today this dual capability is met by the so-called attack fighter.

Foreword

This book is the culmination of over ten years ardent study of United States Navy and Marine Corps aircraft, their history, purpose and development. While basically it is the story of fighter types, it is also the story of the ingenuity and courage of the men and women who produced them and the men who flew and maintained them. No air historian, single-handed, could have produced all the material necessary for the complete story of United States Navy and Marine Fighter aircraft. However, with the aid of those listed below, I have endeavoured to compile the most complete and authentic account of this vast subject.

I am indebted for information, guidance and verification to the Department of the Navy, especially their Bureau of Naval Weapons, where Mr. Lee M. Pearson and Mr. Harold Andrews were of great assistance; L./Cdr. F. A. Prehn, U.S.N., Head, Magazine and Book Branch, Department of the Navy, Office of Information, who afforded great help in compiling much of the illustrative material. In the Federal Bureau of Records, Miss M. E. Lynch, Chief of Reference Branch, made available certain drawings and data of an historical nature. Additional information came from Mr. J. T. Bibby at the Naval Ordnance Test Station, China Lake, and Mr. Royal D. Frey and Mrs. F. R. Biese of the Wright-Patterson Air Force Museum. Further verifications were provided by Mr. C. L. Grahn of the Naval Air Material Centre.

Many photographs were willingly supplied by airframe and engine manufacturers as well as much valuable information. I am also grateful to Mr. M. C. Olbina of the Allison Division, General Motors Corp., Messrs. T. L. Tobin and A. C. Manson of Bell Aerosystems Co., Mr. H. V. Brockin, Jnr., of the Boeing Co., and Mr. F. G. Clark, of Bristol Aircraft Ltd., Mr. C. Welti and Mr. F. J. Bettinger of Convair Division, General Dynamics Co., The Curtiss-Wright Corporation; Messrs. W. Morrison and C. Maynard of Douglas Aircraft Co. Inc.; Mr. T. F. McDonald of Flight Propulsion Division of General Electric Co.; Mr. N. G. MacKennan of Grumman Aircraft Engineering Corp.; Mr. S. A. H. Scuffham of Handley Page Ltd.; Mr. M. V. Mattson of Lockheed Aircraft Corp.; Messrs. W. A. Wheeler, J. O. Buerger and G. H. Hall of North American Aviation Inc.; the Ryan Aeronautical Co.; and Mr. A. L. Schoeni, Chance-Vought Division, Ling-Temco-Vought Inc.

Over-all assistance and encouragement was provided by fellow members of the American Aviation Historical Society (A.A.H.S.), viz. in alphabetical order: Mr. Harold Andrews, Mr. Fred Dickey, Jnr., Mr. Bude Donato, Mr. Harry S. Gann, Jnr., Mr. Herbert Kelley, A./Gy./Sgt. Walter F. Gemeinhart, U.S.M.C., Mr. Clay Jansson, Mr. William T. Larkins, Mr. Edward Maloney, Mr. Bruce Reynolds, Mr. William A. Riley, Mr. James J. Sloan, and Maj. Truman S. Weaver, U.S.A.F., to all of whom I give sincere thanks.

My special thanks are due to early Curtiss test pilot Roland Rohlfs and another early pilot, Maj. Howard Wherle, who provided information and inspiration on the pioneering days of flying. Apart from Mr. Bruce Robertson, who edited the manuscript, and who joins me in thanking all who assisted, I am particularly grateful to his assistants, Mr. M. J. F. Bowyer and Mr. Peter Berry, also Lt.-Col. J. D. Thompson, U.S.A.F.

About thirty of the drawings were prepared by Messrs. J. D. Carrick and Frank Yeoman of the *Harleyford* team. Messrs. Herbert Kelley, E. F. Schmidt, Roblin Lambert and Larry Vossbrink assisted me with several. I prepared the balance of some forty out of the total of seventy.

The assistance of Chester Polek, who kept my filing system organised, is acknowledged with sincere thanks.

Deep appreciation is expressed to my wife, Joan, not only for typing much of the correspondence and original manuscript, but mainly for her patience!

In keeping with previous *Harleyford* publications, my aim was to cover the subject of the title more thoroughly than in any previous work. As a companion volume to *United States Army and Air Force Fighters 1916-1961* this book on United States naval fighters does not deal with extraneous aircraft such as those built for export, or types which were utilised in ways other than for their basic missions. Many United States Navy and Marine Corps fighters *were*, however, called upon for various missions, i.e. scouting, light bombing, attack, photographic, racing and training. This, in itself, is a great tribute to the versatility of the United States naval fighter.

Temple City, California,
October 1962.

PAUL R. MATT

ACKNOWLEDGEMENTS FOR PHOTOGRAPHS

Grateful acknowledgement is given to the Office of Information, Department of the Navy, the Marine Corps Museum, Quantico, and the Air Force Museum, Wright-Patterson Air Force Base, for permission to use photographs from their archives. Many photographs have come from private sources. The following organisations and individuals, listed alphabetically, are gratefully acknowledged. Mr. Harold Andrews, Mr. Roger F. Besecker, The Boeing Company, Mr. Fred C. Dickey, Jr., Mr. Bude Donato, Mr. Harry S. Gann, Jr., A./Gy./Sgt. W. F. Gemeinhardt, U.S.M.C., Mr. Frank C. Hartman, the Imperial War Museum, London; Mr. Clay Jansson, Mr. Herbert Kelley, Mr. William T. Larkins, Herr Heinz J. Nowarra, Messrs. Real Photographs Ltd., Mr. Roland Rohlfs, Mr. James J. Sloan, Mr. Frank Strnad, Major Truman S. Weaver, U.S.A.F., and Mr. Frank Yeoman.

About this Book

By D. A. RUSSELL, M.I.Mech.E,

The two outstanding names of pre-war U.S. Navy fighters were Boeing and Curtiss. This Curtiss F4C-1 Hawk brings out the significant fact that aircraft designated as fighters did not necessarily serve as fighters or in fighting units, as it bears the unit letters of Bombing Squadron VB-1, 'The Red Rippers'.

This is a companion volume to 'United States Army and Air Force Fighters 1916-1961' and, dealing as it does with the Navy and Marine Corps fighters, completes the overall picture of fighter types of aircraft developed and used in the United States of America throughout the last fifty-six years.

The success of the earlier 'Harleyford' book has prompted the Publishers to fulfil the requirements of the many enquiries received, by this further and complementary work dealing with the aircraft used by the Navy and Marine Corps. The fiftieth anniversary of aviation in the United States Navy was celebrated in 1961 and that of the Marine Corps in 1962—the year of publication of this book. In that span of time these two Services have progressed side by side, whilst making their individual contributions to the air power of their nation.

The expansion of both Services has continued throughout the years from 1911-1912 until to-day, (1962), they form the most powerful naval aviation element of any nation of the world. In common with the air components of other countries armed forces their beginnings were humble, and caution and financial restrictions were but two of the obstacles they had to surmount from birth. Until the requirements of the 1914-1918 War became apparent no great enthusiasm was evinced by the authorities for this new arm and their development was, perhaps naturally, retarded by the traditional preference for the more orthodox land and sea services.

Fortunately, however, the importance of aviation and the implications in its adoption to naval uses were not entirely ignored, and certain far-sighted and enthusiastic supporters were successful in persuading the United States Government that naval and marine aviation forces were desirable if not a necessity. As a result, the training of the first pilots and acquisition of aircraft commenced in 1911 for the Navy and in 1912 for the Marines. Thus, were laid the foundations of these two services. At the time of their inception, the need for fighter types of aircraft was not, of course, envisaged. Technical development somewhat naturally, was confined to float planes, flying boats and land planes which might be employed on duties which would assist sea-warfare of that period. Not until 1917, when some of the lessons of the War then being waged had been learned, did it become apparent that the fighter plane must be included in all future requirements for naval and marine aviation.

Fighters at war. The war came shortly after the transition from biplane to monoplane and it brought a temporary stay to bright finishes. The Corsairs shown are of the Second Marine Air Wing on a rocket strike against Japanese positions on Okinawa.

This Boeing (BuAer No. 8637) also bears the boar's head insignia of 'The Red Rippers' and is significant of changing unit designations. It is a fighter by designation—F4B-2, by unit—VF-5, yet it is fitted with bombing racks and bears an 'E', indicative of excellence attained in both bombing and maintenance.

It is from this point that this book takes up the story and describes the long line of fighter aircraft used in the ensuing years until 1962. Every progressive step in naval fighter evolution is described; every design which was accepted for service with the Navy or Marines, or both, is reviewed from its inception. A number of these aircraft found favour with air forces other than American, and these too, are comprehensively dealt with.

The appearance of a new design and its acceptance into use merely hints at the vast amount of research and experimental work involved by the industry and services, and this important facet of aviation progress is described in full, descriptions of both successful and unsuccessful designs being included in this book.

The serious student of the United States Naval and Marine Corps aviation will appreciate the amount of background information that has been incorporated into these chapters. This analytical approach and the explanations of the creation of certain designs (and their eventual acceptance or rejection) enhances the value of the book, with its ten-pages of Data Tables. In this way, it has been possible to incorporate all the salient facts relating to a design.

In addition to the treatment accorded to the individual service aircraft types, many interesting and related subjects are discussed—the racing float-land planes of the early 'twenties—the intriguing 'parasite fighters' which operated from the airships of the United States Navy—and, of course, the continuing story of the aircraft carriers.

The operational activities of Naval and Marine fighters in the 1939-1945 War and also, in the Korean conflict, are thoroughly described; the latter being significant as the first occasion on which jet fighters were flown in action by the United States Air Force.

The markings and colouring systems of the fighters of the two services are of considerable interest, especially so to the model-maker, and these are very fully illustrated and examined in Appendix One; additional information being provided in the form of reproductions of sixty unit insignias. In a similar manner, aircraft carriers are described—with a number of illustrations—in Appendix Two.

The illustrative material in this book consists of the usual detailed photographic coverage as in all 'Harleyford' books. These total some hundreds of photographs including a number of rare and hitherto unpublished ones, whilst no less than seventy three-view tone drawings, all to the popular 1/72 scale, depict a representative selection of the fighters featured in the book and cover the entire period of their history.

A 'Collector's Album' illustrates lesser-used, experimental or rare fighters, United States designs in use by the British Services and a number of 'atmosphere' shots of aircraft operating in the Pacific theatre of War.

As in pre-war days, so post-war, fighters perform dual roles and these Furies are the FJ-4B version with an attack squadron, VA-146. Typical of a basic design that has been developed for both U.S.A.F. and U.S.N. use, the FJ Fury has its counterpart in the F-86 Sabre.

United States Navy and Marine

CONTENTS

		Page
Painting	By J. D. Carrick	Dust Cover
Frontispiece	Early 'thirties	2
Foreword	Paul R. Matt	3
About this Book	D. A. Russell, M.I.Mech.E.	4 and 5
Plate	Early 'sixties	8
Part One	Narrative, Chapters 1 to 15	6 to 136
Part Two	Appendices	137 to 151
Part Three	Pictorial Review	152 to 227
Part Four	Squadron Badges	228 and 229
Part Five	Collector's Album	230 to 235
Part Six	Data Tables	236 to 245
Index and Glossary		246 to 248

PART ONE

Chapter		Page	Chapter		Page
One	Fighting without Fighters	9	Eight	Changing the Fighting Form	59
Two	The Quest for a Fighter	18	Nine	Grummans Forge Ahead	68
Three	A Zest for Speed	25	Ten	Fighters of Fame	80
Four	Framing a Fighting Force	31	Eleven	Jets—The New Generation	95
Five	'Will of the Wasp'	37	Twelve	Finding a Fighter for the 'fifties	103
Six	A Boom in Boeings	47	Thirteen	Korean Kinetics	109
Seven	Parasite Project	54	Fourteen	The Radicals	117
	Fifteen	Fighters in Fine Fettle	121		

PART TWO

		Page
Appendix One	U.S. Navy and Marine Corps Fighter Markings	137 to 145
Appendix Two	Aircraft Carriers of the U.S. Navy	146 to 151

Carrier Photographs—Langley CV-1, Lexington CV-2, Saratoga CV-3, Ranger CV-4, Yorktown CV-5, Enterprise CV-6, Hornet CV-8, Essex CV-9, Yorktown CV-10, Independence CVL-22, Leyte CV-32 148

Carrier Photographs—Antietam CVA-36, Midway CVB-41, Franklin D. Roosevelt CVB-42, Coral Sea CVA-43, Wright CVL-49, Saipan CV-48, Saratoga CV-60, Ranger CVA-61, Independence CV-62, Kittyhawk CVA-63, Constellation CV-64, Enterprise CVA(N)-65 149

Carrier Photographs—Essex CV-9, Kearsarge CVA-33, Forrestal CVA-59, Intrepid CVA-11, Shangri-La CVA-38, Independence CVA-62 150

Carrier Photographs—Enterprise CVAN-65, Ranger CVA-61, Saratoga CVA-60, Intrepid CVA-11, Forrestal CVA-59, Kitty Hawk CVA-63 151

PART THREE
1/72 SCALE THREE-VIEW TONE PAINTINGS

	Page		Page		Page
Curtiss HA	152	Curtiss TS-1	157	Chance Vought FU-1	162
Curtiss 18-T	153	Curtiss-Hall F4C-1	158	Boeing FB-5	163
Hanriot HD-1	154	Curtiss F6C-2	159	Curtiss F6C-4	164
Thomas-Morse MB-3	155	Boeing FB-1	160	Boeing F2B-1	165
Vought VE-7F	156	Wright F3W-1	161	Curtiss F7C-1	166

Corps Fighters 1918-1962

	Page		Page		Page
Eberhart XFG-1	167	Curtiss XF13C-1	185	Chance Vought F4U-5N	205
Boeing F3B-1	168	Grumman F3F-1	186	Grumman F7F-3N	206-7
Hall XFH-1	169	Brewster F2A-1	187	Douglas F3D-2	208
Boeing F4B-1	170	Grumman F3F-2	188	Grumman F9F-2	209
Berliner-Joyce XFJ-1	171	Bell XFL-1	189	McDonnell F2H-2	210
Curtiss F8C-4	172	Grumman F4F-3	190	Convair XFY-1	211
Boeing XF5B-1	173	Grumman XF5F-1	191	Convair XF2Y-1	212
Grumman FF-1	174	Vought-Sikorsky F4U-1D	192	Grumman F9F-8	213
Berliner-Joyce XF3J-1	175	Grumman F6F-3	193	Grumman XF10F-1	214-5
Curtiss F11C-2	176	Curtiss XF14C-2	194	North American FJ-3	216
Boeing F4B-4	177	Grumman F8F-1	195	McDonnell F2H-4	217
Curtiss F9C-2	178	Boeing XF8B-1	196-7	Douglas F4D-1	218
Douglas XFD-1	179	Ryan FR-1	198	Chance Vought F7U-3	219
Curtiss BF2C-1	180	Curtiss XF15C-1	199	North American FJ-4	220
Boeing XF7B-1	181	Ryan XF2R-1	200	Grumman F11F-1	221
Curtiss XF12C-1	182	McDonnell FH-1	201	McDonnell F3H-2N	222-3
Grumman F2F-1	183	Chance Vought XF5U-1	202	Chance Vought F8U-1	224-5
Northrop/Douglas XFT-1	184	North American FJ-1	203	McDonnell F4H-1	226-7
		Chance Vought F6U-1	204		

PART FOUR

A Representative Selection of U.S. Navy and Marine Corps Fighter Unit Badges   228 *and* 229

PART FIVE
A Collector's Album

	Page
From War to War—Curtiss TR-1, Eberhart XFG-1, Boeing F4B-4, Grumman F3F-2, D.H.4B, Naval Aircraft Factory TF-1, Curtiss F6C-3, Boeing F2B-1, Curtiss BF2C-1	230
Racers and chasers—Wright 'Apache', Curtiss land and seaplane racers, Vought UO-1, Vought FU-1, Thomas-Morse MB-7, Curtiss F6C-2, Douglas XFD-1	231
In the Pacific—Grumman F4F and F6F; and Vought F4U aircraft. A selection of photographs showing these aircraft operating in the Pacific	232
'Neath foreign flags—Brewster F2A; Grumman F4F and F6F; Vought F4U. A selection showing these aircraft operating with the R.A.F. and R.N.Z.A.F.	233
Mainly experimental—Curtiss F7C-1, F8C-1/OC-1, XF8C-4, XF11C-3, Grumman F4F-3 Seaplane, Curtiss XF8C-2, OC-2, F8C-5, Grumman XF3F-1, Bell L-39	234
Miscellaneous monoplanes—Ryan FR-1, Grumman XF5F-1, F9F-6, Convair YF2Y-1, Vought F7U-3, XF4U-3B, XF5U-1, Grumman F9F-5, N. American P-51A, P-51H	235

PART SIX
5 Double-Page Data Tables

In these tables are given particulars of the different aircraft types, sub-types and experimental designs which form the subject of this book. This information includes the names of firms who produced these aircraft, crew numbers, aircraft type (monoplane, biplane or triplane), date of delivery, type and horsepower of engine(s), maximum speed, wing span, length, loaded weight, quantity built and their appropriate serial numbers. For every type information is also provided in a 'Remarks' column in which many other items of interest, such as armament details are also described.

Fighting Form—Early 'sixties

Possibly the most formidable naval fighter in the World, the McDonnell F4H-1 Phantom II twin-engined, two-seat all-weather interceptor and attack fighter, represented here by F4H-1 BuAer No. 148375 of VF-114 shown being catapulted from the angled deck of the U.S.S. Coral Sea *(CVA-43) on September 8th, 1961. The attack fighter is somewhat removed from the original VF class concept of fighters to protect the fleet, but once committed, attack—it is said—is the best method of defence.*

PART ONE

CHAPTER ONE

Fighting Without Fighters

The first take-off from a ship, November 14th, 1914. Eugene Ely's Curtiss dipped perilously close to the water before rising and flying to Willoughby Spit for a safe landing.

That the battleship was the capital ship and the backbone of any navy is unquestionable. Battleships were built to fight battleships, although only once in living memory did they clash in force—at Jutland in 1916. Nevertheless, the battleship was a great potential force and nations strove for parity by matching battleship with battleship. There emerged a 'battleship mentality', an outlook by officers so steeped in the order of naval routine that their outlook became conservative. They are sometimes referred to as the 'die-hards'.

This frame of mind was not peculiar to any one Service nor yet to any one country. The British Commander-in-Chief in the Field 1915-18, Earl Haig, after the object lessons of tanks at Cambrai in 1917 and Amiens in 1918, was endeavouring to convince officers in a lecture at the Staff College post-war that the horse still had a part to play on the battlefield in future wars.

The wind of change to the American naval scene came in 1921 when General 'Billy' Mitchell forcibly demonstrated the vulnerability of the battleship to aircraft by sinking German reparation warships. Still the 'die-hards' were not convinced, but the next World War convinced even the most conservative, that the day of the battleship had ended. Their place was taken by the aircraft carrier, which combined striking power with great radius of action.

No longer is the Navy hidebound in its approach. It is even willing to consider and indeed to plan for the missile-ship as the capital ship of the future. But at the moment, the carrier is at its zenith and the world's largest ship is an American aircraft carrier.

In these times of eager acceptance of technological aids, the difficulties of the past may seem difficult to grasp. When U.S. Naval aviation was in its infancy the Navy was dominated by 'battleship admirals' and it cannot but be admitted that on the face of it, such frail structures of wood and canvas, at the mercy of the slightest wind, could hardly be looked upon as a menace, potential or actual, to the monsters with 18 inch steel armour-plate that could steam into the teeth of a gale. It was a long struggle before the aeroplane could prove its worth in U.S. military affairs.

When finally the aeroplane was accepted as an adjunct to the navy it could be foreseen only as the 'eyes of the fleet'; roles other than scouting and patrolling had not at that time been envisaged.

The first few aircraft purchased by the Navy prior to World War I, were obtained merely as 'aeroplanes' not for any specific purpose. Adapting aeroplanes was a new art in itself, and water-based aircraft—seaplanes—were an entirely new venture.

When interest in aviation was shown by the Army and Navy the Marine Corps was not far behind. As in the Navy proper, their activities attracted enthusiastic individuals who came forward to investigate and study aeronautics. These advocates were usually granted special permission by their superiors to pursue this new science but generally this had to be extra-curricular without neglect of their regular duties. Lt. Alfred Austell Cunningham was one early enthusiast. In fact U.S.M.C. aviation can count its beginnings from when he received orders to report to the Navy's Aviation Camp near Annapolis on May 22nd, 1912. When on the following August 1st Cunningham flew solo at the Burgess Company's flying school where he had been sent to receive formal instruction, it could be said that Marine aviation had taken the air. Lt. Cunningham was followed by Lts. Bernard L. Smith and William M.

A-1, the U.S. Navy's first aircraft, a Curtiss pusher. This seaplane was first flown in Navy service on July 1st, 1911, by Glenn Curtiss and tested next day by Lt. Ellyson.

McIlvain, both Marine officers, to form the nucleus of a Marine Corps Aviation Branch.

Together with the first three Navy pilots, Lts. Theodore Ellyson, John Rodgers and John H. Towers, the Marines faced a long, tough struggle promoting aviation in the Naval Services. They had to convince the sceptics that aviation was a necessity to the Service. Their first major point was made on August 30th, 1913, when Admiral Dewey (of Manila fame) signed a report recommending the establishment of an air service—'Suited to the needs of the Navy in War'. This came two years after the Navy received its first aeroplane, ten years after the Wright Brothers' first flight, but fortunately four years before America entered the European war in 1917.

On January 6th, 1914, Lt. John H. Towers, U.S.N., received an unsigned letter suggesting 'a separate Marine Section of the Navy Flying School at the newly established Pensacola Flying School. Later the same year, Lt. Smith, U.S.M.C., came out boldly with the proposal (making Lt. Towers believe it was Smith who had sent the unsigned letter) to make Marine Aviation an organisation separate from the Navy. By the same token it was stipulated that it would be still under direct orders from the Naval Department. The recommendation was indeed farsighted. Lt. Smith could at this early date, see the need for Marine Aviation operating in close conjunction with Marine ground forces. His argument was that this would relieve the Navy of support and other duties.

At a General Board Meeting in 1915, the proposal, being of mutual interest to both services, was accepted. Henceforth, the United States Marine Corps had a separate Aviation Section, under Naval jurisdiction. This unique form of two separate naval air departments, has become accepted in the U.S. Navy.

By that time naval aviation had been in action for the first time, albeit briefly, in the Vera Cruz incident during the Mexican troubles. On May 6th, 1914, a Curtiss AH-3 rising from the water, after being off-loaded from the U.S.S. *Mississippi*, to photograph areas for the subsequent landings, was fired upon from the ground and it was flown back with bullet holes through its fabric.

From the time the Navy purchased its first aeroplane, the Curtiss A-1 on July 1st, 1911, to 1916 when the war in Europe was in its middle year, there were but seventeen aircraft, on the Navy's inventory; fourteen seaplanes of assorted types in various conditions, and three flying boats. All were antiquated by European military standards and were used mainly for training and experiments. Detailed specifications were issued that year for naval seaplanes, but a fighting role was still not envisaged.

The maintenance of neutrality itself involved a policy of rearmament. Provocative acts by German U-boats and the possibility of the war engulfing the United States, led to substantial funds being allotted to the Navy and a proportion was allotted to Naval aviation to enable a new establishment to be introduced and this was being studied when America entered the war on the side of the Allies. Naval aviation, including its new sister service, the Marine Corps Aviation Section, had first to concentrate on the submarine menace in conjunction with the convoying and escorting of troop and suppy vessels and many of the formalities of service organisation went by the board. In April 1917 there were but a handful of men involved in the new organisation; ten U.S.M.C. and thirty-eight Naval aviators—no students, no officer candidates, not even ground administrative officers. The total of 287 naval aviation personnel consisted of the flying personnel and 163 Navy and seventy-six Marine ground crew. Their equipment was fifty-four aircraft (forty-five seaplanes, six flying boats and three land planes) but none classified as ready for combat or even of combat type.

The year 1917 was an historic one for naval aviation; the first Marine Aeronautic Company formed at Philadelphia Navy Yard on April 27th, the first standardised national insignia appeared on naval aircraft in May and the first uniform for aviators (Order Change No. 11) was effective from June 22nd. By that time, the 'First Aeronautic Detachment' consisting of seven pilots and 122 men under Lt. K. Whiting had landed in France, the first United States military unit to land in Europe for the First World War.

First elements of the unit arrived at Bordeaux on June 5th, 1917, in the collier *Jupiter* which later became the U.S.S. *Langley*, the United States first aircraft carrier. Whiting contacted the French naval authorities to arrange for the participation of naval aircraft in operations by taking over bases. One of the first of the twenty-two bases eventually operated in France, the British Isles, Canada, Italy and the Azores, was Le Croisic from where, on November 18th, 1917, the first coastal patrols commenced by the Navy, using French Tellier flying boats.

Later in 1918, a naval air base was opened at Dunkirk from where both French and British aircraft operated. This was perhaps the most active of Naval Stations as the district was repeatedly subjected to air raids and on occasions long-range shelling. The personnel lived in dug-outs. It was here that the need for fighter aircraft to protect patrol aircraft on anti-submarine and convoy duties was most apparent. Lt. Cunningham, following an inspection tour in Europe in late 1917 had clearly seen the need and hurried back to Washington to present a case. Together with Lt. John Towers, he appeared before the General Board of the Navy and advocated as imperative at least four fighter squadrons if air control over the Dunkirk-Calais area was to be maintained.

Indicative of the intensity of air fighting over the area were the operational records supplied by the British for the Royal Naval Air Service squadrons based on Dunkirk which showed in the previous six months, July to December 1917, that 228 tons of bombs were dropped twenty-one aircraft having been lost and of enemy aircraft, they claimed that eighty-four had been destroyed and 252 driven down out of control, during this period.

Towers pointed out that slow lumbering patrol craft were of little use unless they could operate unmolested, but Admiral Sims and his staff were apparently not impressed and, ignoring the advice of his specialists, decided that bombers were more important than fighters. But in any case aircraft of all types were in short supply and the Army appeared to have prior claim on production.

At this time all U.S. military forces in the European Theatre were under the unified command of the Commander of the American Expeditionary Force, and the matter of provision of aircraft to meet the Navy's needs was placed with General J. Pershing himself, who decided that the Navy should undertake the bombing of submarine pens and should be given whatever equipment was necessary to carry out that function. The Navy requested seventy-five bombers and forty single-seat pursuits as escorts. The United States Air Service under the command of Maj. Gen. Benjamin D. Foulois, balked at the request, insisting that landplanes or fighter type aircraft and land targets were in their province, not the Navy's.

General Pershing's decision, however, was final. Reluctantly the Army agreed to furnish some fighters but jibed at supplying any bombers. The Navy waited, but the fighters never did arrive and they were left with the American equipment at hand, plus whatever could be procured from other governments—Capronis from Italy, Handley Pages and de Havillands from Britain, Tellier and Donnet-Denhaut flying boats from France.

What the fighters would have been is anyone's guess since the U.S. Army Air Service itself did not operate one American-built or designed fighter aircraft in action during the 1914-1918 War, but used numbers of French and some British aircraft to equip their pursuit squadrons; not so fortunate was the Navy or Marine Corps who were unable to draw on this source for their fighters, in spite of the failure of their homeland to supply the need. This governmental policy of discrimination in the Army's favour was so discouraging to the Navy that not surprisingly a bitter inter-service dispute arose over the jurisdiction of overall military air control. A dispute that was continued for some years.

Meanwhile the British had already merged their air services into a separate Force and their example gave weight

Amphibious trials by Curtiss with the 'Triad' prior to its acquisition by the Navy as the A-1. Ordered on May 8th, 1911, its career lasted until late in 1913.

to a movement for the United States to do likewise. The Army Air Service was ready to do this and openly advocated such action; any move to escape the overall control of their service by ground officers met with their favour. One cannot but sympathise with the feelings of active fliers who, at this time, found their orders emanating from senior officers who knew nothing of the correct application of aircraft in warfare, and who took no steps to improve their lack of efficiency by acquiring the practical knowledge they lacked. They could not properly appreciate the weapon they had had thrust into their hands, and as a result it was often wielded blindly.

It has been argued since that had a United States Air Force been created in that War, it would have been better equipped to meet the double threat of Germany and Japan in the later war; if the Royal Air Force may be taken as a criterion then it would seem very true, but not so far as naval aviation was concerned. From the leading naval air power in the world in early 1918, naval aviation regressed to an almost fatal degree from the Royal Navy, who gained control, almost too late, shortly before the 1939-45 War. The U.S. Navy fortunately held fast, and fought hard in insisting upon an air fleet of its own, suited to its own particular requirements and operating only under naval direction, factors which, in retrospect, cannot be called unreasonable if the air fleet was to have practical value.

The lack of fighter aircraft by no means implied that the personnel did not fight; many naval airmen were actively engaged by operating with British, French and Italian units. One of the early participants was Ensign Fallon, U.S.N., who was attached to Felixstowe Royal Naval Air Station. On March 12th in F.2A flying boat N4582, accompanied by N4510 another F.2A and 8661 a Curtiss H.12 flying boat, he left Felixstowe shortly after midday for a war patrol to the North Hinder. N4510, however, had engine trouble, and with a spluttering engine turned back for the coast; the other two carried on. When seven miles east of the East Hinder the crews spotted five German seaplanes sitting on the water, evidently awaiting their patrol craft. Immediately the seaplanes took off, rose and came into an attack in formation. The flying boats, with the advantage of height, dived into them to break up their formation and then for thirty minutes they battled. Wireless Operator Gray in the Curtiss shot one two-seater seaplane down which crashed into the sea and another was forced down but later took-off again. Fallon shot the gunner of another two-seater, which dived steeply, and then righted, but broke off the engagement. The wireless operator of his own machine was shot in the neck and collapsed; during a brief lull Fallon was able to render first aid before rushing again to his guns. Both flying boats returned safely to Felixstowe.

Five days later in a formation of five flying boats, Ensign Stephen Potter shot down one of the attackers and was officially credited as being the first American naval aviator to destroy an enemy seaplane.

At home strident steps had been made towards providing the aircraft needed and the Navy, in the face of competition from the Army, had decided to build a Naval Aircraft Factory, to assure in some measure the supply of aircraft of the best quality, to assess costing as a guide in dealing with private manufacturers and to conduct experimental work.

Authorisation to build the Naval Aircraft Factory (N.A.F.), was granted on June 27th, 1917, and Lt.-Comdr. F. G. Coburn of the Construction Corps was ordered to proceed with haste and build the plant. A site was picked on vacant land near the Navy Yard at Philadelphia. Ground was broken on August 10th and just sixty-seven days later, on October 16th, 1917, the first machine tool was installed. Production started the same day on the N.A.F.'s first Navy built plane, a Curtiss H-16 flying boat, A-1049, which first flew on March 27th, 1918.

For the first time a fighter aircraft was a specific naval requirement in mid-1917 and an optimistic specification was issued for a speedy fighter, capable of a good rate of climb and, for the period, with heavy armament. It asked much of a floatplane. The Secretary of the Navy authorised the design for immediate construction and the Curtiss Company responded to the challenge.

In June 1917 the collier Jupiter *brought the first American units to France. Post-war, with the superstructure shown, it became the U.S.S.* Langley *(CV-1), America's first aircraft carrier.*

The 'Kirkham Fighter', known officially as the Curtiss 18-T, was the first U.S. Navy fighter in landplane form. This machine was first flown on July 5th, 1918, and is shown here in its later form with a slight sweepback to the wings and larger radiator.

The Navy's full requirements could not possibly be met and in this first venture, the Navy was forced eventually to acknowledge that 'haste makes waste'. The result was a heavy, bulky seaplane that caused a series of alarming incidents before it was finally made airworthy.

Plans called for 'many of these fighters to achieve air superiority over the Dunkirk area by late 1918' and therefore the nickname, bestowed even before it was completed, was 'Dunkirk Fighter'. The official designation was HA, a Curtiss designation; this caused much ribbing at the Garden City Plant where the jocular interpretation was Ha! Ha! No official explanation has even been given for HA but the letters are significant of Hydro Aeroplane.

Curtiss and the Navy ran into many unforeseen problems with this two-seater naval fighter and more time was consumed than anticipated in completing the prototype. The only power plant available, likely to fulfil the design requirements was the famous (sometimes infamous) Liberty 12 and this heavy engine proved so unsuitable that the Navy never considered it again in fighter design; they were thus spared the series of disappointments the Army was yet to undergo over the Liberty. However, for heavier patrol and bombing aircraft types it proved so suitable that even the Royal Air Force used it in their D.H. 9A bombers in post-war years.

With the first Navy fighter underway, Lts. Cunningham, Towers and Smith saw their dreams coming true. With Curtiss and the Navy forcing the pace the HA was made ready for its first flight on March 21st, 1918, the day the greatest battle of the war opened as Germany flung her troops into a gigantic offensive in an attempt to break the Allies before the full weight of America's strength could arrive on the Western Front. The flight test was made at Port Washington, Long Island, near the Curtiss Plant at Garden City in the hands of Roland Rohlfs; Lt. B. L. Smith of the Marine Corps, who had promoted and designed the HA, was honoured with a place in the rear-gunner's compartment.

Curtiss, at this time, had a hard and fast rule that first flights of any new aircraft should not be witnessed by anyone except engineers and employees directly connected with the project. However, so enthusiastic and confident was the Company about the HA, that they broke their rule and invited government officials to witness the trials. A Navy delegation showed up in full dress, headed by Commander A. C. Read who later achieved fame as commander of the first aircraft to fly across the Atlantic.

The HA slid easily down the slipway into the water where it appeared to sit low in typical Curtiss fashion, with the aft end of the main float almost completely submerged. Curtiss's theory, proven on earlier craft, was that once the aircraft gained speed it would level up into a normal take-off attitude. As the aircraft raced across the water, gathering speed it appeared to suddenly jump into the air. The spectators noted with satisfaction its short take-off run and were suitably impressed—not so the occupants who quickly realised that something was seriously wrong.

After shooting into the air, they were at the mercy of the craft, see-sawing up and down and rolling from side to side as it wallowed along. Smith at the back, pumping fuel for the engine realized that there was trouble, but managed to continue his job to keep it going in spite of being tossed about violently. Both occupants experienced real fear and cold sweat stood out on their faces.

Rohlfs recalled that he had never tired of a flight so fast in all his life. His only thought was to set the machine down before it tore itself apart. They had barely reached 1,000 feet when Rohlfs fought to get the machine into position for a landing into the wind. Tail heavy and with poor directional stability, Rohlfs experimented with the throttle. He waited for the next dive of a series over which he had no control, then closed the throttle and noted that the aircraft steadied, but as speed built up during the next dive, it became unstable again.

Using this knowledge for landing, he throttled down and with a definite flop, but in one piece, the HA landed in Hempstead Harbour whereupon Rohlfs turned to congratulate Smith on manning the pump throughout their very rough ride so successfully.

Thus ended the first flight of the Navy's first fighter. Perhaps Rohlfs and Smith were not greatly surprised when it was found that owing to a serious miscalculation, the wing was about a foot too far forward! Further evaluation pointed to the enlarging of the fin area and the aircraft was put back into the Curtiss plant for alterations, which included this and other small changes.

Rohlfs returned to the fray a few days later. On this second flight Lt. Smith was given a ground view of the

The U.S. Navy's first fighter, the Curtiss HA 'hydro-aeroplane', photographed immediately before the first flight on March 21st, 1918. While the Curtiss Aeroplane & Motor Corporation produced famous trainers such as the JN-4 'Jenny' and reliable flying boats that were used by both the Americans and British, their fighter venture was not a success.

proceedings and Curtiss's chief project engineer, Joe Meade, had the doubtful privilege of occupying the rear seat. Rohlfs attempted a ten minute flight to check climb and stability. After some seven minutes, the combined effect of the summer heat and restricted ventilation owing to the new cowling, the engine overheated. The Liberty misfired and a puff of black smoke blew back at the flyers. Petrol in the carburettor had caught fire.

Rohlfs, at 2,000 feet, cut the engine, and started a dive. As he gained momentum, flames flickered back in long thin jets into the cockpit through small screw holes left in the dividing wall. Rohlfs was forced to leave the cockpit and sit on the turtle deck with only his feet in the cockpit. Even this was almost unbearable. Hooking his feet around the control stick he flew the flaming mass to a rather rough but successful landing on the water. Watching officials agreed that only the sound design and sturdy construction prevented the machine from collapsing.

As it wallowed to a stop, Rohlfs and Meade hastily jumped overboard and swam away. They were picked up almost immediately and looking back, watched the U.S. Navy's first fighter being consumed by fire. In ten minutes it was destroyed, quite beyond repair or even of salvage.

In accordance with the Navy's method of numbering its aircraft numerically as each is scheduled for production, that first HA was A (Airplane)-2278. This system was started with the first aeroplane, the A-1 in 1911 and is still in effect today, although the 'A' prefix was abandoned in the late 1920's. Thus 2278 aircraft of various types had been ordered before this—the first of the Navy's fighters.

Somewhat dejected but undaunted by the misfortune to A-2278, the problem was tackled day and night with a wind-tunnel model to test and prove, or disprove, the efficiency of the HA's design. A report issued on March 23rd, 1918, assessed the original A-2278 as longitudinally unstable and tail heavy, and an increase in the negative setting of the fixed stabiliser was advised. Curtiss incorporated these recommendations in the second and third models that emerged later in the year, but by that time the name, Dunkirk Fighter, was an anachronism.

While they proceeded with the HA 'Dunkirk Fighters' to Navy specifications, the Curtiss Company were also busy

After the HA Dunkirk Fighter's first flight in the form above, it was modified to the form shown here. On its first flight to test the improvements, it caught fire in the air, and the pilot and passenger were forced to abandon it as soon as it had settled back on the water.

on several fighter designs of their own. These dated back to 1916 when the Navy showed interest in the triplane following news of the successful Sopwith Triplane and their attention turned to the L-1 and S-3 Triplane Scouts being tested by the Army Signal Corps, that showed promising performance. The Navy ordered three of these Triplane Tractors and their engineers proceeded to re-design the craft with floats, one main float and wing tip floats, to meet their particular requirements.

Wind tunnel tests, the first undertaken by the Navy, were conducted at the Navy Yard, Washington, D.C., on January 19th, 1917. These proved that the design was inherently stable and three machines, designated L-2 Hydro-triplane tractors, were produced the following March as A291 to A293. Powered by a 90/100 h.p. Curtiss OX engine, they could reach only 90 m.p.h. flat out and the estimated performance with the only other suitable engine, the 120 h.p. Curtiss OXX, promised only a 6 m.p.h. increase in top speed which precluded any idea of using it as a fighter. The L-2s, however, provided useful data in triplane design and having side-by-side seating, with low stalling speed and inherent stability, were considered suitable as trainers. The triplane form had thus come under Navy scrutiny and further development by the Company in consultation with naval officials resulted in a new and more promising triplane the following year.

Employed at the Garden City Plant under W. L. Gilmore their chief engineer, Curtiss had the brilliant young aeronautical engineer Charles B. Kirkham, pioneer aeroplane and motor builder, who worked on a triplane design which was, in fact, the first time an engine and airframe had been designed as an integrated concept. His design was sufficiently convincing for the Navy to place contract No. 37372 of March 30th, 1918, for two triplanes which materialised as the Curtiss 18-T with the naval serials A-3325-6.

Like the HA, the 18-T was conceived as a single en-gined, two-seat fighter, adaptable to sea or land use. The proposed armament was two forward firing machine-guns, synchronised to fire through the propeller arc and two flexible gun mountings in the rear cockpit. The first was towed behind Kirkham's Packard car from Garden City to Mitchel Field, Long Island, for the first flight on July 5th, 1918. After careful checking and preliminary engine warming, Roland Rohlfs, after climbing in for the initial flight, glanced over at Kirkham, and noted that he was calmly talking to field officials paying no heed to the roar of the engine or indeed to the aircraft. Impressed with his apparent confidence, Rohlfs opened up the engine and took to the air after an extremely short run. Kirkham's confidence was justified, the 18-T performed perfectly and was perfectly balanced.

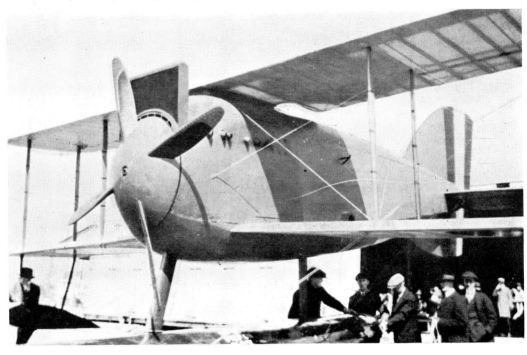

No. A-2278, the Navy's first fighter, the Curtiss HA 'Dunkirk Fighter' had many modifications in its ephemeral existence and the two photographs opposite represent the two extremes in its changing form. Perhaps its failure was due mainly to a dogged persistence, by both the U.S. Navy and the U.S. Army, to adapt the heavy Liberty engine to a fighter design. This photo shows adjustments being made to A-2278.

A fortnight later, after minor modifications to the lower wing, A-3325 was ready for display before officials. This beautifully streamlined aircraft making its official debut was by then known as the 'Kirkham Fighter', after the designer. The powerful new engine, the K-12, also of Kirkham's design was a compact 12 cylinder, 'Vee' type, of aluminium, developing 400 h.p. at 2,500 r.p.m. It had a 4½ inch bore and a stroke of 6 inches. Its dry weight was only 680 pounds in comparison to the 844 pounds of the Liberty engine of approximately equal power and overall size was a few inches less than that of the Liberty 12.

Roland Rohlfs as Curtiss's chief test pilot, again piloted while Lt. C. N. Liqued, U.S.N. Inspector of Naval Aircraft, who had been assigned to the Curtiss Plant, witnessed the flight from the ground and was greatly impressed. It appeared to perform all the evolutions expected of a fighter. Throughout July and August Rohlfs flew nearly five hours in the triplane before official naval trials were held during five days in September at the Curtiss Field at Mineola, Long Island.

On the first day, September 1st, a speed trial was made of 27½ minutes duration. Partly with a 45° cross-wind, and

An American-built fighter did not reach the Western Front during the 1914-1918 War and American pilots used French and British fighters. U.S. Marine Corps pilots flew R.A.F. Sopwith Camels such as that depicted.

partly down-wind, the 18-T achieved 167·1 m.p.h., coming back into the wind 153·1 m.p.h., giving an average of 160·1 m.p.h. Take-off was at 65 m.p.h. and it was airborne in 4 seconds from standstill; landing speed was 55 m.p.h. Official climbing tests could not be made because the side radiators proved to be too small causing the engine to overheat. Four days later, larger radiators were experimentally installed and a brief flight was made to test them out. Mr. Thurston of the Curtiss Company went up on this flight as a flight project engineer. Kirkham reported some loss of speed but next day the official climbing tests were made, during which 12,000 feet was reached in 10 minutes (12,500 feet was the official goal). This time the top speed over a measured course was 150–156 m.p.h. On another climbing test, four teeth of the magneto drive pinion broke and after landing, the engine had to be removed and sent back for overhaul.

By September 19th tests were under way again for speed and climb trials with full military load. No ordnance gear was actually installed, but sand bags to the equivalent weight were distributed in the fuselage. All cockpit instruments were installed bringing the total weight to 2,864 lb. Over a measured course, official speeds were 153·15 and 168·02 (average 160·5 m.p.h.). Loaded to 'overweight' conditions of 2,902 lb., the final official climb to 12,000 feet was accomplished in ten minutes at which height the Kirkham Fighter was still going strong at 78 m.p.h.

As a result of the trials the Navy Board recommended, on September 20th, that the Curtiss 18-T Kirkham Fighter A3325 be accepted by the Navy for payment and the second machine, A-3326, procured on the same contract, also be completed and accepted. Thus the Navy received its first fighter, although not the first naval fighter by design. In spite of outside controversy over a land machine, in the fighter class, in the hands of the Navy, plans proceeded for further testing and experimenting along these lines and the Engineering Division of the Army Air Services were sufficiently impressed with the 18-T's potentialities to purchase two similar models for experiments.

The Kirkham Fighter was indeed of advanced construction for the period. It had a monocoque fuselage of 3-ply birch and redwood veneer. The wings had seven spars each with the leading edges covered with Spanish cedar. The tail surfaces were covered with Spanish cedar and all surfaces given a final covering of linen, doped with a light grey pigmented varnish. Plans called for two forward-firing ·300 Marlin machine-guns and 750 rounds of ammunition plus two free firing ·300 Lewis guns on a Scarff ring on the rear cockpit, with 875 rounds, and this armament was installed for tests.

The Navy Board had several recommendations to make on the 18-T, including conversion to a floatplane for trials and tests with two and four-bladed Charavay propellers, which were built by the Hartzell Walnut Propeller Company. While initial trials had proved that the design was basically sound, difficulties were encountered with the centre of gravity in varying load and flight conditions. To compensate in some degree for this a 5° sweepback was incorporated on all wings shortly before the first two were accepted. In this form there was no wing stagger, whereas in original configuration each wing had been staggered six inches.

With the officially designed HA being far from satisfactory in its present form and with the privately designed 18-T needing further development, it was evident that delays in the delivery of a fighter to naval units had to be accepted. Marine and Navy aviation personnel that could be spared were sent to France in the Spring of 1918 and organised into the Northern Bombing Group. It was planned to introduce the HA fighter later in the year, meanwhile patrols would have to be carried on with the existing American-built DH-4s and borrowed DH-4s and DH-9s.

A Naval Aircraft Base was set up at Eastleigh in Southern England to accept and erect aircraft shipped from the United States, while plans went ahead for airfields to accommodate the Northern Bombing Group in France. The cadre of Marine fighter squadrons actually arrived in France, under Major A. A. Cunningham, being based at Oye and Le Frene. Since no fighter aircraft were available the pilots were rotated with local British squadrons on a scale of three operations per pilot.

The Group was organised at first as part of No. 5 Group, Royal Air Force, based in the Dunkirk area. To escort and protect the bombing and patrol squadrons of the Group were three R.A.F. Squadrons of Sopwith Camels, Nos. 204, 210 and 213 and it was to these three squadrons that the Marine fighter pilots were rotated.

During these closing months of the war, the U.S. Navy's first and only ace of that war, operated. Marine pilots were itching for combat duty in the fast and nimble fighters such as their Allies used. Some pilots, not content with long hours flying patrols, would, when off duty visit nearby British and French airfields to beg flying time in fighters. Such a 'fill-in' was Lt. (j.g.) David Sinton Ingalls who regularly flew Camels on a 'spare-time basis' with No. 213 Squadron, R.A.F.

Born in Cleveland, Ohio, in 1899, where he received his early schooling, Ingalls went on to enter Yale University in 1916. Being interested in aviation he joined the University's flying club, which later became a part of the Naval Reserve. When war broke out, the unit was ordered to active duty with further training at Pensacola. As part of the 3rd Yale Unit, he graduated as Naval Aviator No. 85 in 1917. Later that year his patrol group arrived in France. During off-duty time he got the feel and thrill of flying the speedy little Camel. Then No. 213 Squadron allowed him to fly with them on combat patrols. On July 18th, 1918, he was credited with his first victory, a two-seater Rumpler. When news of this reached Navy officials they instructed young Ingalls to cease this double-time flying and fly regularly with the British as a fighter pilot! His instruction read—Lt. Ingalls, David S. TDY (Temporary duty)—Sqd. 213 R.A.F.

The British respected ability rather than age in a pilot, and Ingalls, a 'teenager' incidentally, certainly had ability. On July 21st, he was allowed to lead a British flight on a strafing mission. So successful was the operation that the Squadron Commander himself appointed him a flight commander in No. 213. On July 24th he shot down an observation balloon and a few days later his second Rumpler. Another enemy machine fell to his guns in the days that followed and on September 24th, while flying over Nieuport, he sent down yet another two-seat Rumpler. For this he received the D.F.C. (British), D.S.C. (American) and was made a member of the French Legion of Honour. With five enemy aircraft shot down to his credit, he became, by American standards of that war, an ace.

Although not one American naval fighter nor yet one American Army fighter, went into action, at least many Navy and Marine pilots saw action in fighters in ways similar to Ingalls. The experiences and knowledge gained by such men as Lt. Cunningham of the U.S.M.C. and Lt. Towers of the U.S.N. and that of Lt. Ingalls, was to lay the groundwork for the future of Naval Aviation.

After participating in raids with R.A.F. squadrons of No. 5 Group, the Northern Bombing Group was at sufficient strength to attempt formation bombing with their own squadrons. On October 14th, 1918, eight D.H.4s of the 9th Marine Squadron bombed an important rail junction at Thielt Ring in Belgium. Observation proved difficult due to mists and an attack by eleven German fighters over the target area. Two of the enemy were shot down but one D.H.4 was missing and the gunner of another was fatally wounded in the action. Had fighters been available as escort the result might well have been different. In the ensuing weeks, the last of the war, several such raids were made as well as supply dropping missions, but the main targets, the submarine bases of Zeebruge and Ostend were evacuated by the Germans. At the time when American naval fighters had been most needed, they had not been available.

In fact the only fighters in the whole of the Navy were three experimental machines of two types detailed earlier in this chapter, the HA and Curtiss 18-T. Three fighters only among the total procurement of aircraft for the Navy in the war, which was as follows:

Service Types procured in America	1,144
Training Types procured in America	1,084
Experimental Types procured in America	36
Ex-Army (155 D.H.4, 144 trainers)	299
Purchased in Britain, France and Italy	142
	2,705

To a large extent the die-hard outlook of the naval staff had been responsible for this state of affairs, and had the Northern Bombing Group operated earlier, before Germany's military strength had been weakened, it might well have been dependent on R.A.F. fighter support to enable operations to have been carried out at all, or to have accepted an exceedingly heavy casualty rate. In convoying and anti-submarine patrol work U.S. Naval Aviation made a substantial contribution to final victory. On many occasions their work was menaced by enemy fighters and they were involved in many actions. The U.S. Navy and Marine Corps fought in the air in the first World War, but they fought without fighters.

The U.S.M.C. operated American-built D.H.4s, supplemented by British D.H.4s and D.H.9s, in the Northern Bombing Group in the closing months of the War. Note United States cockade on upper wings of this DH-4B

CHAPTER TWO

The Quest for a Fighter

One of the six Sopwith Camels, Type 2F.1, which were purchased by the U.S. Navy after the 1914–1918 War. Some of these were used for gun-turret platform take-off experiments and for the subsequent ditching were equipped with hydrovane and flotation gear.

So the war ended with two promising naval fighters on the stocks, but none that had reached operational status. The prospects for development were by no means clear with the economies that logically followed the vast expenditure of war; the air services however had proved their worth and both Naval and Marine Aviation was likely to play an important part in the future, but whether or not they would remain an integral part of the Navy was soon to be debated.

With limited funds and the need to implement a new peacetime establishment, the policy for 1919 was the development of existing types that showed promise and the study of others, to decide which was best suited to naval needs. Much had been gained from the Allies, and there was much more to learn and assess.

A preliminary programme issued by the naval staff on March 13th was indicative of the new conceptions of the roles of naval aviation. Provision was made for the Marine Corps to operate both land and seaplanes and the term ' ship-board fighter ' was first used. In connection with this the Secretary of the Navy authorised on July 1st, the erection of launching platforms on the turrets of eight battleships.

Shipboard fighters could well be considered expendable items under certain fleet conditions, and thus cost, together with lightness, speed and manoeuvrability, was a prime consideration for the type requirement. No American aircraft, not even at design stage, could meet these conditions and the Navy turned to the war surpluses of her Allies. Two Sopwith Pups, originally sent to the U.S.A. for design study, were taken over for a mere $2,147 apiece. About the same time six Sopwith Camels were acquired from Britain, who, in January 1919, had a total of thirty-two B.R.-engined Camels aboard warships other than carriers and were thus much experienced in their use from platforms.

On March 9th, 1919, Lt. Cdr. E. O. McDonnell made the first turret platform take-off from a United States warship, flying a Camel from the U.S.S. *Texas* at anchor in Guantanamo Bay, Cuba. In this connection it is interesting to reflect that the first-ever take-off by an aeroplane from a ship was made by the American aviator Eugene B. Ely in a Curtiss biplane from a quarter-deck superstructure on the U.S.S. *Birmingham* moored in Chesapeake Bay. The

A French Hanriot HD.1, assembled by the N.A.F., on a United States battleship. It was necessary for the 130 h.p. Clerget to be fully revved and for the ship to steam into the wind for the Hanriot to take-off from its wooden platform on a forward turret.

significance of the turret platforms was that they were operationally expedient and did not greatly interfere with the normal functioning of the ship.

Fighters for other ventures were acquired from various sources during 1919. Ten Standard E.1 (M-Defense) biplanes came from the Army and were turned over to the Marine Corps to acquire experience with fighter types. Twelve Nieuport 28s also went into service to help bolster the fighter complement and were used mainly for practical experience in formation flying and combat tactics.

Probably the best known and most widely used of the foreign fighters were the Hanriot HD-1s of which ten were purchased from France at $5,282 each. They were received in their shipping crates at the Naval Aircraft Factory where they were assembled and modifications were made including a revised rudder and heavier tail skid. At first they were used as fighter trainers and had two ·300 forward firing machine-guns installed. Like some of the other surplus equipment, they were used in the early 'twenties in catapult tests.

Two S.E.5As were picked up in excellent condition but most of the other surplus aircraft were purchased as they stood and had to be overhauled to get them into flying trim. Apart from that, they were apparently not even re-painted in many cases retaining their original markings.

Not only from the Allies, but from the former enemy came aircraft for evaluation. A particularly rare specimen received by the Navy in mid-1919 was a Dornier D.I. Only four examples of this revolutionary design had been built. A conventional biplane of all-metal construction, it featured cantilever wings with no interplane struts. Originally it had been completed for the third fighter aircraft contest for the German Air Service held at Adlershof in June 1918. On July 3rd, D.I.D2085/18 was flown by Hauptmann Reinhardt (Manfred von Richthofen's successor to the command of the famous ' Circus ') who was killed when the machine crashed after the wings had collapsed. That completely ruined its chances in the competition but other models were built and proved satisfactory with reinforced wings. The Armistice stopped further development, and Dornier himself fled to Switzerland.

After the Armistice three D-Is were found intact and two were shipped for study in America and in mid-1919 the Army Air Service and the U.S. Navy each had one. The only remaining D-1 was put into the Dornier museum. It was destroyed by bombing in the Second World War.

Ground evaluation of the Navy's D-I, assigned No. A-6058, was made mainly by N.A.F. officers who found interest in the constructional methods employed by Dornier and particularly with the cantilever wing. The D-I was not flown and the effect of adding interplane struts was the subject of a paper study only, on metal construction.

Such experiments as these were consistent with the post-war charter of the Naval Aircraft Factory. There was a considerable run-down of personnel after the war and production was no longer a major concern; instead part of the establishment was used for storage and there was concentration on design and experimental work which, in succeeding years, performed a vital function in the development of Navy/Marine fighters. It became in fact the development centre of naval aviation.

Aircraft were still to be produced at the Factory mainly trainers or batches of urgently required operational types, but at no time did the Navy ever discourage private manufacture; on the contrary industrial firms were very actively encouraged and the Factory had the skilled personnel to

A German biplane in which both the U.S. Army and Navy were sufficiently interested to acquire examples was the Zeppelin (Lindau) Dornier D-I. This biplane did not find favour in the German Air Service, but its metal construction and strutless wings evoked considerable interest.

assess the operational worth of the considerable enterprise shown by the American aircraft industry as well as costing for naval purchasing.

On the fighter aspect, their immediate post-war problems were the evaluation of the Allied and enemy aircraft mentioned, the continual development of the two experimental fighter types, the HA and Curtiss 18-T, and a new and original design of their own for a fighter.

With the promise of years of peace there was time to assess the programmes so hastily embarked upon in the war and development of the HA and Curtiss 18-T continued. When A-4110, the revised HA, emerged, Rohlfs was again at the controls for the initial flight in late 1918 at Port Washington. This time it performed beautifully and the trials that followed led to only minor modifications.

During an early flight on A-4110, Rohlfs had the bottom of the wooden float shear off on take-off. Although he got the craft off the water, he was unable to hold it in the air. As the HA hit the water, it nosed over into a complete somersault. Lt. A. Stengle of the Navy who had gone along as observer, jumped into the water, but Rohlfs stayed with the machine and clambering over the ship like a squirrel, ended up sitting on top of the underside of the main float! Neither Stengle nor Rohlfs were hurt, Rohlfs didn't even get wet, while the wreckage floated and simplified salvage.

The last of the three HA 'Dunkirk Fighters' for the Navy. Planned armament of two ·300 forward-firing Marlins and twin Lewis guns on a Scarff mounting in the rear fuselage, was heavier than the armament of most fighters for the next twenty years.

Repaired, the HA was again set afloat at Port Washington for Rohlfs to make the final trials in the presence of naval officials while Lt. Bradley of the Navy was elected to occupy the observer's seat. On the final climb of the trial programme, Rohlfs, making a grand flourish, dived steeply before hauling back on the stick to start the final climb. He cut it too fine and a float struck the water making the machine bounce into the air. After re-gaining control, Rohlfs turned to see how Bradley had fared and was horrified to find that he had disappeared! Immediately he banked round to search the area where he imagined his passenger had been thrown, but could see only the patch of foam where the machine itself had struck. As an icy chill came over him he felt a light touch on his shoulder. There was Bradley, bleeding but grinning broadly. The seat and floorboard had collapsed plunging him into the deep fuselage.

Immediately Rohlfs set-to for an emergency landing, but Bradley intervened and insisted on the completion of the trials, while he braced himself for support inside his cockpit. Persuaded, Rohlfs attempted the climb, but the machine only made some 5,000 feet. Sensing some trouble after the impact on the water, Rohlfs throttled back for a glide down, when Bradley suggested a spin. Rohlfs balked at that but he did promise a loop if Bradley could secure himself sufficiently. The first attempt was sloppy and it embarrassed Rohlfs. Again he tried, but this time the HA remained inverted! The whole structure vibrated and the wings swung inches fore and aft, first one then the other, and flapped until finally it nosed down and completed its odd loop. Carefully Rohlfs brought the machine to land.

Examination showed that the float struts had been stoved into the fuselage and rigging wires were dangling loosely. Only the rugged construction of the HA had saved the occupants from more serious injury.

A-4110's sister-ship, No. A-4111, was delivered while tests on the first of the two were still underway. It had an increase in wing area by 100 square feet and subsequently the top wing was raised above the fuselage on cabane struts. This greatly improved handling. In spite of their discouraging outward appearance, by the unusual rigging characteristics of dihedral on the upper wing and anhedral on the lower, a great deal of attention was nevertheless given to careful streamlining and robust construction.

Like the 18-T, the HA's were to have two synchronised forward-firing Marlin machine-guns and two free firing Lewis guns on a Scarff mounting in the rear cockpit. Possibly due to the marked performance of the Bristol Fighter in Britain, the Navy, as well as the U.S. Army, was thinking in terms of two-seat fighters. This concept was, however, soon dropped by the Navy in favour of single seat fighters, and was not revived until the 'thirties. Thought was given to adapting the HA to an observation role but rigging problems and difficulties with the water-cooled engine led to them being scrapped.

Curtiss were sufficiently interested to produce a single-seat landplane version for the civil market. This model, fitted with a Liberty engine driving a four-bladed Hartzell propeller met with only limited success, although the U.S. Air Mail Service did purchase three as HA Mailplanes.

The Curtiss 18-T triplane, sometimes referred to as the Wasp, was also put through its paces in 1919 by both the Navy and Curtiss. As a landplane, it was a foregone conclusion that it would have a better performance than the HA floatplanes, but the astonishing average speed of 160 m.p.h. was obtained with a useful load of 1,076 lb. With Rohlfs at the controls and a passenger, 13,000 feet was reached in ten minutes at its gross weight.

Rohlfs was sufficiently impressed to talk the Curtiss Company into attempting a world altitude record, then held by Lt. Casale of the French Army at 33,000 feet. The first 18-T, A-3325, had been handed over to the Navy as early as September 1st, 1918, but A-3326 was still being tested by the firm. Since no hand-over for the latter had been effected, the firm were free to use this aircraft and being in limbo, so to speak, it bore no markings.

From March 1919 onwards Rohlfs made a series of attempts after initial alterations to the aircraft, which was further modified including new wings of increased span to 40 feet 7½ inches to assist lift. On his first attempt the oxygen equipment failed at 26,000 and he was forced to descend. In a series of practice flights and unofficial attempts three engines were burnt out before the oiling system could be made to work satisfactorily in the prolonged climb, which caused one postponement after another.

On July 30th, 1919, Rolhfs set an American altitude record at 30,400 feet and on September 14th a height of 34,200 feet was attained but was not considered official. However, four days later, in the presence of Government, Aero Club, Curtiss and Navy officials, he left Roosevelt Field, Mineola, L.I., at 12.06 p.m. and by the time he had landed at 1.54 p.m. a new world's altitude record had been established at 34,910 feet. No supercharger or special fuels were used with the K-12 engine and the only major innovation was controllable shutters on the side of the radiators.

The Kirkham Triplane fighters were without a doubt exceptionally clean, well streamlined and speedy craft for their day. The Navy sometimes referred to their two 'Wasps' as 18-T-1 and 18-T-2 to differentiate between the first and the second. A total of seven 18-T triplane and 18-B biplane models were built by Curtiss, two 18-Ts each for the Navy and the Army Signal Corps, two 18-Bs for the Signal Corps and one 18-B for the Bolivian Government. The biplane version preceded the triplane designs in construction, the 'T' models being basically a 'B' with a set of triplane wings. The original 18-T was Curtiss design 15, and the second 18-T and 18-B model was design 15A.

Early in 1919 the Naval Aircraft Factory engineers had on the boards a flying boat fighter design, looking much like a miniature of the NC boat that achieved fame that year as the first aircraft to cross the Atlantic. Since the large boat had proved successful, it was probably considered that a scaled-down version would have similar attributes. The design evolved as the TF Fighter, and A-5576 and 5577, the first of four ordered, emerged from the N.A.F. in 1920. It proved a reasonably sound design, but as a fighter it failed. Cooling of its two 300 h.p. Hispano engines proved a problem, manoeuvrability suffered because of the large and heavy hull, and a poor performance precluded its use in a fighting role.

A further two were planned; one with 230 h.p. Packard 1-A-825 engines installed was built, but the additional weight of over 1,000 pounds that the Packards entailed, jeopardised performance even more and since a cost of $30,000 each was evoked for these fighters, additional expenditure was not considered justified. The other was cancelled. Nevertheless, the TF has a place in American aeronautical history as the first twin-engined fighter.

Re-organisation of the aviation services was the keynote of the immediate post-war period and for a time it seemed as if America might follow the example of Britain and create a separate service. Vociferous in his defence of the Navy retaining control of its own air units was the Assistant Secretary of the Navy, whose name was to become a household word on both sides of the Atlantic, Franklin D. Roosevelt. He pointed out that to separate the command of naval aircraft from the navy it is intended to serve, bordered on the ridiculous. And when the example of Britain was quoted, Roosevelt pointed out that the Royal Air Force was a separate service in name alone and that the Royal Navy controlled its naval aircraft. This was in fact substantially true for early 1919, for on December 31st, 1918, the Royal Navy had 478 aeroplanes, 1,221 seaplanes and 105 airships in commission under its direct control. Not until 1919 did the R.A.F. assume full control.

The Naval staff may not have been fully alive to the importance of their naval branch, but at least they now regarded it as an essential adjunct to the Navy and they were not going to hand over control to an outside body without a struggle. In an attempt to show that their own house was in order and to promote the status of aviation, a Bill was raised to form a Bureau of Naval Aeronautics charged with the design, construction and administration of all Navy and Marine Corps aircraft. This became law on July 12th, 1921, and naval aviation was placed on a par with the Bureau of Ordnance and Engineering.

The assent of Congress was no doubt influenced by pending developments. Two new phases were being introduced into naval aviation, one proved a gross failure, the other a marked success. The first affected fighters to a very limited degree, the success of the second was to lead to the fighter being ordered in greater quantities than any other naval aircraft type.

The United States Marine Corps already had an association with the famous Sopwith Camel in France during the war, an association with the equally famous British S.E.5A evidently came after the War by this photograph of an S.E.5A at the U.S.M.C. Quantico depot.

Not only in America, but in Britain and Germany the development of large airships was attended by disaster, but in 1919 both Britain and America still had high hopes of lighter than air (LTA in U.S. naval parlance) craft. Construction of the *U.S.S. Shenandoah* (ZR-1) began at the Naval Aircraft Factory in August 1919; although final assembly did not take place unil 1923 at Lakehurst. Since the British, for reasons of economy, were seeking to sell some of their large rigids, the R38 was acquired as the Navy's ZR-2. It had not passed unnoticed that fighter protection could be given to airships by the carriage of underslung fighters, from experiments carried out by the British with a Sopwith Camel underslung from the R23. However, on its fourth trial flight over Hull in England on August 24th, 1921, a searing flame broke the envelope asunder and the R38 fell into the waters of the Humber.

The other innovation, again following British practice, was the acquisition of an aircraft carrier. This was achieved by converting the collier *Jupiter* which was commissioned on March 20th, 1921, at Norfolk Virginia, as the U.S.S. *Langley* (CV-1). Of 11,500 tons, her flight deck was 64 feet wide by 534 feet in length; small by today's standard, but adequate for the period. For over two years she was used to qualify Naval aviators in the fleet for carrier duty, and this was successfully done without a single failure, or even injury, during training. CV-1 (Carrier Aircraft-One), affectionately named the 'Covered Wagon', proved her usefulness in many ways over the years following Lt. Comdr. G. de Chevaliers' first landing in an Aeromarine trainer on October 26th, 1922.

Like the Engineering Division of the U.S. Army Air Service, the Naval Aircraft Factory was interested in welded metal construction and from the Army's war reparation surpluses, six unarmed Fokker D.VIIs and three of its two-seat version, the C.I, were purchased; these became BuAer Nos. A-5843-8 and A-5887-9 and they were left in their original Army olive drab colouring. After both static and flying tests the value of a trussed steel fuselage was apparent and various components including wing sections were ordered from Charles Ward Hall, a Curtiss engineer.

After the Fokkers served their purpose at the Factory they went over to the Marine Corps, and two were flown at Quantico. As proficiency trainers they served a useful purpose particularly to trainee fighter pilots and were retained in service for three years. The D.VIIs had 350 h.p. Packard 1-A-1237 engines while the C-1s retained the original German 245 h.p. B.M.W. engines.

The main strength of the fighter arm of the U.S. Navy in 1920 was the eight Nieuport 28s acquired in 1919 which constituted Combat Squadron Three. This was the sole fighter squadron and their aircraft were the only fighters to be fully armed.

Tests concluded at the Navy Laboratory on March 15th, 1921, proved the high strength of the chrome-vanadium steel alloy developed there and further tests proved it satisfactory for aircraft. Studies were also conducted on aluminium and aluminium alloys as structural parts as apart from exterior coverings. Salt water was apt to play havoc with aluminium, and even chrome-vanadium; constant research and development in this sphere led later to chrome-molybdenum steels which were much more suited to naval requirements.

The Army, now fairly well supplied with fighters, was in a position to dispose of some of its older types. When the Thomas Morse MB-3As became available in 1922 the Army discarded some of the earlier MB-3s with which the Marines saw their opportunity to organise a fighter unit. Together with eleven MB-3s they acquired some ex-Army DH-4s and Curtiss JN-4s which served as interim equipment until more suitable aircraft could be obtained. At least they kept units up to their full complement. Thus another fighter type, although unsuited to naval needs, went on the fighter type list. The MB-3s with their original Hispano-Suiza engines and armament were in a very poor condition when the Marines received them, and although reconditioned they were left in their Army olive drab. However, to distinguish them from the standard Army finish, the engine cowling was painted grey and the U.S.M.C. insignia was added to the fuselage side, while on the rudder the BuAer Nos. A-6060-6070 were marked.

In July 1920 the mainstay of the Navy's fighters were the twelve Nieuport 28s that had been acquired in mid-1919. Some, such as the example shown, were later used as racers. In the background can be seen S.E.5As and a Standard-built H.P. O/400.

First of the American fighters to match European contemporaries—the Thomas Morse MB-3. Eleven of these biplanes were transferred from the Army to the Marine Corps in 1922. They were used by 'F' Flight of the Third Air Squadron at Quantico.

The Marines took over the MB-3s from the Army at Gerstner Field, Louisiana. They went first to Norfolk and then to Quantico where they were used to form 'F' Flight of the Third Air Squadron. This was the first true fighter unit of the Marine Corps and it gave the Marines the opportunity to become proficient in fighter tactics and it provided the nucleus for expansion later. During manoeuvres in rough weather General Rogers flying A-6062 was forced to make his first parachute jump. In a summing up of the Thomas Morse MB-3, they were said to be 'fast, tricky and as tiring as hell to fly'.

But at last the Marines had a true fighter; they also acquired by devious means, a single example of another Thomas Morse type, the MB-7. Only two were ever produced. The first, originally produced for the Army, was entered in the 1921 National Air Races at Omaha. During trials it showed a top speed of 180 m.p.h., but a fuel pump failure caused a forced landing which resulted in slight injury to the pilot and serious damage to the machine. A bystander negligently dropped a match on the remains and fire completed the destruction of the aircraft.

The Army retained interest in the design, which was the first monoplane built by the Thomas Morse concern, and purchased the only other example of the type constructed for possible use as a racer. It was entered in the 1922 Pulitzer Race in Detroit, but it proved unsuccessful and the Army offered it that same year to the Navy who in turn passed it on to the Marine Corps. There being no racing category in the Naval services, it was listed as a fighter, in spite of the fact that it was never armed and was used mainly for pilots to gain flying hours. Two years later, in 1924, it was still serving as a general utility, test-bed and proficiency trainer aircraft.

It was a trainer that provided the Navy with its first carrier borne fighter; this was a Vought, a name that was already familiar to the Navy. Vought had been building aircraft since 1917. Maj. Howard Wehrle, late of the Air Branch, U.S. Signal Corps, remembers him as a young man in 1917 bending intently over a drawing board at work. To the Major he did not then appear to have any special talent, but he was busy on the seventh, and the most successful of his designs to date, as the Major could later testify as he flew the VE-7.

The VE-7 appeared to be a refined version of the popular and widely used Curtiss JN4 'Jenny' and proved to be reliable and manoeuvrable. The Army had seen it as a successor to the Jenny during the war and ordered over 1,500 but this order was drastically cut and only twenty went to the Army. The Navy saw possibilities in the VE-7 and acquired A-5661 in May 1920 and ferried it from Mitchel Field, N.Y., to the Naval Air Station at Anacostia. This was the first of some 128 VE-7s procured from the Chance Vought Aeroplane Company and several more were built by the Naval Aircraft Factory.

Designed around the Wright-Martin Hispano engine, its performance was comparable to that of its contemporary fighters and although it had been obtained as a trainer it became apparent that it had capabilities in scouting and observation roles. So versatile was the little craft that orders mounted beyond the capacity of the Lewis and Vought Corporation on Long Island. After negotiations the Naval Aircraft Factory turned to VE-7 production and while Lewis & Vought built the first 16 as VE-7 trainers followed by three VE-7S, 'scouting' versions, the N.A.F. turned out ten trainers, and produced A-5691 as a VE-7GF, 'ground fighter' possibly with an eye on future use of such an aircraft by the Marine Corps in close support of ground troops. VE-7SF and VE-7F versions followed.

The VE-7SF—scout-fighter was no more than a reclassified VE-7. It had the same appearance and capability as the trainer, except provision for forward-firing machine-guns. Basically it was for a scouting or observation role, but could be converted or pressed into use as a fighter should the need arise. The VE-7F fighter on the other hand, was officially listed as a ship-board fighter; in this case the front cockpit was covered over with a sheet of flat aluminium. With a boosted 190 h.p. Wright-Martin Hispano E-2 engine it had a top speed of 121 m.p.h. and could climb to 10,000 feet in just under twelve minutes.

Through all the VE-7 series the basic configuration and overall dimensions was unchanged. For new fighter pilots no conversion training was necessary from trainer to fighter. The Navy had thus effected great economies, by maintaining the cadre of a fighting force, in which tactical training could be fulfilled at the expense of fighters with a performance far below that of other nations. A policy justified at the time. In mid-1922 total fighter strength of the U.S. Navy was: MB-3, eleven; MB-7, one; Fokker D.VII, two; Fokker C-1, two; TS-1, two; Curtiss 18-T, two.

The appearance of the VE-7F heralded the use of standard colouring for Navy/Marine fighter aircraft. These were overall silver with the top surface of the upper wing and tailplane painted high visibility chrome-yellow —sometimes referred to as orange-yellow.

The VE-7 series pioneered carrier operation to the Navy. With their reliability and low landing speed and equipped with arrester hooks they proved to the Fleet the practicability of using the carrier as a floating airfield. It was in fact a VE-7SF that made the first take-off from an American carrier when Lt. V. C. Griffin rose from the U.S.S. *Langley* on October 17th, 1922.

However, the VE-7s had one major drawback—an in-line engine. This proved a constant source of trouble relative to the earlier foreign purchases, the Hanriots, Nieuports and Camels with their air-cooled engines, dispensing with the need for radiators and the cumbersome equipment associated with liquid cooling. Early in 1919 the Lawrance Aero Engine Corporation of New York, following the trend from rotary to radial engines, had a promising nine-cylinder radial engine of 220 h.p. at 1,800 r.p.m. for the relatively light weight of 442 pounds. They received $100,000 from the Navy to perfect the design and supply examples. A fifty hour test was satisfactorily completed on January 7th, 1922, on a Model J-1, the forerunner of a series that the Navy came to adopt as standard.

At the N.A.F. work was put in hand to design a fleet fighter utilising the promising J-1 engine and this evolved as the TS-1, the first American fighter designed specifically for carrier operations. The first, A-6248 built under contract by Curtiss, was delivered to Anacostia on May 9th, 1922, and further trials were made with N.A.F. designed and built twin wooden floats. As a floatplane it was tried with both USA-27 and cambered N-15 wings. A unique feature was an auxiliary fuel tank of 10 gallons capacity fitted in the lower-wing centre section, which could be jettisoned.

After further trials with A-6248 it was decided to put the TS-1 in production in its landplane form and Curtiss was awarded a contract for twenty-two in late 1921 while the N.A.F. took on production of five. Service trials aboard the U.S.S. *Langley* were conducted by the Navy's first fighter squadron, recently re-designated VF-1.

The re-designation had come about by a re-organisation of squadrons for specific roles and Fighting Plane Squadrons 1 and 2 (VF-1 and VF-2) were formed from Combat Squadrons Nos. 4 and 3 respectively on June 17th, 1922. Both had an optimistic establishment of 18 single-seat fighters authorised for fighting duties with the battle fleet, to be based on battleships with a shore base for replacements and stores at Naval Air Station, San Diego. At the same time VF-3, a new squadron, was created for operation from U.S.S. *Langley*.

As far as actual strength went at the time, VF-1 had fourteen VE-7SFs in commission while VF-2, which as Combat Squadron Three has only recently belied its title with nothing more formidable than ' Jennies ' (five Curtiss JN-4Hs)—had eight VE-7SFs, one VE-7 and a VE-7GF. Later VF-2 achieved a strength of twelve VE-7SFs while VF-1 changed over to the TS-1s.

Evaluating the J-1 engine (which was becoming known as the Wright J-1 following that Corporation's take-over of the Lawrance concern), VF-1 reported that it gave little trouble and was easily serviced and developed 200/220 h.p. at 1,800/2,200 r.p.m. respectively. They found that their TS-1s had a service ceiling of 16,250 feet and could climb to 5,400 feet in five minutes. VF-1 also operated their twin-float TS-1s from battleships in accordance with their planned duties.

The TS-1 could be said to be the Navy's first practical fleet fighter after a quest extending over several years, but a new phase was coming into fighter design and the TS-1 was to play an important part. The N.A.F. followed up the production of the five TS-1s with two further versions, the TS-2 and TS-3 featuring the 210 h.p. Aeromarine U-8-D and 180 h.p. Wright E-2 engines respectively. The return to water-cooled engines may appear a retrograde step but the clue to the policy was in the only other major modification from the basic TS-1 design—the substitution of the original USA-27 airfoil by R.A.F. 15. The TS-2 and TS-3 were being groomed for air racing and this was designed to play its part in naval fighter development.

An adapted fighter, described as a refined version of the Curtiss JN-4 Jenny, was the Vought VE-7 convertable single-seat/two-seat biplane. The example shown here is a VE-7SF model attached to the U.S.S. Langley (CV-1) and carries the appropriate diagonal red, white and blue fuselage stripes.

CHAPTER THREE

A Zest for Speed

A later view of the first landplane fighter delivered to the U.S. Navy (see photograph on page 13). With the Scarff mounting removed and the rear cockpit covered, this machine, A-3325, was used extensively for racing.

About this time, when aviation meets, airshows and races were coming into prominence, the air services too seemed to get racing fever. The Curtiss Marine Trophy Race for seaplanes was of special interest to the Navy and this was held as a regular event of the National Air Races until Glenn H. Curtiss died in 1930. The Pulitzer Races and the Schneider Cup Trophy also attracted both the Navy and Marine Corps. In all these events, there was nothing in the regulations to preclude participation by any of the Services.

Both the U.S. Army and Navy were quick to take advantage of this opportunity. It provided an outlet for healthy inter-services rivalry, and as always, with a target in view, both manufacturers and purchasers were anxious to produce the very best in design and performance, and new methods and equipment frequently made their first appearance on military racing aircraft.

Early in 1921 the Navy had made a contract with Curtiss to build two racers for the Pulitzer Races that year and although the design was not finalised until June, both A-6080 and A-6081 were delivered in August. The Navy actually withdrew from the race but A-6081 won at an average speed of 176·7 m.p.h. as a Curtiss entry with Bert Acosta at the controls. Next year skin type radiators added 10 m.p.h. to their speed and with other modifications to the cooling system they became CR-2s in succession to the original CR-1 (Curtiss Racer One).

Aircraft were specially built by both the Army and Navy and competition was keen. To allay the suspicions of tax-payers the object of the races was often quoted for the advancement of aviation or to keep aviation alive and to stimulate new and better designs. This was largely true and it did provide testing facilities for new designs, particularly with fighters where speed was a major factor. Far from being dare-devil circus stunts they were test-beds for future fighter design.

The Navy put many of its racing aircraft under the category of fighters, although it is doubtful that there was ever any intention of using them as such in service. Undoubtedly a wealth of technical knowledge could be gained from racing aircraft and in retrospect the British attributed their success in the Battle of Britain partly to an aircraft design with a lineage stretching back to Schneider Cup Trophy participation. At the time there was no evidence to support this view and since expenditure on a seemingly sporting project might well receive investigation it was a wise move to classify the aircraft as fighters.

For racing some of the TS series were re-designated TR. While the TR-2 and TR-3 did not achieve distinction, one TR-1 floatplane (A-6303 originally a Curtiss-built TS-1) won the 1922 Curtiss Marine Trophy Race in the hands of Lt. A. W. Gorton at Detroit on October 8th. Using a standard J-1 engine it achieved an average speed of 112·6 m.p.h. over the 160 mile course and the three take-offs and landings stipulated in the race regulations were without any untoward incident.

One TR-3, A-6447, was modified as a 'speed special' seaplane racer and became the TR-3A. The wing area was reduced, and a Wright E-4 water-cooled engine with wing radiators was installed. It was scheduled as entrant No. 3 in the 1923 Schneider Trophy Race in England but it was grounded with mechanical trouble.

It was that year's Schneider Trophy event that brought international fame to Curtiss Racers. Since it was a marine event the two Curtisses re-designated yet again as CR-3s, were fitted with floats and the additional weight was

The first U.S. Navy fighter to go into squadron service, the Curtiss TS-1. This fighter also marked the change from in-line to radial engines. The example shown is of Sqd. VF-1 and it bears a red tail indicative of the U.S.N.'s first carrier, the Langley.

compensated in performance by a Curtiss D-12 engine of increased power, giving 450 h.p. at 2,300 r.p.m., driving a Curtiss Reed metal propeller. Fuel tanks were placed in the floats, an innovation to be adopted later by the British Supermarine racers. But in this event, the Supermarine Sea Lion III designed by R. J. Mitchell who, years later designed the Spitfire, was completely outclassed. On its home waters on September 22nd the Supermarine came in third at 157·16 m.p.h. while the CR-3 A-6081 piloted by Lt. David Rittenhouse was declared the winner at 177·38 m.p.h. and A-6080 in the hands of Lt. Rutledge Irvine came in second. *Flight*, the British aeronautical magazine, stated that the Americans deserved to win as they considered that the Curtiss Racers were 'extraordinary fine pieces of design' and that they were handled with consummate skill.

Still flying in 1923 were the veteran racers, indeed the Navy's first fighters, the 18-Ts. Both A-3325 and A-3326 had been Navy entrants in the 1920 Pulitzer Race and both had retired with engine trouble. Now, two years later both were entered as float-planes for the Curtiss Marine Trophy and trouble with the engines again spoilt their chances. Lt. L. H. Sanderson, U.S.M.C., assigned to A-3326 (Entrant No. 4) did not manage to start and Lt. Rutledge Irvine had to pull out with an overheated engine after completing seven of the eight twenty-eight mile laps.

The Curtiss 18-Ts made a final showing on October 3rd at the 1923 Air Races. A-3325, then conspicuously painted with a deep orange fuselage and cream wings, was taken up for a trial run prior to the main event. Dipping low from the spectators' view, it disappeared behind some trees and crashed. It was a complete wreck but fortunately the pilot escaped with nothing worse than cuts and bruises. A-3326, the other 18-T, was then grounded and later scrapped. Mechanics, who had dubbed them 'Whistling Rufuses', were not sorry to see them go as they had been difficult to keep in rig.

Curtiss undoubtedly led the field in racing and marine aircraft at this time and while the veteran 18-Ts were still in the running, new types, in conjunction with both the Army and Navy, were being developed. Other racing types ostensibly, and perhaps optimistically, classed as fighters were the BR-1 and BR-2 Booth 'Beeline Racers' and Wright NW-1 and NW-2. The latter were known as the Navy-Wrights and one of these, A-6544, might well have participated in the 1922 Schneider Trophy, had not a mishap occurred during a practice in English waters.

Two new Wright Racers acquired in 1923 came under

The success of the Curtiss Racers was largely due to the small frontal area presented by the 400 h.p. Curtiss D-12 engines. Drag was further reduced by fitting wing skin radiators in which form this racer became the Curtiss CR-2.

the new naval aircraft designations which were standardised by an edict from the Bureau of Aeronautics dated August 29th, 1922, whereby in a letter/figure/letter combination the initial letter would signify the design number. However, on March 10th, 1923, the order of the letters was reversed and the system basically the same as in use today, was introduced. The new Wrights as the F2W had a self-explanatory designation as 2nd Fighter design by Wright.

In spite of the 'F' of their designation, these Wright F2Ws were used purely as racers. The first, A-6743, was originally entered in the 1923 St. Louis Races as a landplane, but subsequently became Entrant No. 8 as a twin floatplane in the Pulitzer Races that same year. Piloted by Lt. L. H. Sanderson, it won third place at 230·6 m.p.h. However, after the race, coming into land, it ran out of fuel, stalled and crashed. Although completely wrecked Sanderson came out unhurt.

The second F2W, A-6744, was also entered in the 1923 Pulitzer Races as a floatplane. As Entrant No. 4 it averaged 229·99 m.p.h. with Lt. S. W. Callaway at the controls. This aircraft was later considerably modified with new wings and re-engined to become the F2W-2, the original version having been the F2W-1. This illustrates the use of the final hyphenated figure in U.S. naval aircraft designations as model numbers, corresponding to the mark numbers of British aircraft.

At the other end of the scale from highest speed, was the lowest speed at which an aircraft could remain airborne —that is the stalling speed. As a general rule, the higher the design speed the higher is the stalling speed and consequently landing has to be effected at a higher speed. With the Army the disadvantage of high speed landing in reducing the safety factor and necessitating longer landing runs and therefore larger airfields, was acceptable if a considerably greater operational speed was to be obtained, but for carrier operation, high speed landings presented grave hazards.

Since the war surplus acquisitions, the U.S. Navy had been reluctant to purchase outside America, but tests in England of the Handley Page slotted-wing with leading edge slats and slotted, drooping ailerons, showed great improvement in stall prevention, slow landing speeds, shorter take-off runs and improved lateral control at slow speeds.

To evaluate this innovation, the U.S. Navy had ordered three shipboard fighters from Handley Page in 1922 incorporating the new slotted wing. This resulted in the H.P.21 which was known to the U.S. Navy as the S.B.24 or HPS-1 (Handley Page Shipboard—One).

An ingenious low-wing monoplane of clean design, its fuselage was plywood covered, giving an exceptionally smooth surface. The open cockpit was well forward affording very good visibility. The wheeled landing gear for carrier operation was rubber shock absorbing, much like a Fokker design, and it had added scissor-type rear brace suspension. The Gwynne-built B.R.2 engine was fully enclosed in a streamlined aluminium cowl and the control lines on a push rod system were all neatly faired in. Armament was not supplied but fittings and sundry items were incorporated to accept two ·300 Marlin guns; the occasion for fitting them, however, did not arise.

In order to study low speed landing devices, the U.S. Navy purchased this Handley Page H.P.21. As the HPS-1, it was assigned BuAer. No. 6402, but this number was apparently not applied. Full trials were not made and the project was dropped.

The first of the HPS-1s ordered was built at Cricklewood in 1923 and shipped to America for trials. With a top speed of 145 m.p.h. it could, by utilising its wing devices as flaps, land at the astounding low speed of 44 m.p.h. U.S. Navy trials proved it quite satisfactory, but no carrier landings were attempted and for some inexplicable reason the remaining two were cancelled and the HPS-1 was scrapped. Not only that but the project of evaluating the slotted wing was shelved.

It might be expected that somewhat greater interest would have been shown in this design by the air services in the country of its origin but this was not so. While Britain did adopt the slotted wing for many R.A.F. aircraft no H.P.21 orders were received by Handley Page for either R.A.F. or Royal Navy use; thus this interesting design fared no better at home than abroad and was soon forgotten by all concerned.

To return to the racers, Curtiss Racers had been hitting the headlines since 1921 and in November 1923, Lt. Williams in a Curtiss R3C-1, wrested the World's Air Speed Record back from France. However, these racers did not affect fighter development to the degree anticipated. It was quite evident that they would not be sufficiently manoeuvrable for fleet use and their in-line engines had already proved troublesome. Looking ahead in this respect, the Navy had their Aeronautical Engine Testing Laboratory at the Washington Navy Yard moved to the Naval Aircraft Factory, thus allowing experimental work on airframes, and engines to be co-ordinated at one location under unified control. Since the general trend, even before the move in January 1924, was for a metal airframe mated to a radial engine, the famous CR (Curtiss Racer) series did not find favour as potential fighters.

It followed that the success of the TS fighters, led to the Navy approaching Charles Ward Hall again, this time for the design of an all-metal version of the TS to the N.A.F.'s basic design, using the J-1 engine for power. Detail specifications were set out in April 1923 and two machines evolved—A-6689 for flight and fleet service evaluation and A-6690 for static test and construction detail examination.

Official trials at Curtiss Field, Mineola, N.Y., commenced on September 4th, 1924, with Lt. Cuddihy, U.S.N. A full load was carried, except for ammunition, but the

made; no drastic changes were suggested, but one item was the workmanship and finish which was not considered first class. All in all it was considered satisfactory as an experimental project but not suitable for production as a fighter type.

The F4C-1 was a subject much discussed at a time when metal construction was coming in favour by both the military authorities and commercial organisations. A comparison between the original TS and its metal counterpart, the F4C, showed the latter to be stronger and lighter and therefore faster and more manoeuvrable. The fuselage of the F4C weighed only 164·53 pounds to the TS's 358·25 pounds and the wings were correspondingly lighter. Only the Curtiss Reed metal propeller of the F4C weighed more, actually 27 pounds more, than the propeller of the TS. Two forward-firing ·300, 1917 Pattern, Marlin guns were fitted with provision for 250 rounds per gun, and trials with this armament proved satisfactory. With a capacity of 50 gallons of fuel and 4·5 gallons of oil it had a cruising range of 525 miles or 350 miles at 125 m.p.h. with full combat load.

The F4C, the first of the Curtiss fighters to be desig-

The Navy's zest for speed led to several racers being ordered that were optimistically classified as fighters. This Wright NW-1, known as the Navy-Wright, existed first in sesquiplane form with Lamblin radiators as shown here. Later it was modified to a biplane floatplane.

1,656 pounds loaded weight did include a pigeon and its box! Our feathered friends had proved useful for stranded airmen during the 1914-1918 War and were still carried as a safety measure. The F4C-1, as the metal TS machine was known under the new system of designation, achieved a top speed of 130-132 m.p.h. on its second test.

Later that same day Lt. Oftsie took over and ascended in A-6689 for a rate of climb test; this, its third test, was documented as Test 'C'. To reach 10,000 feet the F4C-1 took 8·35 minutes and the ceiling was recorded at 18,400 feet. For Test G, manoeuvrability, also tested that day, the two pilots reported that it handled well in both rough and smooth air, but was slightly nose heavy at full throttle; stability was stated to be excellent and control good.

In spite of this largely favourable report on the little F4C, twenty-two recommendations for improvement were

nated under the new system, marked a turning point in development. It's constructional features set a new style that had one great drawback—its cost precluded the possibility of production.

Naval funds were not even sufficient to carry the development of the F4C further and after proving that it had all the characteristics of a successful naval fighter, it was dropped. This was not so unreasonable as these bare words may sound. Expenditure on the Services must, in wise statesmanship, be linked to the political situation and America, pursuing an isolationist policy, sequestered by two giant oceans, saw little need to put dollars into metal airframes. The single F4C-1 built to fly, was however utilised to the full; it went to VF-1 for service trials and for most of 1926 they operated it on a test basis alongside their seventeen TS-1 aircraft. It apparently

The Navy-Wright monoplane as shown on page 30 modified to biplane floatplane form, photographed at the Naval Aircraft Factory on June 8th, 1923. It was sent to England for the Schneider Trophy but a mishap during a test flight precluded it from participating.

remained with the squadron until mid-1927 when they were re-equipped.

Another study in metal construction reached the Navy in a round-about fashion, and was evaluated under conditions which in present-day parlance would be called security. This was the Wright-Dornier Falke parasol monoplane which owed much of its origin to the Dornier D.1 design already familiar to the Navy. It was truly all-metal with metal covering as well as structure and had been built in Switzerland because of the restrictions on aircraft construction in Germany after the war. Imported as a private venture by the Wright Aeronautical Corporation, a 300 h.p. Wright-Martin H-3 engine was installed. It was hopefully loaned for both Army and Navy evaluation.

For some reason the Navy designated it WP-1 suggesting Wright-Patrol type, but it was tested as a VF class aircraft—perhaps an Army designation as Wright Pursuit was intended. Its one great attribute at a time when there was a zest for speed was its maximum speed of 162 m.p.h., but it lacked manoeuvrability and gave a poor sphere of visibility to the pilot; thus it proved unsuited to naval needs as it failed in precisely the requirements necessary for a good carrier aircraft.

Perhaps it was too early to expect metal construction to catch on and it was not until Anthony Fokker started building in the United States that metal construction was adopted throughout America. For the time being the Navy turned to a proven type to meet their next requirement—a fighter type suitable for catapult launching.

Up to this time aircraft allotted to catapults on ships of the fleet, were on more of an experimental basis than service status. In any case priority was given to observation, scouting and patrol types; but in 1923 the turn of the fighter came—perhaps partly came would be more correct

An improved Navy-Wright, the F2W-2 pictured on October 10th, 1924. This was one of the first naval aircraft to come under the present system of U.S. naval aircraft nomenclature, although the 'F' classification to racing aircraft was something of a misnomer.

The U.S. Navy won the Schneider Trophy Race for 1923, when Lt. David Rittenhouse was declared the winner at 177·38 m.p.h. The aircraft, depicted here, was a CR-2 modified to CR-3 standard as a floatplane. Wing skin radiators were used and the floats were utilised to carry fuel.

for the requirement, somewhat half-heartedly as far as the fighter concept went, was for a two-seat fighter/observation type that could serve as interim equipment aboard battleships and cruisers.

The TS-1s with their twin-pontoon floats were not considered suitable for catapult launching but the proven Vought VE-7 series showed more promise and one specially modified and stressed for this was designated the VE-9. Vought followed this up by a pure fighter project, with the qualities of the VE-9, and eighteen were ordered as the UF-1 (Vought Fighter One) under the old nomenclature. (This had no direct connection with the Vought FU-1 fighter that appeared a few years later.) Delay in delivery was occasioned by a changeover before completion from an Aeromarine U-873 to a Lawrence J-1 engine.

As fighters the UF-1s came and went in quick succession. Although reliable and standing up well to operations from land, water or carrier, as well as catapult launchings, speed among other factors rendered them unsuited to a VF role and the UF-1s were redesignated UO-1 as observation aircraft. Their value in this role is evinced by the fact that apart from the eighteen original conversions, over 400 UO-1s were eventually procured by the Navy.

Naval aviation in 1923 had all the prestige that racing aircraft could bestow, by not only holding their own in competition with the Army, but they had achieved a World Air Speed record and resounding success in the premier international event—the Schneider Trophy Cup which would run until the Cup itself was won, as M. Jacques Schneider had decreed, by the country who won the event three times in succession. Turning from prestige to practice, a much different picture is presented. In the Fleet exercises that same year single aircraft had to rate as squadrons and when ' Black Fleet ' ' attacked ' the Panama Canal by catapulting aircraft from two warships representing enemies, there were no fighters from the defending ' Blue Fleet ' to intercept. Racing aircraft and a wealth of detail on their capabilities hit the headlines but the fleet exercises were barely mentioned and those details were not for publication. Racers provided a good façade.

The influence of Charles Ward Hall in designing an all-metal version of the Navy's first fighter, the Curtiss TS-1, resulted in this F4C-1 biplane being generally known as the Curtiss-Hall. The model shown is the first of the two built.

CHAPTER FOUR

Framing a Fighting Force

A policy for the role of naval aviation as a whole, let alone a policy for fighters, was formulated in the mid-'twenties. Board followed board and report followed report and the greatest stimulus to fighter development was the final recommendation in 1926 that adequate funds be made available.

The status of naval aviation was clearly set out in General Order No. 132 of July 7th, 1924, after the Army and Navy authorities had mutually agreed upon an air policy. The basic concept for the Navy stated that aircraft would operate from mobile floating bases or from naval stations on shore in co-operation with the fleet —as an arm of the fleet; for overseas scouting; against enemy establishments on shore when such operations are conducted in co-operation with other types of Naval forces, or alone when their mission is primarily naval; to protect coastal sea communications by reconnaissance and patrol, convoy, attacks on enemy submarines, *aircraft*, or surface vessels; in co-operation with the Army against enemy vessels engaged in attacks on the coast.

redundant and it was decided to convert these to meet the requirement.

The new carriers were several years in the building, but construction received a stimulus after the Fleet Exercises of 1925 when the *Langley* participated for the first time in the U.S. Navy's annual war game and much impressed the Commander-in-Chief, Admiral R. E. Coontz.

The year the new carriers *Saratoga* (CV-3) and *Lexington* (CV-2) were launched, 1925, brought changes in the acquisition of aircraft following the 'Lampert Report'. This was the report to Congress of a special committee under Representative Florian Lampert of Wisconsin to make 'inquiry into the operation of the United States Air Services'. Not only the Army and Navy, but the state of the aeronautical industry was carefully examined. Hitherto, under ancient laws the Navy had been bound to accept the lowest tender and this was now changed. A series of grievances were aired as Board followed Board in these formative years of the mid-'twenties. It was recommended that the Army and Navy should survey and condemn as unneces-

The first of the Navy's famous Boeing fighters, the FB-1. This series paralleled the Army's PW-9s. These machines were not fitted for carrier operation and were handed over to the Marine Corps. The machine depicted is of Marine Fighter Squadron One. The aircraft is Bu. No. A-6893 which was the last production FB-1. In the background is A-6888, the Squadron Commander's aircraft of the same unit.

Marine Corps Aviation received a little different treatment. The functions normally assigned to Army aircraft would be performed by Marine aircraft when the operations were in connection with an advance base in which operations of the Army were not represented and under certain conditions, in the national interest for military security, it could be placed under Army jurisdiction.

With land based aircraft there was no limit, except funds for a fighter complement, but with the fleet, fighters were strictly limited. Floatplane fighters were limited in performance and could only take-off from calm waters unless catapulted, and since it was not normal for more than two aircraft to be carried aboard, with patrol and observation a prime necessity, few fighters operated at sea except from the single carrier in commission. However the U.S.S. *Langley* was to be supplemented by two carriers that had been planned in 1920. Later, under the Washington Disarmament Treaties, two battle cruiser hulls were declared

sary old equipment, and embark upon a new five-year programme for which not less than $10,000,000 should be spent annually on aviation by each of the two services. With some reservations and amendments this was accepted and whereas in the early 'twenties fighters had been dealt with mainly in ones and twos, in the late 'twenties they appeared in their tens and twenties. The significant point here, is that the expansion and efficiency of any service is not due in the main to the dedication of its officers, or to the ingenuity of the supporting industry, but to the willingness of their government to vote sufficient funds.

An indication of the trend of development can be gained by flicking over the pages of drawings and observing the change from in-line to radial engines 1918-1939 and the switch from biplane to monoplane configuration. But one important factor remained unchanged—armament. Perhaps this was the most important factor involved in development, for primarily the fighter is a flying gun platform.

One of three FB-1s fitted as convertible floatplane/landplane. In this form it was known as the FB-3, but the machine depicted reverted to FB-1 standard in 1928. The inscription by the cockpit relates to crew, fuel, armament and equipment weights.

Both in the Army and Navy, fighter armament had been standardised at two forward-firing machine guns synchronised to fire through the propeller arc. Normally provision was made for the installation of one ·300 and one ·50 machine gun. In this respect America led the world as ·50 was a larger calibre than in general use elsewhere, but while two ·50 guns would have given an even greater firepower, their weight* was a prime consideration and hence the mixed armament was a compromise. Events were to prove this armament inadequate, but not until 1937, after the firepower of the naval fighters had remained unchanged for nearly twenty years, were any real steps taken to remedy this failing. Perhaps the fact that aircraft armament seemed to be a matter poised somewhere between the spheres of the Bureau of Aeronautics and Bureau of Ordnance, was the reason that it received scant attention.

A new weapon would have involved millions of dollars to make it effective by the manufacture, proofing and deployment of ammunition in adequate quantities to meet contingencies. It can therefore be appreciated that while the large stocks of ·300 ammunition remained from the war and with ·50 calibre ammunition being currently produced, there was a reluctance to effect costly changes.

* $2 \times ·300$ m.gs. = 48 lb.; $2 \times ·50$ m.gs. = 124 lb.

Apart from armament, the mid-'twenties were a turning-point. From the out-dated TS-1s and converted VE-7s the Navy turned to Curtiss Hawk and Boeing fighters, already being developed for the Army. These were equal to the fighters of any other nation.

As related in some detail in our companion work, *United States Army and Air Force Fighters 1916-1961*, there was great rivalry between the Curtiss and Boeing firm with their respective designs under the Army designations PW-8 and PW-9. Even before the fillip of increased funds, the Bureau of Aeronautics had been watching the Army's evaluation and its chief, Rear Admiral W. A. Moffett, reported to the Secretary of the Navy on November 20th, 1924, that he was interested in getting better fighters for the Navy and had been watching the Army's Curtiss PW-8A/Boeing PW-9 comparative tests. Naval pilots had been permitted to fly both the aircraft by the Army as a matter of courtesy—and possibly as a matter of good business for if the Navy were to order too, the unit price per aircraft might be less. The Secretary of War astutely communicated with the Secretary of the Navy to the effect that any order be jointly made into a single contract to effect such savings.

The Navy favoured the Boeing. Lieutenant R. A. Oftsie after testing both came to the conclusion that the

The one and only Boeing FB-4, a radial-engined version of the FB-1. This particular airframe started life as an FB-1 and was later modified to become the FB-6. All the FB series had metal fuselages with wooden wings, both being fabric covered.

Boeing had a slight edge over the Curtiss on controllability, manoeuvrability and maintenance, a definite superiority in vision, but slightly inferior qualities in take-off and landing. Under the points system used in assessing types, it was 410 : 365 in favour of the Boeing. Thus, the Navy ordered fourteen as the FB-1 (Boeing Fighter Type One).

It has been generally thought that the FB-1 to FB-4 series developed successively after delivery of the first of the series, but when the first order for fourteen was placed, it was envisaged that ten would be standard service versions, two equipped for carrier landings aboard the *Langley* (FB-2) and two for engine development with the 510 h.p. Packard IA-1500 (FB-3) and 450 h.p. Wright P-1 (FB-4).

sions to 3 feet 3 inches and by a balanced rudder similar to its PW-9D counterpart with the Army.

In general the FB series fuselages were built mainly of welded steel tubing, while the wing had a wooden framework; both were fabric covered except for aluminium sheeting for cowlings forward and fore part of the fuselage decking. They were the first naval fighters to employ oleo shock absorbing landing gears of Boeing design. Armament was to the standard mentioned previously.

Outwardly the FB series matched the Army's PW-9 and PW-9D models, but inwardly they differed by a cockpit designed to meet naval requirements and certain special anti-corrosive materials. In spite of the addition of naval equip-

Production version of the FB-3 and the most widely used of the FB series, the FB-5, which had a revised cowling and tail. This was in fact the first true carrier-borne fighter to be built by Boeing. The first was delivered early in 1927.

The two FB-2s were nothing more than FB-1s delivered by Boeing with fittings for a carrier hook, hoisting lugs and an improved undercarriage. They reached VF-2 on December 25th, 1925, for trials aboard the U.S.S. *Langley* which proved them suitable for fleet use, apart from minor modifications recommended by VF-2 which were incorporated on the following production batch—the FB-5s.

Apart from the two engine development projects the standard power unit was a version of the famous Curtiss D-12 series, and so the Navy reverted to liquid-cooled engines for a period. Tests with the projected alternative in-line Packard (FB-3) were held up as A-6897 crashed before delivery. Two more aircraft were ordered and fitted with Packards, but these later reverted to standard FB-1s.

Tests with a radial engine, at first proved no more promising. The last of the FB-1s delivered, A-6896, had a Wright P-1 radial installed which, being air-cooled, dispensed with the heavy cooling system of the Curtiss engine and saved all round some 200 lb. Its gross weight was in fact 2,817 lb. as a landplane and 3,237 lb. as a floatplane. On the other hand, the large frontal area presented by the engine reduced the aircraft's speed by some 7 m.p.h. and to a smaller degree adversely affected ceiling and range. With the Wright engine installed, A-6896 was designated the FB-4.

The first major production variant was the FB-5 of which twenty-seven were delivered in the first half of 1927. This version differed in configuration, by having the stagger of the lower wing set from the 8 $\frac{13}{16}$ inches of the earlier ver-

ment, their performance with full load surpassed their Army sister ships. The FB-5 topped 168·6 m.p.h. and as a float-plane this top speed was reduced by only 3 m.p.h.; all of the FB-3 to 6 models were convertible as twin-float seaplanes.

The first of the series, the ten FB-1s, were all assigned to the Marine Corps for two reasons; firstly there was a need to bolster the newly-formed Marine Fighter Squadrons and secondly the aircraft were not rigged or strengthened for carrier or catapult operation—at that time the Marines did not operate from carriers. With a speed comparable to their contemporaries in Britain and France and their landing speed of 57·5 m.p.h. permitting operation from small airfields, they served the Marines well. They could climb to 5,000 feet in 2$\frac{1}{2}$ minutes and their service ceiling was 21,200 feet. An internal fuel capacity of 112 gallons gave them a cruising range of 500 miles at 110 m.p.h.

Six FB-1s went to VF-1M the Marines Fighting Plane Squadron One. This unit was part of the Marines Aviation Group at Quantico and on July 1st, 1925, Fighting Plane Squadron Two was added. Two months later with the formation of the Second Aviation Group at San Diego, Fighting Squadron Three was added. As with the Navy the short title form VF was used, but with an 'M' suffix to distinguish U.S.M.C. from U.S.N. squadrons. While the six FB-1s went to VF-1M, the remaining four were divided between VF-2M and VF-3M. These squadrons were re-designated on July 1st, 1927 when VF-1M, VF-2M and VF-3M became VF-8M, VF-9M and VF-10M respectively

First of the famous Curtiss Hawks in naval service. This aircraft was virtually the same as the Army's Curtiss P-1 Hawks and the particular example shown was the first of nine delivered to the U.S. Navy as Curtiss F6C-1s in 1925.

and the possibility of confusing them with the Navy squadrons was eliminated. Nine of the ten FB-1s went with VF-10M to China in September 1927 while the tenth stayed at San Diego with VO-1M. They were recalled to the States in mid-1928 and the final assignment of those remaining, was to VF-6M at San Diego. By mid-1930 all had been condemned and destroyed.

While the FB-1s went only to the Marine Corps, the FB-5, the main production version, went first to the U.S.N. and later to the U.S.M.C. They were considered bouncing, tricky machines to fly. With extra heavy members to accommodate a carrier hook they operated from the U.S.S. *Langley* and although convertible to twin-float seaplanes they were only used in that configuration to test their serviceability. Their internal fuel capacity was only 100 gallons which allowed a range of 325 miles, but an underbelly auxiliary tank could increase the range to 530 miles when occasion demanded. They could climb to 5,000 feet in 3·2 minutes, and had a service ceiling of 22,000 feet which was reduced to 17,800 feet as a seaplane.

The Boeings had been in competition with the Curtiss Hawk series for both Army and Navy use. In early 1925 the Navy decided to order nine Curtiss aircraft identical to the Army's P-1s except for cockpit layout and the finish of silver, light grey and yellow to naval specifications. It was the first Curtiss fighter to be ordered in quantity by the Navy. Strangely, it was designated the F6C-1; there had been an F4C-1 following the TS-1 to 3 series but no F5C-1. It has been represented that F5C-1 as a designation was omitted to avoid confusion with the F5L flying boats then still in service. In any case they were often referred to as Hawks.

Soon after the first Hawks reached VF-2 in late September 1925 the Commanding Officer was deputed to

A Curtiss F6C-1 used by the commanding officer of Squadron VF-2 and tested in comparison with a Boeing FB-1. These aircraft were delivered with floatplane conversion kits but were rarely used as seaplanes. In 1927 a few machines of this model reached the U.S.M.C.

A production batch of Hawks, fitted for carrier operations, were designated F6C-3 in 1927. Some of these were passed to the United States Marine Corps and the example shown at Quantico in 1929 was used by Squadron VF-4M. The inscription by the cockpit reads PILOT—LT. W. O. BRICE.

compare one with an FB-1. His report, written early in 1926, stated that the FB-1 was more sensitive on the controls and slightly superior in manoeuvrability; he considered that the F6C-1 had poor upward and forward visibility. These were matters that had been thrashed out even before the Boeings or Curtisses were ordered for the Navy, but the Bureau of Aeronautics was anxious to procure both types to encourage development. As it was the Curtiss D-12 gave a top speed of 164 m.p.h. and allowed a cruising speed of 110 m.p.h. for a 355 mile range. Its ceiling was 21,700 feet and it could climb to 5,000 feet in three minutes.

Development of the new Curtiss series proceeded with that of the Boeings. The initial batch were not delivered as convertible to floatplanes, nor did they have carrier equipment, but four were modified as F6C-2s by strengthening the rear fuselage and modifying the undercarriage to a normal axle type to facilitate hooking up with transverse deck-lines on the U.S.S. *Langley*.

They proved very adaptable and more were ordered to incorporate the recommendations of VF-2. These were the F6C-3s which had a strengthened fuselage and a new landing gear with oleo shock absorbers and smaller radiator; all were delivered with floatplane conversion kits following special trials of an F6C-1 with twin-floats. It was intended that they should be capable of operating from catapults, but for this they proved unsuitable. Nevertheless they proved to be the most popular of the Hawk series although their top speed and rate of climb was slightly inferior to the initial batch. Their cruising range was in the order of 350 miles but an auxiliary tank could extend the range by 300

Curtiss F6C-3 biplanes, like most of the series, were convertible to floatplane form. This aircraft is shown with its Curtiss Marine Trophy number in May 1930, an event which it won at a speed of 164 m.p.h. This machine was, soon after, extensively modified as shown overleaf.

miles. They provided the initial equipment for the newly-formed VF-5, commissioned in January 1927 under Lt. Cdr. O. B. Hardison, which used them until 1929.

The F6C series made an important contribution to tactics—by demonstrating dive-bombing. This was not an innovation, for dive-bombing had been practised by aircraft in the 1914-1918 War, but the U.S. Navy and Marine Corps appear to be the first service to develop a dive-bombing technique. This first came into prominence in October 1926 during routine flight manoeuvres. The Pacific Fleet, coming in from San Pedro received a wireless warning that aircraft would make a mock attack on their ships; they were even advised of the time.

Presumably the Pacific Fleet expected a conventional 'fly-over' and were not prepared for what happened in effect. VF-2 took off from San Diego in the F6C-2 Hawks and climbed to 12,000 feet. Below they spotted the Fleet ploughing serenely on towards their base. After waiting until the time previously advised and then taking up position, Lt. Cdr. F. D. Wagner led his pilots down towards the Fleet in an almost vertical dive. In spite of the warning, complete surprise was achieved. So impressed was the Admiral, that he admitted that under the circumstances no defence was possible and the 'attack' had succeeded.

As well as tactically the F6C series achieved some fame as racers: a float-equipped F6C-1 piloted by Lt. T. P. Jeter won the 1926 Curtiss Marine Trophy held on May 14th, and a 'hotted up' F6C-3, A-7147, operated by the Marines with special floats, proved a winner in a literal sense. Moreover it was a consistent winner and Major Charles A. Lutz, U.S.M.C., Lt. W. C. Tomlinson, U.S.N., and Capt. Arthur H. Page, U.S.M.C., successively won the 1928, 1929 and 1930 Curtiss Marine Trophy Races in this machine.

After winning the 1930 race A-7147 was re-designed and entered for the Thompson Trophy Races the following September. It emerged as a parasol monoplane powered by a specially tuned supercharged 600 h.p. Curtiss Conqueror. With faired undercarriage, streamlined wheel spats and and overall high gloss dark blue finish, it was hard to realise that it was in fact the A-7147 airframe.

Prior to the race, Capt. Arthur Page, U.S.M.C., estimated its top speed in the region of 210 m.p.h., with the reasonable landing speed of 63 m.p.h. Its service ceiling was 23,400 feet and rate of climb 1,725 feet per minute. Piloting A-7147 in the actual event at Chicago, Capt. Page gained an early lead and was increasing that lead in the seventeenth lap of the twenty-lap course race, when he was overcome by carbon-monoxide fumes and crashed to his death.

Converted from the Curtiss F6C-3 biplane shown on page 35, this extensively modified airframe was re-designated XF6C-6 and became known as the Page Racer. A Curtiss Conqueror engine gave it a speed of well over 200 m.p.h. It crashed in the 1930 Thompson Trophy race.

Almost simultaneously, VF-5 were attacking the Atlantic Fleet. These two demonstrations of dive-bombing by fighters was to have a profound effect on future tactics and the Hawks have been linked with the origin of dive-bombing. Major Ross R. Rowell who commanded a dive-bombing unit of the Marines, had seen it practised by the U.S. Army at Kelly Field in 1923; the Army when asked about it stated that they had first seen it in Europe. In any case glide bombing, loosely called dive-bombing, had been practised by the 4th Marine Squadron in Haiti in 1919.

Perhaps the most significant fact is that at the Cleveland Air Races the Navy/Marines put on a dive-bombing display that profoundly impressed Ernst Udet, who later fostered the idea of the 'Stuka' and was appointed Inspector of Fighters and Dive Bombers to the *Luftwaffe* in 1936.

Races were still maintaining U.S. Naval prestige in the air, but no longer was it a facade as the framework of the build-up was evident. By April 1926 there were 393 trained pilots in the Navy and 59 in the Marine Corps. The combined fighter strength was 135. At sea the U.S.S. *Langley* had from early 1925 been assigned to the Fleet and was no longer regarded as experimental; while the *Lexington* and *Saratoga* were fitted out and would accommodate fighter squadrons. There were also plans for a further carrier for the 'thirties (U.S.S. *Ranger*—CV-4).

If many of the fighters were obsolete, the new Curtiss and Boeing types were the equal of any other fighter of the period and a new engine which was to become a common factor of both the later Boeing and Curtiss types was being ordered in quantity.

CHAPTER FIVE

'Will of the Wasp'

An interim fighter, the Vought FU-1, of which the majority of the twenty built went to Squadron VF-2 and the example shown bears the markings of that squadron. The single float, to facilitate the standard method of catapult launching, was typical of the period.

Of the various facets in the development of naval fighters in the United States during the late 'twenties, there is little doubt that the most significant was the engine that powered them and, for the first time, gave them parity with fighters of any other nation of the world. By name that engine was the Pratt & Whitney 'Wasp'.

The Wasp was conceived by George Mead and refined by Andrew Van Dean Willgoos in a frame garage belonging to the company of Pratt & Whitney, newly formed by Frederick Brant Rentschler, who had, until 1924, been president of the Wright Aeronautical Corporation.

Convinced that the radial engine showed more promise in the future, the Bureau of Aeronautics encouraged development of new fighters from both Boeing and Curtiss to be powered by the promising Wasp engine. The particular need was for single-seat fighters to operate as a landplane or floatplane, and in the latter form to have a central float with two stabilising floats supported from the wings to facilitate catapult launching. But by the time they evolved it became policy to carry only scouting and observation aircraft aboard warships, restricting fighters to land or carrier operation, where catapult launching was not required.

In the interim, a new fighter was quickly produced by adapting an observation type, the Vought UO-1. After modifying this new observation type to a single-seater the designation changed to UO-3 and in October 1926 again to FU-1 by which time twenty were under construction. Basically they were the same as the Vought series of observation aircraft, but with diving speeds up to 200 m.p.h. permitted and no restrictions on aerobatics. Construction was conventional with a fabric covered metal tube fuselage and wooden wings. Power was supplied by a radial engine, a Wright J-5, which, developing 220 h.p. at 1,800 r.p.m., was low-powered for the period; however it had a Root Model 3 supercharger to enhance its performance.

The bulk of FU-1s went to Squadron VF-2B (B suffix for Battle Force). Between October 1927 and June 1928 they operated from the twelve battleships of the Battle Fleet and as such were the last catapult fighters. VF-2B then became shore-based at San Diego and wheels replaced floats on their aircraft. Their poor performance led to their early relegation to a training role for a new lease of life.

The FU-1s had been particularly unpopular for carrier landings because of the poor visibility from the cockpit. This was remedied by adding another cockpit which was easily facilitated since the basic design was that of a two-seater. Dual controls were fitted and with this modification they became FU-2s.

As interim fighters until more powerful equipment was available and as fighter trainers they served until 1931 when they were re-assigned as training or utility aircraft. The West Coast fighter squadrons had one each allotted and three went to the utility Squadron VJ-1B. By then, after their inauspicious start, they had at least achieved a reputation for ease of handling and durability. The last one in service, A-7378, was written off in May 1932.

A contemporary of the first FU-1, and indeed in competition with it, was the Wright F3W-1 Apache designed for fleet use by catapult launching and to have a convertible wheel/float undercarriage. Of Wright design too, the engine was the 200 h.p. Wright Whirlwind which, reliable as it was, could not produce a performance to match current observation types, let alone fighters. Discussion between the firm and the naval authorities led to a change-over to the 325 h.p. Wright Simoon with an extremely low power/weight ratio, but this was still at the testing stage. Hastily installed in the Apache at the Wright company's field, it gave the machine a most promising performance, but owing to technical difficulties at this stage of its development, a production schedule could not be guaranteed for the near future.

Pratt and Whitney, however, with their Wasp, could offer the navy an engine of the same weight and practically the same fuel consumption, that could give 400 h.p. Thus,

The Pratt & Whitney Wasp engine was introduced on the Wright F3W-1 Apache, which, retrospectively, became the XF3W-1. The single example, A-7223, attained a world's altitude record, and was also converted to a single-float seaplane.

instead of becoming a prototype for a new Wright series, the F3W-1 became a test-bed for the Wasp, the first of which had been delivered to the Navy on Christmas Eve 1925.

The proving of the Wasp was a highlight in engine development. First proof came on the warm, humid morning of May 5th, 1926, after the F3W-1 had taken off from the staked-out flying field on a mud flat along the Anacostia River—the Anacostia Naval Air Station of 1926. To Lt. C. C. Champion, Jr., the Bureau of Aeronautics duty test pilot, fell the privilege of conducting the test and it evoked from him a report couched in glowing words of praise. Tests were continued, the F3W-1 Apache was purchased, and contracts were placed for the Wasp.

Exactly a year after the Wasp's convincing performance, the single Apache, rigged as a floatplane, with a Wasp having an experimental N.A.C.A. supercharger, achieved a world altitude record in its class of 33,455 feet. Converted back to a landplane, and still in the hands of Lt. Champion, it broke the world's altitude record on July 25th, 1927, by reaching 38,419 feet. Incidentally this record stood for two years, until May 8th, 1929, in fact, when the same machine reached 39,140 feet with Lt. Apollo Soucek at the controls. Not content with that, Soucek later lifted the altitude record for floatplanes and also his own landplane record by reaching 43,166 feet, still using the same Apache but with the Wasp supercharged to give 450 h.p.

Since it was obvious from the initial trials of the Wasp that it was a winner, both Curtiss and Boeing commenced designs around the new power plant. In conjunction with naval engineers, the Wasp was also adapted to existing designs and the single FB-4, A-6896 test-bed for the Wright P-1, was tried with a Wasp as the FB-6.

Curtiss adapting their proven Hawk design to the Wasp received an order for thirty-one, with purchase subject to its first meeting the performance predicted by the firm. The first example soon proved itself at Anacostia and the Navy gave the go-ahead for the balance. Produced as F6C-4s, they were basically the same as earlier Hawks except for the engine and Hamilton propeller, a re-designed wing structure and rear fuselage, that hardly affected outward appearance.

Rear Admiral Joseph M. Reeves had ordered a concentration of the Navy's battle squadrons at San Diego in October 1926 to explore new tactics and obtain a general exchange of views. At that time the F6C-4 Hawk with the first production Wasp engine installed was at Anacostia. Lt. Ralph Ofstie was assigned to fly it over to demonstrate it at San Diego to give the Navy a preview of their new equipment. After having the engine checked at the Hartford, Connecticut, plant of Pratt & Whitney, he set out on October 6th, flying the southern route with three overnight stops.

Not to be outdone, Boeing had their new fighter there, too. The new XF2B-1, A-7358, had just arrived from the company's Seattle, Washington, plant. Both the F6C-4 and XF2B-1 were pitted, in competition, against the service type fighters then in use. They completely outclassed the liquid-cooled Boeing FBs and F6C-1/3 Hawks. Admiral Reeves gave instructions that every pilot on the station could have fifteen minutes flying time in either the new Curtiss or Boeing. Most of these flights turned into mock combats between the liquid-cooled service types and the two air-cooled radial-engined newcomers. After three days of hectic flying, the Boeing had 100 engine and airframe hours and the utilisation of the Curtiss was about half that, yet both were in fine fettle with nothing more required than routine maintenance. The prototypes had passed what would in later years be described as intensive flying trials.

Following these gruelling exercises Lt. Ofstie took the original F6C-4 on to Seattle for a conference with the Boeing Airplane Company. Then, following the air mail route through Salt Lake City, he returned to Washington. Of this trip, covering some 7,000 transcontinental miles, Lt. Ofstie's flight report read: "New P. & W. engine proved to be remarkably satisfactory for this service from time of leaving Washington until return. The engine had about 75 flying hours and only work done was removal and inspection of spark plugs at San Diego. Nothing else was required beyond filling the fuel tanks and cranking up the engine." It endorsed the Navy's action in having already placed a production order.

Both the Marine Corps and Navy liked the Hawks. They were easy to fly, manoeuvrable and had a reasonable all-round performance. Aileron control was most efficient right down to the stall—an important factor in carrier operations. The landing speed of the F6C-4 was 57 m.p.h. It could climb to 10,000 feet in six minutes and had a service ceiling of 22,900 feet. In the Navy it served VF-2 with replacements as necessary, while the bulk went to Marine Corps squadrons VF-8M, VF-9M and VF-10M.

Following Army practice, the Navy commenced using the prefix 'X' to aircraft description to denote experimental models and this applied to F6C-4 A-7403 when it was used as a test-bed for further development as the XF6C-5. As such, it had a 525 h.p. Pratt & Whitney R-1690 Hornet engine installed which gave slightly increased range. During tests both the split axle type of undercarriage of the F6C-1 and straight axle type of the F6C-4 were used. Experiments were also conducted with spinners, which gave a slight increase in speed, but not sufficient to compensate for the maintenance difficulties they engendered.

Further modifications came to A-7403 as the XF6C-7 (XF6C-6 had been allotted to a special racing version). In this case the inverted air-cooled Ranger V-770 engine had the advantage of the low drag presented by the small frontal area of an in-line engine, without the weight and intricate plumbing of the usual liquid cooled system. Unfortunately its performance was below an acceptable standard. This proved to be the last of F6C series.

By mid-October 1926 the concentration at North Island, San Diego, was over and the prima donna, the Boeing XF2B-1, was turned over to Lt. Compo for a further three weeks of intensive official flying trials. It had yet to prove itself sufficiently to warrant a production order. The Test Board ran speed runs at different r.p.m. to determine fuel and oil consumption. Climbing tests were made and Lt. Compo also took the XF2B-1 aboard the U.S.S. *Langley* for deck landings and take-offs.

Tests continued into February. Impressive as its performance was, all did not go well. There were engine troubles at high altitudes and the landing wheels were found to be improperly aligned. During armament trials the blast tube of the ·50 machine-gun fell off, and was struck by bullets which ricocheted into the engine. The armament included the carriage, as necessary, of five bombs in racks under the fuselage and this arrangement was also tested.

VF-1B was the first to receive the production F2B-1s in December 1927 and following deliveries were to VF-6B the next June. They were used exclusively by the Navy; VF-1 and VF-6 both retained theirs until well into 1930 and additionally VF-2 operated them for the brief period, June to December 1930.

The Boeing F2Bs were one of the few naval types that became well known to the public. This was largely due to Lt. D. W. 'Tommy' Tomlinson who formed, unofficially and informally, a team consisting of himself, and Lts. W. V. 'Bill' Davis and A. P. 'Putt' Storrs. They operated together, in 1928, for less than twelve months as the first, it not the most famous, of service aerobatic teams. They achieved a remarkable precision in formation flying and kept their service in the public eye at a time when the Navy was withdrawing from racing events.

From the U.S.S. *Saratoga*, where the three team members were with VF-6 (later re-designated VB-2B as a light bombing squadron) on routine fleet service, they never lost an opportunity to thrill a crowd, just for the fun of it, at their ports of call. Lt. Tomlinson, who had a flair for showmanship, was twice court martialled, was grounded off and on, but kept asking for more! He was irked by the fact that the Army had a stunt team 'The Three Musketeers', on a more official basis. Knowing that his Boeing F2Bs outclassed their Boeing PW-9s, he wished to press the point for the prestige of his Service.

It was during the inauguration of Lindbergh Field at San Diego that they were afforded their first opportunity to perform publicly as a team. Rolls, loops and Immelmann turns in formation impressed the public and indeed the Navy too. The Press were evidently over-impressed, for the newspapers headlined them as 'The Suicide Trio'! The 'Three Musketeers' were getting away with such stunts, although the performance of their aircraft did not permit the manoeuvres of their naval counterparts. In an attempt to be on par with the Army team, 'Putt' Storrs suggested the name 'Three Sea Hawks'—and it stuck.

On Saturday, September 8th, 1928, at Mines Field, Los Angeles, the team first made headlines as a precision stunt team. The National Air Races were being held and

Marrying the Pratt & Whitney Wasp to a Boeing FB-4, produced the single FB-6. The P. & W. company started business in 1925 and nine years later became a wholly owned subsidiary of the United Aircraft Corporation. From 1936 P. & W. have been a division of that Corporation.

The first production naval fighter to utilise the Pratt & Whitney Wasp engine, the Curtiss F6C-4. Of the thirty-one produced, eight went to Marine Squadron VF-10M and the example illustrated bears the winged devil insignia of that unit on the fin.

Admiral Reeves permitted naval representation by VB-2B. There followed one of the most spectacular performances in formation and aerobatic flying ever witnessed.

Loops, Immelmann turns, up-side-down 'Vee' formations, looping four times in succession with the first at a mere fifty or so feet off the ground after take off, stalling and spinning in 'Vee' formation and pulling out at between fifty and a hundred feet above ground, successive looping and then down to a formation landing. The crowds were thrilled with aerobatics performed at such low altitudes.

Perhaps the most spectacular item in the repertoire of the 'Three Sea Hawks' was their inverted flying, which, together with the bunt, was forbidden by naval orders up to 1928. However, with the new air-cooled radial engines this was possible as only carburation was likely to be affected and in this respect, the Stromberg Carburetor Company together with Jimmy Doolittle of the Army Air Corps and Lt. Alford Williams of the Navy overcame this difficulty.

The 'Three Sea Hawks' did try the stunt of performing 'roped-together' with tapes from wing-tip to wing-tip but this cramped their style for such manoeuvres as spinning. This stunt was however demonstrated a year later by the F2B-1s of VF-1 at the 1929 Cleveland Air Races, and while it was hailed with great applause from the public, the tapes limited the range of aerobatics.

Two promising fighters, Boeing's Model 74 and Curtiss's Model 43, sponsored by their respective companies, were delivered to the Navy for evaluation in 1927. Both had been designed from the outset for carrier operation, with convertible wheel/float undercarriages and to utilise the Pratt & Whitney 'Wasp' engine.

The Boeing 74, designated the XF3B-1, arrived at Anacostia in March 1927. It was seen to bear resemblance to the earlier F2B series as it featured tapered upper wings, straight lower wings and tail surfaces similar to its immediate predecessors with oleo shock absorbers similar to the FB series. Its performance of 156-7 m.p.h. maximum with a 2,000 feet per minute rate of climb and a ceiling of 21,300 feet was sufficiently impressive for the Bureau of Aeronautics to simultaneously purchase the prototype and place an order for seventy-three production models on June 30th, 1927.

Tests of the XF3B-1 had not been without snags; lateral stability in particular was poor. The production F3B-1 models had re-designed tail surfaces, a $6\frac{1}{2}°$ sweepback

The Curtiss XF7C-1. The first of the Seahawks, which was the first fighter for the Navy by Curtiss that was designed from the inception as a VF type. It was convertible for use as a floatplane and was stressed for launching by catapult.

on the top wing of increased span, while the fuselage was shortened. Corrugated dural sheets were an innovation for covering certain areas and metal control surfaces were used by Boeing for the first time. Improvements were made to the cabane strut arrangement and cockpit, and a completely new oleo shock absorbing system was contained within the undercarriage struts. Two ·300 Browning machine-guns were normally fitted in the nose cowling, firing between the cylinders and synchronised to fire through the propeller arc.

Like the prototype, the first production F3Bs were tested at Anacostia. This time lateral stability was found to be satisfactory and landing characteristics, a point of criticism on the ' X ' model, were greatly improved. The first service models reached VF-1B on October 27th, 1927, and with this squadron they served aboard the U.S.S. Saratoga for the greater part of 1928. There were also some of the earlier F2Bs aboard the Saratoga and these actually had the edge over the F3Bs for speed, but the latter were more manoeuvrable and robust, and in fact great favourites with Navy personnel. VF-2 received F3Bs in June of 1928 and they also served in VF-3, VF-5 and VF-6.

The final deliveries, in January 1929, went to VF-6 and a

and duralumin tubing except for the wings which were of spruce conforming to Curtiss C-72 airfoil. The airframe was fabric covered.

With the recommended modifications incorporated, orders were placed for sixteen production machines. The first three were subjected to rigid testing and further modifications resulted; spinners were removed and the tripod type undercarriage was changed for an oleo-leg type similar to the Boeing F3B. A short chord N.A.C.A. cowling was fitted over the engine. On A-7655, the second production model, tests of Reed tandem propellers were made while on the third, A-7656, the effectiveness of leading edge wing slots was tested. Experiments with auxiliary fuel tanks was among a number of innovations for which the first F7C-1s were ' guinea pigs '. By mid-1928 the trials were completed and production models went into service—exclusively to the Marines with whom they served for some five years.

1928 was a significant year in U.S. naval aviation. The U.S.S. Saratoga (CV-3) and U.S.S. Lexington (CV-2) had been commissioned late in 1927, on November 16th and December 14th respectively, and the first take-off from the Lexington was by Lt. A. M. Pride in a Vought UO-1 on

The first fully aerobatic VF type, the Boeing F2B of which this is A7385 the prototype (XF2B-1). This design marked the first use by Boeing of bolted duralumin in place of welded steel tubing or wood. Provision was made for the carriage of five 25 lb. bombs.

few years later, the last one built, A-7763, became the only one of the type to reach the Marines with whom it served as a utility aircraft for some two years. Throughout the active life of the F3Bs, a number of service and individual modifications were made including the introduction of wheel spats and N.A.C.A. short chord ring cowlings.

Competitive with the F3B at the design stage was, as mentioned, the Curtiss Model 43. Construction of this model had commenced in 1926 and while it was designed from the beginning as a naval fighter, the theme was seemingly that of a scaled-down and refined Falcon (a successful observation aircraft that the firm were turning out for the Army). It first flew on the last day of February 1927 and although purchased in June by the Navy, at the same time as the Boeing, it was not until late August that, as the XF7C-1, it reached Anacostia for testing which followed the same routine as for the Boeing. Modifications advised after testing included a new landing gear and propeller.

By looks alone, the Curtiss with a beautifully streamlined exterior and its engine cowled to line up with a large propeller spinner, appeared the better aircraft, but this was not confirmed by tests. In general construction it was of steel

January 5th, 1928. Six days later Cdr. Marc A. Mitscher took off in a similar type from the Saratoga. That same month the U.S. Navy's airship U.S.S. Los Angeles (ZR-3) successfully transferred passengers to the U.S.S. Saratoga at sea and took on fuel and supplies. At this time the U.S.S. Langley (CV-1) still had bow and stern catapults fitted, although little used. Their removal later in the year showed that no compromise was intended in operations from carriers. Experiments in radio telephony were being carried out to increase the tactical efficiency of fighters and R/T sets were first utilised by a formation of Boeing F3Bs in the last month of the year.

As far as fighter strength went, it was a case of a boom for Boeings. The actual establishment at the beginning of 1928 with planned changes throughout the year, was:

VF-1B 18 Boeing F3B in service
VF-2B 12 Vought FU (replacement by 12 Boeing F3B planned)
VF-3B 6 F6C-4 to be given up for 18 Boeing F3B.
VF-4B 18 Boeing F2B in service
VF-5B 18 Curtiss F6C-3 in service
VF-6B In process of increasing establishment to 18 Boeing FB-5.

In the bewildering changes that took place in squadron reorganisation, the personnel of the old VF-1 became VF-4, while their FB-5s went to VF-6; the reconstituted VF-1 with new F3B aircraft re-formed with the personnel of VF-4.

Two companies at this time challenged the Boeing/Curtiss monopoly with experimental fighter types. These were known as the Eberhart XFG-1 'Comanche' and the Hall XFH-1.

The Eberhart Steel Products Company had erected SE-5s for the Army in the early 'twenties, but were not otherwise familiar with airframe construction. They formed an aircraft division as the Eberhart Aeroplane and Motor Company in 1925. Two years later they submitted an ambitious design for a naval shipboard fighter. On a 'contractor owned' basis, a contract was signed to test the Eberhart by Navy officials. Known by the firm's name of Comanche, it appeared without any official designation marked, or any national insignia and serial markings applied. Only latterly were red, white and blue rudder stripes applied by the contractor to denote a military type of aircraft. For accounting work it was known officially as the XFG-1 and the BuAer No. A-7944 was assigned, but not marked on the aircraft in the usual way.

Submitted as a convertible land/seaplane fighter, it was fully rigged for carrier service or catapult launching in either version. During initial trials by the Company, prior to delivery for official tests, it was slightly damaged and was returned to the factory where repairs were overtaken by considerable refinements resulting from test evaluation. The main change was an increase in the span of the top mainplane by adding 18 in. to each wing.

The modified Comanche flew, in landplane form, to Anacostia for test on June 26th, 1926. After demonstration by the contractor's pilot and stress analysis approval by the Bureau, it was taken over three days later by Lt. E. W. Rounds of the Flight Test Section. Trials commenced on July 11th, 1927, but had to be discontinued on the 26th, due to a failure of the tailskid assembly. The XFG-1 was then rigged as a floatplane and trials proceeded until August 10th. At this stage many discrepancies showed up and it was necessary for the contractor to make various and repeated changes. Re-converted, further trials ensued of the landplane version until completed on August 19th, by which time seventy distinct recommendations for improvement were made ranging from re-design of the fuselage to streamlining the compass, and from modifying the tail structure to sealing oil holes in the fuselage. According to N.A.C.A. Report 15, the greatest problem concerned the poor aerodynamic characteristics, particularly lateral stability.

Nevertheless, the Comanche had many qualities of a fighter type. The wing arrangement was rather unusual. The top plane had a 7° sweepback on the outer sections and a 5° forward sweep of the lower wing, an arrangement designed to give good visibility to the pilot. The gap was 4 ft. 3½ in. and the maximum stagger 3 ft. 3 in. Neither wing had dihedral and they were set at 3/4° incidence. Provisions were made for standard ·300 machine guns, but armament trials were never run. Power from the P. & W. R-1340D engine rated at 425 h.p. at 1,900 r.p.m. gave a top speed of 153·9 m.p.h. and since the figure for the seaplane version given at 155·5 is faster, it was evidently attained with some change in equipment.

Inevitably the floats reduced general performance and the respective comparative figures at land/sea versions were

An aircraft that existed in several forms, the Curtiss XF6C-5, is shown in its final shape under that designation, with a P. & W. Hornet engine. This airframe had an F6C-1 type undercarriage, Wasp engine, larger propeller and spinner in its various forms.

12,100/10,600 feet climb in ten minutes and 20,400/18,700 feet service ceiling for 2,938/3,208 lb. gross weight.

Fuselage and empennage construction of welded steel tubes and wings represented methods in advance of the period, but unfortunately, the production difficulties were great and the construction came in for criticism. The Navy did purchase the Comanche for $19,000 and the airframe was used to study, test, and evaluate the all-steel constructional methods employed by the Eberhart Company who were currently engaged on supplying the aircraft industry with metal components.

It was interest in metal construction that induced the Bureau on November 1st, 1927, to send to Charles Ward Hall, who had been responsible for the successful F4C-1 specification for an all-metal fighter that would have all the capabilities of a carrier-based fighter, plus two innovations: Specification No. 388A stated that the fuselage should be of all-aluminium alloy, monocoque type, construction and water-tight, permitting the aircraft to be landed safely in the water and float upright at a reasonable angle of trim; Specification No. 422B stated that the wheel-type undercarriage should be jettisonable for emergency landings on the water. The designation XFH-1, was given and BuAer No. 8009 assigned.

Leader of the famous aerobatic team, the Three Sea Hawks, seated in the cockpit of his Boeing F2B-1 of Squadron VB-2, which has a white tail indicative of the U.S.S. Saratoga. A shot taken during the National Air Races, Los Angeles, in September 1928.

It was June 18th, 1929, before Hall's design materialised at Anacostia, having been sent by freight. Erection was accomplished by the Hall Company's representatives but prior to flying an inspection revealed weaknesses and some structural changes were effected. Struts were substituted for wires bracing the tail surfaces and additional wire bracing to the cabane was made.

Even before the Navy requested the contractor to demonstrate the machine, it was evident that it would not meet the exacting standards of a fighting type. Trial flights were unsatisfactory and buffeting damaged the machine.

In mid-September when the contractor considered that the demonstration had been fully carried out, the Navy contended that it had not been demonstrated satisfactorily. However, official trials did get underway later in the month, but a series of mishaps ensued. First the rudder was found to be out of balance and then the rudder pedals sheared off! During October, while in a vertical power dive to a speed of about 200 m.p.h., the rear spar of the upper wing buckled and partially froze the aileron control. The XFH-1 was then put aside until a new, and modified, wing was completed. It was January 1930 before it flew again and the next month it was flown to Hampton Roads for arresting gear trials.

On February 18th, 1930, the water-tight fuselage was unintentionally demonstrated. An engine failure caused the aircraft to land in the water—with wheels attached. It was reported that after first nosing up, and the pilot having climbed out onto the aft turtledeck, the fuselage settled to a position about 40° down by the nose. The engine was submerged to about the middle of the top cylinder and there was about two feet freeboard at the cockpit cut out. It floated until hoisted from the water some forty minutes later when the fuselage was about a quarter full of sea water and a small amount had seeped into both wings.

The Hall XFH-1 faded out in March 1930. It was evident that a watertight fuselage was possible but not practical, because construction and maintenance was too complicated for service fighters. In any case the aircraft had proved sluggish on controls and was tail heavy with engine on—nose heavy with engine off. A speed of 152·6 m.p.h. has been obtained, but presumably this was without full equipment as the service cruising speed was only 105 m.p.h. It took ten minutes to climb 2,530 feet and had a service ceiling of 14,250 feet. Like the Eberhart, it had a Wasp engine and featured a sweep back on the upper wing with a forward sweep on the lower wing which allowed good visibility, particularly when landing aboard a carrier.

The Hall XFH-1 did mark a stage in the transition from wood to all-metal construction which became the official policy from May 1927 when the Naval Aircraft Factory had promulgated a report advocating the use of aluminium with protective anodic treatment.

The first of a famous series, the XF3B-1 of which the production form was considerably revised. Boeing introduced corrugated dural sheets for control surfaces on this machine which are apparent on the fin and rudder.

With 'Wasps' coming off the assembly line at a rate of fifteen a month in early 1927, several aircraft firms were designing fighters around this outstanding engine. Boeing had a new model, Curtiss was developing a shipboard fighter, while the Navy was showing interest in developing a high performance two-seat fighter and notified the industry accordingly. For fleet use, the two-seat fighter was conceived primarily for air defence but could also be utilised as a light bomber.

A contract was signed on June 30th, 1927, with Curtiss for an experimental two-seat defence fighter, the XF8C-1, based on the Army's O-1 'Falcon' series of observation aircraft. A similar contract was made with Chance Vought.

The Curtiss did not match up to expectations. While it had possibilities, the legacy of its basic observation type design resulted in an inordinately large machine. The two pre-production XF8C-1s and four production F8C-1s were relegated to observation types and turned over to the Marines in the spring of 1928. The original model, A-7671, was re-designated OC-1, while the second, A-7672, was modified to become the XOC-3. The few production models became OC-1s or OC-2s with Marine observation squadrons based at San Diego. They were relegated to utility duties in 1935 and finally disappeared from service in February 1936.

Contemporary with the Curtiss F8C-1 at the design stage was the Vought XF2U-1, but construction, although conventional, was subject to long delays due to changes in design and preoccupation of the firm with production of the famous O2U series of observation aircraft for the Navy and export orders. Its lines were cleaner than its counterpart by Curtiss and it met its original performance specification without difficulty, but by the time it made its first flight, two years after the order was placed, it was already outdated by improved models in the Curtiss F8C series.

A feature of the XF2U-1 was its long-chord N.A.C.A. cowling. Mindful that this might obstruct the pilot's view in carrier operations, trials were conducted on land on a simulated carrier deck at Norfolk. However, after several routine landings and take-offs, it was proved that it did not adversely affect visibility, and an appreciation of the aerodynamic efficiency of N.A.C.A. cowling was made and noted for future use. In January 1930 the XF2U-1 was returned to Anacostia for further flight tests and next month it was assigned to the N.A.F. for general use by the station. After being badly damaged on March 6th, 1931, in a landing accident while coming into Norfolk on a ferrying hop, it was declared beyond repair and later scrapped.

Some confusion has resulted from the Curtiss F8C series. While the original F8C-1 models were really naval versions of the Army O-1 Falcons, the XF8C-2 model

The first carrier-based fighter to be produced in quantity was the Boeing F3B-1 of 1928. A total of seventy-three were built and saw service aboard the U.S.S. Lexington and Saratoga. They were the first Boeing type to feature all-metal control surfaces. The U.S.M.C. used only A-7763, the last one built.

became the prototype of the famed 'Helldiver' series. The 'two' model was much different from the 'one'. Seemingly a somewhat scaled down version, the second of the series was more suited as a fighter, being more compact and of cleaner design. Only A-7673 was built and this became the first to bear the well-remembered name of 'Helldiver'— a name that stuck after test-pilots had assessed its diving capabilities. In spite of the fact that the XF8C-2 crashed during final terminal velocity dive tests, the Navy was satisfied that a new vertical power dive could be performed with stresses well within the safety factors.

The next model in the series by designation, the XF8C-3, was applied to another Falcon type, basically the same as the F8C-1 but with increased armament by forward-firing ·300 machine-guns placed in the lower wing to fire outside the propeller arc in addition to the standard armament, and increased bomb load on under-wing racks.

Although twenty-one F8C-3s were ordered and

All sixteen Curtiss F7C-1 Seahawks went to the Marine Corps. This was yet another design for which the power was provided by the Pratt & Whitney Wasp engine. Armament, when fitted, was the standard two forward-firing machine-guns.

delivered, they were so unsuited to a fighting role that almost immediately they were redesignated OC-1 or OC-2 for observation units. This concluded the development of the 'Falcon' line of the F8C series, while the 'Helldiver' line went into a succession of models. As remarked, F8C-1 and F8C-3 had been taken up by the Falcons and F8C-2 by the original Helldiver, so that when a further experimental Helldiver was ordered with an improved Wasp engine, incorporating refinements, it became the XF8C-4.

The first twenty-five F8C-4 production models reached the Navy in mid-1930 and by the end of the year, the 'High Hatters', Squadron VF-1, had been equipped. Their rugged construction and flying characteristics soon won approval and their limitations were overlooked. Optimistically, a further order was placed for sixty-three—a considerable order for peacetime. These, as the F8C-5, were to use the Wasp R-1340-88 engine with collector ring exhaust system and improved cooling louvres under a Townend ring cowling. A detail improvement was a pneumatic tail-wheel. The F8C-5 prototype was fully rigged for carrier use and showed an increase in performance sufficient to justify production.

It was with VF-1B at sea aboard the U.S.S. *Saratoga* that the Navy had misgivings about their new F8C-4 Helldiver. They found that they could not keep up with their single-seat contemporaries. Since the two-seat fighter concept envisaged a mutual defence system in mixed formations, it meant reducing the speed of the formation to that of the slower aircraft—the Curtisses. In fact, it had all the disadvantages of a bomber without the bomb-carrying capacity. Being unsuited for Fleet use, action was taken to adjust the current order and the first twenty of the sixty-three ordered were delivered without arresting hooks and sundry carrier equipment; the remainder were delivered as observation aircraft (O2C-1).

The F8C-4, had in fact a performance well below that of the XF8C-2 on which it had been based. They took two minutes extra to the XF8C-2's four minutes to gain 5,000 feet and their service ceiling was several thousand feet below that of their prototype. This was largely due to increased service gear that was added, including bomb racks and also a fuel capacity increase to a give a 450 mile range. The tanks themselves, although mounted internally, formed part of the external contour of the fuselage with one of their sides. As a two-seat fighter, the firepower was no improvement on 1918 standards. Two Lewis machine-guns were on a Scarff mounting in the rear cockpit, while two ·300 Browning machine-guns were forward-firing; the latter were mounted in the upper wing to fire outside the area of the propeller arc and so dispensed with synchronising gear.

By the end of 1931 the F8C-4 had been withdrawn from first-line fleet service and the remainder of the F8C series were being re-designated for observation units. The Helldivers had short lives as fighters, but as observation aircraft, capable of light dive-bombing, they made a name. The F8C-5s went to the Marines around the middle of 1930 and the sequence of events followed the same pattern. By 1931 the Marines' Helldivers were serving as O2C-1s

A watertight all-metal fuselage and a jettisonable undercarriage were unorthodox features of the Hall XFH-1 which was eventually considered impracticable for production. It is shown here in its modified form after initial tests.

A fighter designation, F8C-1, on the fin that belies the observation role of the squadron marking. The fact is that this Curtiss Falcon, ordered as a two-seat fighter, was relegated to an observation role with the redesignation OC-1. It was in competition with the Vought design shown below.

and as late as 1937, the U.S.M.C. still had an O2C-1 (ex-F8C-5) on their active list.

The Navy did not abandon their requirement for a two-seat fighter. They modified their conception in order to make such a VF type practical by giving up any idea of the machine performing a dual role. By reducing the bomb-carrying capacity and increasing the power, utilising innovations such as slots, slats and flaps and the new Frise-type ailerons, they considered that there was a good chance of such a type achieving parity with the single-seaters. The advantage of this design would be the increased firepower from the rear cockpit that could be used defensively in mixed fighter formations. Attack from behind was the 'bogey' of pilots in both World Wars and here, in the interim years, attempts were being made to find a counter to that threat.

Interest was being shown in two new power units. Pratt & Whitney had a couple of new engines on their test stands; the geared and supercharged R-1535 and a new twin-row cylinder design, the R-1830. Wright Aeronautical was also on the market with radial engines, namely the Wright R-1820 and the twin-row R-1510. Both the twin-row engines were boosting the output to over the 500 h.p. mark and they were noted for use with the F8C airframe.

A study for adapting wing flap and slot devices and Frise ailerons came first. The last two production F8C-4 models, A-8446-7, went back to the Curtiss factory for these features to be incorporated, and returned for trials as the XF8C-6. The same 450 h.p. Wasp R-1340-88 engine was retained and from test results, the maximum speed remained about the same, but the landing speed reduced to 50 m.p.h. represented a 10 m.p.h. improvement over the earlier models while the service ceiling went up to the 20,000 feet mark. The Frise type ailerons improved control and all-round the innovations showed a marked improvement and eventually became standard features in later fighters.

An additional Helldiver, A-8845, was ordered as the XF8C-7 to evaluate the new 575 h.p. Wright R-1820E nine-cylinder Cyclone. Basically it was the same as earlier models but it was greatly streamlined, having a N.A.C.A. cowl and large streamlined wheel spats. A new feature was enclosed cockpits and both had sliding hoods with an aft turtle deck faired down to the tail fin. Initial tests late in 1930 showed a considerable increase in top speed and at 178·5 m.p.h. it was some 35 m.p.h. faster than the earlier F8C-5 production Helldiver. The service ceiling of the version was over 20,000 feet and it climbed to 10,000 feet in 7·2 minutes. After tests it was taken over as a command transport and sporting a blue, yellow and silver paint scheme it was used by senior officers and the Secretary of the Navy.

Further tests of the Cyclone were made in three additional Helldivers that were ordered. These delivered by Curtiss in November 1931 differed from the XF8C-7 by the absence of streamlined spats and became XF8C-8s, the last of the series. This brought the Wright Cyclone into focus and so threatened the monopoly so far enjoyed by Pratt & Whitney. Competition was a healthy state for industry and of great importance to fighter development.

Another design to the Navy's two-seat fighter concept, the Vought XF2U-1, which was in competition with the Curtiss F8C-1 above. Changing requirements led to the type being abandoned. Surprisingly, a single Lewis gun is mounted in the rear when much earlier types had provision for two.

CHAPTER SIX

A Boom in Boeings

Of the Navy and Marine Corps aircraft that came and went in the 'thirties possibly the Boeing F4B series were the most outstanding. The Boeing Airplane Company, encouraged by orders, were all out to capture the market for fighters and, as related, the Curtiss Company were their only serious competitors. Two prototype naval shipboard aircraft, their Models 83 and 89, fully rigged for carrier service, emerged from the Boeing works in mid-1928. After the Model 83 had taken to the air on June 25th tests showed a top speed of 168·8 m.p.h., an initial rate of climb of 2,920 feet per minute with a ceiling of 29,600 feet.

latter part of June 1928. It was delivered to the Navy as the XF4B-1 at the races, just three days after its initial flight at the Boeing Field. The sister ship, Model 89, received the same designation.

With Lt. Thomas P. Jeter, U.S.N., at the controls, the XF4B-1 was entered in the Aero Digest Trophy Race for military pursuit aircraft. Over the 120-mile course the little biplane averaged 172·6 m.p.h., winning by a wide margin over Lt. Edgar A. Cruise who took second plane in a standard Boeing averaging 159 m.p.h. The best the Air Corps could do was 147 m.p.h. in a Curtiss P-1D 'Hawk',

The start of the Boeing era with that firm's Model 83 which became the XF4B-1. Evaluation of this aircraft by the Navy led to the Army's P-12s and the Navy's F4Bs. This machine made its debut in the 1928 National Air Races at Los Angeles, where its fine performance led to the Army placing the first firm order.

With performance figures superior to existing fighters anywhere, Boeing felt that their future was assured. A second fighter, their Model 89, identical in most respects to the 83 except for a bomb rack fitted under the fuselage between the undercarriage legs, first flew a few weeks later on August 7th, and gave a similar performance. These two aircraft were the prototypes of the Navy's F4B and the Army's P-12 series which, together with export versions, ran to a production figure of 586. Subsequent models varied little from the prototypes. The wingspan remained constant and the fuselage length varied only a matter of inches.

The construction throughout deviated from the conventional only by the use of square tubes of bolted aluminium for the fuselage framework, while the wings were built as a one-piece unit, of spruce and mahogany framework conforming to the newly-developed Boeing 106 aerofoil. Early production aircraft were fabric covered, but later models had semi-monocoque all-metal fuselages. Corrugated aluminium, as used on earlier F3B-1 fighters, was employed on some surfaces of the F4Bs. Here was a trim, compact aircraft mating the new 500 h.p. Pratt & Whitney R-1340B Wasp to a strong light airframe.

The Model 83 was flown down from Seattle to make its debut as 'The New Boeing Naval Fighter' at the National Air Races held at Mines Field, Los Angeles, during the

with a water-cooled Curtiss D-12 engine that had powered so many successful racers of the past. As a further demonstration of the XF4B-1, Lt. M. T. Seligman, U.S.N., challenged the Air Corps to a 'rate of climb' competition. Seligman in the XF4B-1 reached the specified 10,000 feet and was touching down to a three point landing in 5·92 minutes before the Hawk had reached the required height!

Ironically, the Navy missed their chance with this new Boeing fighter. Although the prototype was designed for the Navy and loaned to the Navy for evaluation, they delayed placing production orders due, partly, to the time necessary to conduct carrier evaluation trials before accepting any type for service. Meanwhile, its introduction at the National Air Races made such a profound impression on the Army, that General James E. Fechet of the Army Air Corps placed a verbal order with Boeing officials on the spot and confirmed this in writing from Washington, D.C., later. Thus the Air Corps had prior call on this fighter and as the P-12 first deliveries reached them on February 27th, 1929. They were identical to the models 83 and 89 (XF4B-1s) but omitted the arresting hook and other naval equipment.

The two XF4B-1s in Boeing silver and blue markings were tested by both the Air Corps and the Navy. After exhaustive tests from which it came through with an enthus-

Representative of the first production F4Bs, an F4B-1. Their standard finish was: top surface of upper-wing orange-yellow dope with all other fabric surfaces aluminium colour; metal surfaces were navy grey enamel except for the corrugated metal control surfaces in orange-yellow enamel. The aircraft depicted is seen carrying the markings of Squadron VF-5.

iastic report that criticised little other than lateral control, the Navy ordered twenty-seven as the F4B-1.

The two original XF4Bs were re-conditioned by Boeing and brought up to production standard. They were allotted BuAer Nos. A-8128 and A-8129 and the twenty-seven production aircraft followed on at A-8130 to 8156. The only apparent difference between A-8128 and A-8129 was that the wheels on the latter were moved forward four inches. These two, together with A-8133, were subjected to a series of tests, the results of which led to subsequent orders. A-8128 was used for performance tests including those with the carriage of a 500 lb. bomb. Equipment was tested on A-8129. Final recommendations for service included the removal of the individual cylinder streamline cowlings as these did not appear to give any advantage in performance, but they did make difficulties for inspecting the plugs.

After first deliveries in June 1929 the F4B-1 went into service as a fighter bomber the following August with VB-1B (Light Bombing Squadron One of the Battle Fleet). Later in 1930, VF-5 (re-designated VF-5B) equipped with F4B-1s and had them for about a year, and VF-2 operated them from mid-1932 to early 1934.

The Boeing P-12 and F4B series proved to be the mainstay of the Air Corps, Navy and Marine fighter units for the next five years. But even before a contract was raised to procure a second batch, a new design, Boeing's 205, was being offered to the Navy. As a company-owned and sponsored machine, it bore the civil registration X271V on the rudder and wings. A similar model, as the Boeing 202, was delivered to the U.S. Army with the registration X270V. The Model 205 was procured by the Navy on May 10th, 1930, as the XF5B-1. It represented a completely new development being a high wing, strut braced, monoplane.

Apart from the general configuration, many features resembled closely the F4B-1, including the overall dimensions and plan form. In construction however it differed considerably. The wing, both structure and covering, was 17ST dural with conventional spar and rib arrangement conforming to Boeing 106 High Speed aerofoil. No flotation gear was provided, but it was intended that the wing itself would be water-tight and provide flotation. The spaces between the spars and in the leading edge forward of the front spar, were divided into water-tight compartments to effect this, but the whole design was abandoned before it was put

Basically an Army XP-15 with an arrester hook, the Boeing XF5B-1 was the first monoplane fighter tested by the U.S. Navy. It is shown here under the civil registration X271V. The wing was intended to be watertight to provide flotation.

A cowled engine and Frise ailerons distinguished the F4B-2 from the original production batch. Many of these aircraft were later fitted with tail units similar to the F4B-4. Racks were fitted for the carriage of four 116 lb. bombs under the wings. Note the efficiency 'E' on this aircraft of Squadron VF-5.

to the test. The under surface of the wing had rib webs which extended ⅜ in. below the wing contour, while the top surface was reasonably smooth.

The fuselage was of semi-monocoque construction with the forward area of welded steel tubing, with dural cowlings which were removable for servicing the power unit, fuel and armament installations. Standard arresting gear was installed, brakes were fitted on the main wheels and the tail wheel could swivel. Standard armament was employed, ·300 or ·50 calibre, synchronised machine-guns.

Two engines were tested on the XF5B-1, the Pratt & Whitney R-1340-B and 1340-C. A 10 : 1 blower system was used and engine compression ratios were 5·25:1 and 6:1 respectively. The output of each engine varied from 450 h.p. at 2,100 r.p.m. at 8,000 feet to 485 h.p. at 2,200 r.p.m. at 8,000 feet. During trials, flight performance figures were obtained under four separate conditions, plus comparison tests with an XF4B-1 and a service model F4B-1. In addition, two tests were made with the engine uncowled, while the last two were made with two different types of ring cowlings; Townend and N.A.C.A./Boeing. Two different ground-adjustable propellers were used; they were 2-bladed, all-steel Hamilton Standard airscrews of 9 ft. and 8 ft. 8 in. diameter. Direct drive was employed throughout the tests.

The XF5B-1 was delivered by air from Seattle, Washington, to Anacostia on February 14th, 1930. Initial testing was by Lts. Ofstie, Trapnell, Jeter, and MacComsey at Anacostia and carrier suitability trials followed by Lts. Pride and Pihl at Hampton Roads, N.A.S., between February 19th and June 11th, 1930. Maximum speeds at Anacostia proved to be from 167·5 to 183 m.p.h. at 2,100 and 2,200 r.p.m. at critical altitude of 8,000 ft.; the higher figure being accomplished with a Townend ring fitted.

The XF5B-1 was tested with gross weights of 2,808 to 2,848 lb. Because of the varying factors involved and several sets of performance figures tabulated, those quoted need qualification. The initial rate of climb averaged 1,780 feet per minute and the best recording was 1,850 feet per minute. A climb to 15,000-16,000 feet took 10 minutes and the service ceiling varied from 24,700 to 27,100 feet. Landing speeds of 70 m.p.h. and stalling speeds of 66 m.p.h. were fairly constant throughout. Manoeuvrability was considered very good and directional stability varied; it could be flown hands off, but not feet off; it had a pronounced tendency to skid, but not to slip.

During deck trials ten landings and take-offs were made under various conditions with full load. Take-off was considered excessively long, but landings good and the aircraft was deemed acceptable for carrier use, with changes and modifications—not least of which was to the tail wheel assembly which had failed three times. Directional instability during early trials necessitated an increase in size of the

The first Berliner-Joyce Naval fighter, the XFJ-1, shown at Anacostia in 1930 shortly before being sent back to the firm at Dundalk, Maryland, for repairs following damage in a landing mishap which buckled the undercarriage and lower wing.

The U.S. Navy's second Bristol Bulldog, a Mk.IIA. Official reports on this British fighter included these words: Load factor materially less than American practice, difficult ground handling, narrow landing gear, no brakes, engine very smooth, good control and visibility.

fin to more than double its original area. The general view was that, although it had a good turn of speed, it was not suitable as a service type in its present form but that certain of its innovations should be incorporated on newer VF types. The Army testing its sister ship as the XP-15 came to about the same conclusion before their machine crashed. It seemed that neither service was yet ready to accept the monoplane for service use.

Static tests of the airframe in 1932 sealed the fate of the Boeing XF5B-1. Tested to destruction, the design showed that it was capable of withstanding a 32 per cent increase over the original design load. Recommendations were made for future designs regarding strengths and weights.

While engineers went to work to incorporate many of the XF5B design features on future VF aircraft, the Navy ordered an additional forty-six improved F4B types, as the F4B-2, under two separate contracts. Since it was basically a refined version of the original there was no experimental model or prototype built. The first was flight tested at the Boeing plant, and the first production models were delivered to the Navy on January 2nd, 1931. This corresponded to the Air Corps P-12C to which, externally, it appeared the same, but the Navy model was fully stressed for 2,800 lb. gross weight and had flotation and arresting gear installed.

The performance of the F4B-2 with its top speed of 186 m.p.h. was actually faster than that attained by the XF5B-1 monoplane and 20 m.p.h. faster than the first production model. Power was supplied by a Pratt & Whitney R-1340C Wasp engine cowled in a Townend ring which was then introduced as standard equipment. The tail skid was replaced by a fully swivelling tail wheel. In addition to the standard two machine-guns, provision was made for carrying four 116 lb. bombs. All the F4B-2s went to Navy squadrons. VF-6 received a full complement by June 1931, and VF-2 and VF-5 by late 1932. When the last F4B-2 rolled off the lines in May 1931 it was replaced by a model further improved, the Boeing 235 as the Navy's F4B-3. With such a successful basic design, capable of further development, it was difficult for a newcomer to compete in this field—but attempts were made.

The Berliner-Joyce Company submitted their first fighter design to the Navy in 1929 based on the Bureau of

Third version of the Boeing F4Bs, the F4B-3 with metal-covered fuselage, which supplanted the Curtiss Helldiver during 1931-32. The example shown, the first of twenty-one built, was used in a light bombing role with Marine Squadron VB-4M.

Aeronautics Designs, Nos. 92, 93 and 95. These called for an all-metal biplane, with a dural monocoque fuselage, featuring an underslung lower wing with the upper wing mounted directly to the upper portion of the fuselage.

The Berliner-Joyce project was accepted in May 1929 and a year later, as the XFJ-1, this aircraft was delivered to Anacostia for trials. With a standard 450 h.p. Pratt & Whitney R-1340C Wasp this small biplane weighed 2,046 lb. empty, with a gross weight of 2,797 lb. Its service ceiling proved to be 23,800 feet and 10,000 feet was attained in nine minutes. The range when cruising on the internal fuel tankage of 91 gallons was 400 miles and with an auxiliary tank this could be extended to 715 miles. In November 1930 it was returned to the firm's plant at Dundalk, Maryland, for

of the bottom wing. It remained at Anacostia for some years and among the items it tested were the new Goodyear ' balloon ' tyres which later came into general use.

In 1931 the Naval Aircraft Factory had a design on their own boards which showed some promise as a fighter type, but due to BuAer specifications issued to the industry, showing greater promise, the N.A.F. concept was abandoned at the initial design stage in September 1931, but not before it had been designated XFN-1 with the BuAer No. A-8978 allotted.

The U.S. Forces over the years have purchased few aircraft from other governments. Inevitably, at times they lagged behind development in Europe and conversely, at times, they led the world, but in general only the American

A standard production F4B-3 of Squadron VF-1 as the Top Hat insignia and unit markings indicate. This version had the highest service ceiling of the series at 27,500 ft. with a climb to 12,500 ft in 10 minutes. The top speed was 187 m.p.h. This aircraft, the eighth of the batch of twenty-one, was flown by the commander of VF-1.

repairs following damage to the undercarriage, lower wing and engine mounting in a landing mishap, and for aerodynamic improvements resulting from test recommendations up to that time.

A year after initial delivery, the Berliner-Joyce XFJ-1 returned from repair re-worked as the XFJ-2. Virtually it was a different type, although the BuAer number, A-8288, bore witness to the original airframe. The absence of the pronounced dihedral on the upper wing, a large spinner and Townend ring fitted, together with extremely large streamlined wheel spats over high pressure tyres, combined to disguise the direct relationship between the XFJ-1 and XFJ-2. A 500 h.p. Pratt & Whitney R-1340-92 engine gave promise of a better performance.

Trials were resumed and as far as performance figures went, were satisfactory. It was 16 m.p.h. faster in its new form and the service ceiling rose to 24,500 feet with 14,300 feet reached in ten minutes. Landing speed was about the same at 65 m.p.h. Visibility from both versions had been good, especially for landing. The drawback, however, was stability. It was a confirmed ' ground-looper '. In spite of repeated attempts to make the XFJ-2 more docile at take-off and landing it failed as a service type mainly on that score and the idea of accepting it as a fighter was abandoned. Minor problems were also raised with maintenance of the undercarriage and general handling due to the low position

industry could best meet the specific requirements of its country's forces. The last time the U.S. Navy purchased a fighter from Europe was in 1929 when the serviceability, metal construction and promising performance of the Bristol Bulldog impressed the Bureau. The prototype had flown as early as May 1927. The production version, the Bulldog II, went into R.A.F. service in May of 1929 and made its public debut at the Hendon Air Display in the summer of that year. The Bulldog, destined to be a long-lived and colourful fighter, found favour not only in the R.A.F. but with no less than eight other national air forces.

The Navy ordered a single Bulldog II in October 1929 and the Bristol Aeroplane Company earmarked their No. 7358 from the assembly line. Powered by a Bristol Jupiter VIIF engine (No. J.7811) and equipped with a Marconi transmitter and receiver, oxygen, heating, lighting and signalling lamps and bomb rack –but no armament, it was cleared for a flight test at Filton on October 10th, 1929. After trials it was dismantled, crated, and shipped to Anacostia where the main interest centred on its high tensile steel construction. However, during official trials, it crashed as a result of aileron failure from flutter during a terminal velocity dive.

A replacement was ordered in February 1930 and Bristol aircraft No. 7398 was assigned to the U.S. Navy. This was a Mk. IIA with strengthened, wider track undercarriage and detail modifications including reinforced aileron spars to correct the mishap to the first. This time the Bristol came through its trials satisfactorily but no designation was assigned and no production was envisaged. The main purpose of the acquisition was stress analysis of Bristol's metal construction and after flight tests the airframe was tested to destruction. The results were notified to Bristol and the R.A.F. so that both services benefited mutually from the tests.

The Boeing was still the most potent and useful fighter in naval service and their Model 218, registered as a civil aircraft, X66W, was the test vehicle for both the Navy's F4B-3 and the Army's P-12E. An order was placed for twenty-one of this latest version in April 1931. New features included the incorporation of a combination truss-braced all-metal monocoque fuselage as a result of an evaluation of the XF5B-1. A large turtle back and headrest was employed, to overcome a complaint by pilots of the earlier Boeing fighters. A 500 h.p. Pratt & Whitney R-1340D engine gave a top speed of 187 m.p.h. and heights of 27,500

Power was supplied by a Pratt & Whitney R-1340-16 rated at 550 h.p. at 6,000 feet. The normal gross weight was 3,107 lb. with a maximum of 3,539 lb. that allowed for standard armament of one ·300 and one ·50 machine-gun and wing racks for two 116 lb. bombs. These models had greatly strengthened wings, thereby allowing dive-bombing to become not only practical, but standard practice. The top speed of 184 m.p.h. was accomplished at 6,000 feet; the landing speed was 62.5 m.p.h. in normal weight conditions. A service ceiling of 26,900 feet was attained, and the range on the 110 gallon internal fuel tankage was 350 miles, extended to 700 miles with an under-belly auxiliary tank of 55 gallons.

The first F4B-4s were delivered in July 1932 and during the following year became standard equipment in both Navy and Marine Corps units. The last of the 'fabled fours' was delivered late in February 1933. When fighter squadrons were re-assigned to light bombing roles, the Boeings were equally suited to that task, particularly dive-

Final production version of the famous 'Boeing bipes', the F4B-4 with a larger fin and headrest than its predecessors. It was ordered in greater quantity than any previous U.S. naval fighter type. The example shown is serving in a bombing squadron. The fully painted cowling, and the fuselage band together with the individual number '7', identify this as the third section leader's aircraft.

feet were possible giving the Navy and Marines their first experience of high flying.

Late 1931 and early 1932 deliveries of the F4B-3 were mainly to supplant the Curtiss F8C-4 Helldivers of VF-1B, the only squadron to receive a full complement of this model. After using them for some eighteen months, they passed to the Marine 'Red Devils', VF-10M, who, re-designated VB-4M, used them as light bombers.

The classic aircraft of the early 'thirties was undoubtedly the next of the Boeing series, the F4B-4. Its ruggedness, combined with an unequalled all-round performance, made these Boeing fighters a favourite with both pilots and mechanics alike. Foreign orders for quantities proved their superiority to the world. The new, and as it proved, final production version had more fin area and a larger headrest of which the fairing contained a rubber life-raft and emergency supplies. As Boeing Model 235, the Navy ordered a total of 92 on three separate contracts, making the largest order so far placed by the Navy for a VF class aircraft.

bombing. Of the twenty-four assigned to the Marines all but one, that reached VJ-6M, initially went to VF-9M. Later they were re-assigned and divided up between squadrons at San Diego and Quantico. With Navy squadrons they had a long life: VF-2 used them from November 1934 to June 1935; VF-3 received their first ones in October 1932 and used them until they were replaced by Grumman F2F-1s in June 1935; VF-6 utilised their consignment from October 1932 until as late as June 1936. When VF-8 was formed in June 1937, their initial aircraft were old, service-weary, F4B-4s. They served until early 1938 when replaced by Grumman F3F-2s.

The U.S.M.C. actually built an F4B-4 (A-9719) from spare parts at Quantico in 1938. But, inevitably, the 'Boeing bipes' were relegated to training roles, utility flying and reserve units—and finally the scrap-heap.

The views of an F4B-4 pilot, H. L. Kelley, are of considerable interest: 'As a student flyer training at Pensacola, Florida, in 1937 I flew the Boeing F4B-4, and as I

An F4B-4, BuAer No. 9012, in immaculate finish at Anacostia Naval Air Station. This shows to good effect the underslung fuel tank that, added to the internal capacity of 110 gallons, gave a 700-mile range and a maximum speed of 184 m.p.h.

remember the first impressions, it seemed to be a very small 'plane. This is possible because up to that time I had been flying various types such as the SU-2, PM-2, PD-2, SBU-1, O2U-3 and the TG-1 all of them observation and patrol types. When I first sat in the cockpit it seemed as though I could reach out and touch the ailerons and then reach back and wiggle the rudder with my hand; this gives one an idea of how small the 'plane seemed. After the cockpit checkout where one had to show that one could touch all of the controls blindfolded at will, and a few "skull sessions" about flying it, we were assigned a flight billet. Taxying the airplane was very easy, and since the view to the front was impaired by the cowl and motor, we had to turn from side to side. The first take-off was a real surprise, because when the throttle was pushed to the stop, the 'plane seemed to jump down the runway, and with a few bumps was into the air. Our first flight was to be around the field, under 5,000 feet. We were to go up and practise a few stalls until we thought we had the feel of it, try some turns and that sort of thing for the first half hour, then half an hour of "touch and go" landings. The ship was very easy on the controls, in fact it was a bit difficult to get used to the speed with which it reacted. Stalls were quite straightforward with no fall off on either wing, and one had rudder, elevator and aileron control at all times, in or out of the stall. Spins were not permitted until later. My first landing was made under power because I felt that this was the better way to find the ground. When the wheels had grounded I pulled the throttle all the way back and let the tail drop. Then, pushing the throttle forward against the stop I was off again. Sometime around the thirtieth hour, we were to go through regular stunt check, so the hours up to that time were spent on the "Syllabus". I found that the 'plane would do all the stunts well, even when the controls were mishandled—by this I mean that it would even take plenty of mishandling. We did not always try to do the normal kinds of stunts with the plane. I know of no time when I did not have control of the plane when I wanted it, no matter what my position or situation. My only criticism was the seat, after about one hour of sitting in the straight backed seat my back felt like it would break. I often wondered why the designers did not put a little more slant in it.'

The final chapter in the F4B-4s life did not come until much later. In a way it could be said they were re-introduced into service, for twenty-three of the Army's counterpart of the F4B series, the P-12 series, were taken over by the Navy and stripped of all armament and non-essential equipment, were converted to radio-controlled target drones as F4B-4s. Painted all-silver, with a large distinguishing 'D' for 'Drone', some were still being shot at as late as 1942. The 'fabulous fours' were introduced to the Navy at a time when there was a boom in Boeings and they went out under the guns of the Navy—so it could be said that they came in with a boom and out with a bang!

An F4B-4 BuAer No. 8918 of Squadron VF-3 which received their first machines in October 1932 and used them until replaced by Grumman F2F-1s in June 1935. The F4B-4 marked the peak in the era of Boeing biplane fighters.

CHAPTER SEVEN

The Parasite Project

The unlucky contender to BuAer Design No. 96, the Fokker XFA-1 ('A' for Atlantic Aircraft the designing company's former name). This is shown at Anacostia after initial modifications which included increasing the fin and rudder area.

A fighter concept new to the U.S. Navy was introduced in the early 'thirties that linked heavier-than-air fighters with lighter-than-air transports. The success of the Zeppelin built for the U.S. Navy as their *Los Angeles* (ZR-3) led to orders being placed in October 1928 with the Goodyear Zeppelin Corporation at Akron, Ohio, for two rigid airships of 6,500,000 cu. ft. The first of these, the U.S.S. *Akron* (ZRS-4), made its maiden flight on September 23rd, 1931, to Lakehurst, where it was commissioned the next month.

U.S. military policy was purely defensive in the interim years and the primary role of the U.S. Navy's new airships was that of ocean-going coastal defence scouts. They were regarded as forerunners of a fleet that could perform various roles and fighters carried aboard could be used for attack, defence of the airship itself, and greatly increase their scouting range.

The carriage of fighters by airships was not new. It had been tried by both British and Germans during the 1914-18 War and in 1926 two Gloster Grebes had been launched from the British airship R33 flying at 2,000 feet. In America, the U.S. Army had made successful 'hook-ons' using a Sperry Messenger with the airships D-3 and TC-3 during 1923-24. The new American concept envisaged a hook-on and take-off from airships by means of a trapeze swinging below the airship which could also hoist the aircraft into a hangar deck. Naval initial trials included hooking on and off a Vought UO-1 to the U.S.S. *Los Angeles* and experiments with a glider in 1929 and 1930.

The airships were large enough to house several aircraft although the overall dimensions prevented the planes from being very large. Size was not the only problem; weight is a prime factor in any airborne vehicle and this was doubly important for an aircraft that was to be carried within an aircraft. These fighters would have to be unique, diminutive and light, stable and manoeuvrable.

The detailed specification was issued on May 10th, 1930, by the Bureau of Aeronautics as their Design No. 96. Some of the requirements set forth were: normal gross weight of 2,263 lb., of which 654 lb. would be useful load; overall dimensions to be 25 ft. 6 in. wingspan, 20 ft. 3 in. length and just over 7 ft. in height; the top wing, of Boeing 106 airfoil, would be gulled into the fuselage; an 8 ft. diameter propeller was to be used, 26 in. × 4 in. tyres with wheel brakes, flotation gear and 'cobra' type arresting hook installed; armament specified was two ·300 M-2 Browning fixed-guns with 600 r.p.g. The required performance was a top speed of 175·2 m.p.h. at sea level and 61·5 m.p.h. stalling speed; service ceiling was to be 23,900 feet and 14,400 feet reached in ten minutes from sea level; the endurance was 1·21 hours at full power and 4·85 hours at 70 m.p.h.

Only two firms presented bids for this fighter requirement. General Aviation Corporation, formerly Atlantic Aircraft, the continuation of the original Fokker Aircraft Corporation of America, and the famous Curtiss Aeroplane and Motor Corporation. Although the airship fighters would evolve from the design requirements, they were

The original form of the Curtiss XF9C-1 to BuAer Specification No. 96 which called for a compact shipboard fighter without mention of an eventual requirement for a parasite fighter operating from an airship.

The successful completion of the parasite fighter project was not achieved without much experiment in the years 1931 and 1932. This illustration shows a stage in the development of the aircraft hook and comparison with the standard type, seen below, reveals several differences. Additionally, there were several other forms of the hook, each designed to eliminate problems met with in earlier types, until a satisfactory form was evolved.

classed as single-seater carrier fighters, VF class. The contracts were awarded on June 30th, 1930, for the Curtiss XF9C-1 and the Fokker XFA-1, BuAer Nos. A-8731 and A-8732 respectively.

The Curtiss XF9C-1 made its maiden flight on February 12th, 1931, and was flown to Anacostia where from March 31st to June 30th, 1931, the flight tests were conducted. Reports gave no indication that it was to be used in any other category than for shipboard service. At a gross weight of 2,502 lb., a top speed of 176·5 m.p.h. was attained with stalling and landing speeds, 60 and 63 m.p.h. respectively. The service ceiling was 22,600 feet. However, poor visibility, dangerous spins, failures of the tail wheel assembly directional instability and poor landing characteristics were unpromising items in the initial reports. The faults were not so much with the manufacturer as with the severe limitations imposed by the Design 96 concept set down by the Bureau. The XF9C-1 was returned to the Garden City Plant for modification.

The original XF9C-1 failed to receive a contract as a normal production carrier-based fighter and there are no apparent records to substantiate the steps that led to modifying the Curtiss version of Design 96 into the airship fighter model. At the conclusion of the XF9C-1 official test flights, the recommendations for improvement were incorporated in a new version, classified as the XF9C-2.

Its purpose revealed. The second form of the Curtiss Sparrowhawk, the F9C-2, with the hook-on unit for attachment to an airship. No. 9056 shown was the first of six of its type. This concept was first tried by the Germans during 1917-1918 and a Sopwith Camel was successfully dropped from the R.23 in England in late 1918.

Typical of the F9C-2 Sparrowhawks which were planned for attachment to the U.S.S. Akron and Macon for Atlantic and Pacific coast defence patrol respectively. Both airships were fated to crash and when the project was abandoned, only three aircraft were left. These were re-designated XF9C-2 and sent to San Diego.

On November 1st, 1931, the new version was complete with a blue and silver finish, carrying the civil registration X-986M. It was similar to the XF9C-1, but more streamlined. Thin cantilever, P-6E type, landing gear was employed, with spats and struts well faired. Trials led to a pre-production version as the F9C-2, which was in fact the original XF9C-2 modified and with naval markings replacing its former colour scheme. From May to July of 1932 trials were run which were more to the point of assessing its capability as an airship defender. It has been reported that the F9C-2 differed from the XF9C-1 and XF9C-2 mainly in having a split axle tripod gear and the installation of the airship hook-on mechanism. These bare facts, which seemingly affect only a change of undercarriage and fitting of a hook, do not do justice to the series of trials aimed at ironing out the kinks of this unique fighter before adapting it to its intended new role and the ' hook ' too, had involved much experiment with trapeze contraptions on the airships.

Due to reorganisation of the Fokker concern under new management as the General Aircraft Corporation at this time, some difficulties were encountered in getting the XFA-1 built and not until March 5th, 1932, was it delivered by air from the Baltimore plant to Anacostia. Due to the rigid specification, it looked almost the twin of the Curtiss F9C. The contractor's demonstrations were satisfactorily carried out that month and by the 14th all specifications except spins and dives had been met. A slight mishap occurred on the 14th when the tail wheel assembly was damaged, but repairs were effected in time for the Navy to continue their trials that same month. It was tested with both its initial two-blade, and a new three-blade, propeller, but the machine was, *inter alia*, longitudinally unstable and, like the Curtiss, was returned to its manufacturer for modifications.

Mid-June the XFA-1 returned to Anacostia for the effects of the changes to be determined. The rudder and tailplane area had been increased, fairings were added, the large spinner and wheel spats were removed and spring type, adjustable trailing edge tabs were fitted on the elevators. These changes, however, had little or no effect on the unsatisfactory flight characteristics. It was now excessively tail heavy with power on, and nose heavy with the power off. After a month it was again returned to the manufacturer for further aerodynamic changes and in early August it was back at Anacostia for balance and stability trials. The Testing Board was not at all pleased with the plane and requested the contractor to conduct the spinning tests. Three spins were made over the Naval Proving Grounds at Dahlgren, Virginia, and each terminated in a flat, almost uncontrollable spin.

After this, further demonstrations were considered dangerous and the trials were abandoned before dives and carrier qualifications had been conducted. In all a total of forty-three flights had been made with 21·25 hours flying time. Armament tests were conducted on the ground, during which fifteen stoppages were recorded, in firing 200 r.p.g., due to excessively sharp bends in the feed chute. The proximity of the fuel filler neck to the blast tubes was considered a fire hazard, but recommendations on this and other aspects were not taken up as it was considered that little would be gained by continuing the project.

The XFA-1 had been developed to an exacting specification for a particular purpose and credit is due to the designers for original and ingenious features built into the aircraft. Its top speed of 169·9 m.p.h. at gross weight of 2,525 lb., stalling speed of 64 m.p.h., ceiling of 20,200 feet and 11,500 feet climb in ten minutes, was not inauspicious. The real reason for its rejection at that stage is contained in the official report—" inasmuch as another airplane of similar basic design has reached production, the need for the XFA-1 has vanished'. That ' airplane ' was the Curtiss F9C Sparrowhawk.

It has been represented that neither the Fokker XFA-1 nor the Curtiss XF9C-1 were intended for ' parasite fighters ' as neither had hooks, nor were tests conducted that in anyway suggested their future role. If that was so, the severe limitations in design to which they were conditioned by the Bureau of Aeronautics' specification are so highly illogical, as to be completely out of character with that august establishment. This just does not appear feasible and the possible explanation, buried in documents not yet declassified, is that at the inception of the idea, it was highly secret, whereas flight reports, graded lower at confidential—and which might at discretion be shown to manufacturers—could not bear an indication, at that time, of their ultimate and highly secret role.

The Fokker XFA and Curtiss XF9C-1 bore a remarkable similarity. During trials their relative performance figures were often compared and the Curtiss machine came out on top every time. However, for some unknown reason the government furnished a 423 h.p. Wright R-975-C engine for the XF9C-1 that must have made a difference in performance over the 403 h.p. R-985 of the XFA-1.

On June 16th, 1932, the pre-production F9C-2 made the first hook-on to the U.S.S. *Akron*. By July, 104 hook-ons had been successfully made and five additional F9C-2s had been ordered. As production models they were constructed at the Curtiss Buffalo plant whereas prototypes, pre-production and experimental work came from the division at Garden City. An order was also placed with Curtiss to modify the original XF9C-1 to F9C-2 production configuration. This was done with the exception of the empennage which remained the same; possibly because of this it was very unstable and was withdrawn from service later.

During 1931, the Akron Unit formed with an authorised establishment of one VO and four VF type aircraft. With successful hook-ons accomplished the Unit had expanded by the end of 1932 to a complement of six F9C-2, one XF9C-1 (modified), one F6C-3 Hawk and three N2Y-1 trainers. The production F9C-2s were considerably modified from their original form with additional rudder area, simplified strutting and cut-outs on the outboard side of wheel spats to facilitate maintenance.

To fit the plans for four Sparrowhawks to each airship, an additional F9C-2 was built to make up the eight. But for some reason this Sparrowhawk was not fitted with an airship hook and was designated XF9C-2. It was delivered to Lakehurst N.A.S. on September 21st, 1932. Initially all the Sparrowhawks were assigned to the *Akron* pending the commissioning of the *Macon*. It was intended that *Akron* would operate out of Lakehurst on Atlantic patrol, and *Macon*, having taken over her aircraft, would operate from Moffett Field, California, on Pacific patrol.

Then came the first blow. On April 4th, 1933, while the *Macon* was being prepared for its maiden flight later that month, the *Akron* crashed in a storm off Barnegat Light, New Jersey. The bad weather at Lakehurst had grounded the Sparrowhawks assigned for the flight that day but the *Akron* itself with seventy-three of her personnel had perished among them the Chief of the Bureau of Aeronautics, Rear Admiral William A. Moffett, who was America's greatest exponent of the potentialities of the airship.

What was the Akron Unit, subsequently transferred to

Sparrowhawks of the U.S.S. Macon. *It was planned that four of these aircraft would be allotted per airship and provision was made by their intricate trapeze mechanism for handing them up into the bowels of the airship. These three aircraft formed the famous 'Red, White and Blue' flight. The aircraft with the white band was BuAer No. 9057. In this photograph which was taken on July 6th, 1933, the blue-banded aircraft is in the foreground.*

A rare cine-shot of a once-only experiment over San Francisco on October 12th, 1934, when a Sparrowhawk minus its undercarriage, and with a large underslung fuel tank, was hooked-off from the U.S.S. Macon *during trials to increase the range of the F9C-2s.*

the West Coast and re-organised with six F9C-2, one XF9C-2 one Loening OL-8, one Vought O2U-1 and three N2Y-1s. The original XF9C-1 did not go to the new unit; it was scrapped at this time due to its instability. Several aircraft including observation types and even a Curtiss F6C-3 Hawk served with airship units. These did not have a hook-on mechanism installed, having been assigned as utility craft. They relieved the Sparrowhawks of flying other than to perform their primary function. The Consolidated N2Y-1 trainers, however, did have hook-on units; being small aircraft they were suited to training for the operation of parasite fighters.

By 1934 the operations of the Sparrowhawks to and from the *Macon* became almost routine. A minor change was small fins added to the forward edge of the wheel spats so that arresting wires would not slip over the spats during carrier landings. During the 1934 operations the commander of the Macon fighter unit, Lt. H. B. Miller, secured additional range for his F9C-2 by an auxiliary fuel tank under the fuselage, and for release and return operations he arranged for the complete removal of the undercarrage to give a better performance. This was demonstrated during an air show at San Francisco, when the *Macon* passed over the field and dropped a wheelless F9C-2 which flew over low. Spectators regarded this as something of a stunt!

The Sparrowhawks were extremely tricky aircraft to fly and land and the ' hook-up ' gear increased their inherent instability, but in expert hands they could be made to perform their design duties. Their gross weight was 2,770 lb. of which 681 lb. was useful load. A top speed of 176·5 m.p.h. was attained and they stalled at 63 m.p.h. The service ceiling was 19,200 feet, at which altitude a maximum speed of 140 m.p.h. was attained, and 11,300 feet could be reached in ten minutes. Normal range was 200 miles, but with landing gear removed and an auxiliary fuel tank attached, a four hour endurance or 300-mile radius of action was possible. Experiments with detachable undercarriages to increase range were underway when disaster again intervened.

The *Macon* went down in the Pacific Ocean on February 12th, 1935. A structural failure occurred after a severe buffeting in gusts of wind off Point Sur, California. Four of the Sparrowhawks were aboard and although valiant attempts were made to launch them, the twisting of the big airship as she floundered in the storm, jammed the launching rails and they were lost with the *Macon*. Fortunately there were only two fatalities, but this was more than a set-back, it was the end of rigid airship operations on a large scale by the U.S. Navy and, indeed, of America. Of the giants, only ZR-3, the U.S.S. *Los Angeles*, survived. After eight years of service and over 5,000 hours in the air, this airship was decommissioned at Lakehurst on June 30th, 1932, and stored until dismantled and scrapped in 1939.

The three remaining F9C-2s were designated XF9C-2 and sent to North Island, San Diego. By the end of 1935, Sparrowhawks 9057 and 9060 were both scrapped and the sole remaining Sparrowhawk, 9264, was flown to Anacostia for use as a general utility aircraft. The hook-on mechanism and all carrier gear was removed and the aircraft was given a buffed natural aluminium finish. Later it was transferred to Norfolk and in 1939 it was presented to the National Air Museum, Smithsonian Institution, where it is on permanent display today. Before it was shown to the public it was repainted and the serial number 9056 was loosely applied. Latterly a carrier hook, hook-on mechanism and an auxiliary fuel tank have been specially made from original drawings, and physically it is now truly representative of the days of the Sparrowhawk and hey-day of the airship.

The XF9C-2 in civil guise. This became standard F9C-2 in naval colours. After the crash of the Macon *in which four Sparrowhawks were destroyed this aircraft and two other surviving models were re-designated XF9C-2 and sent to San Diego Naval Air Station.*

CHAPTER EIGHT

Changing the Fighting Form

The Pratt & Whitney Twin-Wasps and the Wright Cyclones were nearing the production stage by early 1930 and thoughts were turning toward building airframes around these more powerful engines. The ultimate in biplane design had almost been reached but further improvements were possible from streamlining, more powerful engines, more efficient propellers including controllable pitch, and retractable undercarriages. In the early 'thirties a spate of new designs reached the Navy and Marines for evaluation. Boeing continued to push the F4B series, Curtiss showed that the Hawk series were still capable of development and Grumman commenced their long association with the United States Navy as producers of naval fighters.

The original 'Fi-Fi', the XFF-1, first of the famous Grumman fighters for the U.S. Navy and the first production VF type to feature a retractable undercarriage. This aircraft brought the Wright Cyclone engine into prominence; an R-1820-E model was fitted to this machine but in production aircraft, delivered from May 1933, the model R-1830-78 of 700 h.p. was installed. The FF-1 proved its versatility when it became a successful VF, VB and VS type.

LeRoy Grumman, who had worked for Grover Loening since 1923 had perfected the retractable landing gear of the Loening amphibious aircraft. He then went into business himself in 1929 and having submitted a preliminary layout for a two-seat fighter, received, on April 2nd, 1931, a contract for one experimental machine to be designated XFF-1. The all-metal, stressed skin fuselage, retractable landing gear and enclosed cockpits of the aircraft provided, perhaps, the greatest step forward in refinement since the late 'twenties. Completed at what was then a small plant at Bethpage, Long Island, the XFF-1 reached the Navy late in 1931 and made its first flight on the penultimate day of the year. The rugged construction that characterised Grumman designs was evident from the inception and the prototype went through its series of trials with a minimum of recommendations for modifications.

The XFF-1 employed a 600-625 h.p. Wright R-1820E engine which gave a top speed of 195 m.p.h. at sea level at its gross loading of 3,933 lb. The service ceiling was 23,600 feet and on an internal fuel capacity of 120 gallons, a range of 800 miles was possible. Only in climbing ability was the XFF-1 mediocre in that it took ten minutes to reach 10,000 feet. This performance made a profound impression on naval officials, as this two-seater not only equalled, but in some cases surpassed the performance of the single-seat fighters then in service. A production order was placed in December for twenty-seven FF-1s. Apart from detail refinements, no basic exterior change was advised except for the accommodation of a more powerful Wright engine which could be installed without extensive modifications.

Another two-seat fighter was the subject of a contract awarded to Berliner-Joyce on June 30th, 1931. Originally a Pratt & Whitney single-row Hornet was the intended power plant, but since a Wright twin-row was available, this was fitted when the XF2J-1 arrived for test at Anacostia two years later. It featured a gull-type upper-wing, faired into the fuselage and unlike the company's first naval fighter, the lower wing was raised to fit directly, and conventionally, into the lower fuselage section. Its top speed of 193 m.p.h., service ceiling of 21,500 feet and climb to 10,000 feet in 7·5 minutes at a gross weight of 4,520 lb., would have been considered good when it was first conceived, but now that the FF-1 had come on the scene it was outclassed. It was a stable aircraft but rather sluggish on the controls and had poor visibility. An enclosed sliding canopy was belatedly fitted, but with the decision not to put the new Wright 14-cylinder twin-row radial in production this project was brought to a standstill.

The Curtiss Company were trying to make a come-back with a fighter, and a final model was planned to the Helldiver series. The XF10C-1 evolved in 1931 as an improved Helldiver, based on the XF8C-8, with the undercarriage modified to a faired single-strut type with large streamlined spats, and the rudder was revised. However, before it was flown, it was re-designated XS3C-1 as an experimental Scouting type.

The Curtiss Helldiver models were apparently unsuited as fighters, so Curtiss re-introduced the Hawk in 1932 to show that its basic design was still capable of development. A contract was awarded on April 16th, 1932, for one prototype, designated XF11C-1. This was a completely re-

A second attempt at producing a naval fighter by Berliner-Joyce resulted in the XF2J-1 biplane with a 'gulled' upper wing. When originally delivered for test the cockpits were open as shown here.

designed 'Hawk' cleaned up for naval operations, built around the 600 h.p. twin-row Wright Cyclone. It presented a well proportioned, streamlined appearance. The wings were of metal similar to the Army's new XP-22, while the fuselage was based on the Army's P-6E Hawks. At one time it was fitted with an experimental cowling with nose slots designed to increase air velocity to facilitate cooling during low-speed, high power operations, when loaded as a fighter-bomber. Trials proved it still had the much envied characteristics of the old Hawks—ease of handling and responsive to the controls, especially the ailerons, right down to the stall. A top speed of 203 m.p.h. at 6,000 feet and a service ceiling of 23,500 feet was claimed. The landing speed was 67 m.p.h. and range, on the 94 gallons internal fuel capacity, was 525 miles.

Light dive bombing was becoming more practical with the Naval Services at this time and the single-seat fighter was being utilised more and more in a dual role. Fitted with bomb racks, several VF machines were tested as BF (Bomber Fighter) types. The cantilever landing gear strut arrangement made it particularly suited for carrying a bomb or auxiliary fuel tank between the undercarriage legs and bomb racks were also placed under the lower wings. As the XF11C-2 performed this dual role quite satisfactorily, it was re-designated XBFC-1 in 1933.

By now the new Curtiss biplanes were known as 'Goshawks'—for which the dictionary gives—'genus of the large short-winged hawks, noted for their powerful flight, activity and courage'. Chronologically, the XF11C-2 came before the XF11C-1. This was because the former was first flight tested on March 25th, 1932, as a contractor owned aircraft, and its initial success led to a contract for both an XF11C-1 with a 600 h.p. Wright and an XF11C-2 with a 700 h.p. Wright, being placed the next month. The XF11C-2 was received first, and differed from the XF11C-1, which was delivered soon after, by smaller wheel spats with balloon tyres and a narrow chord engine cowling. In performance it had, not surprising in view of the extra power, the edge over the other prototype.

The XF11C-1 was retained for further experiments while the XF11C-2 was tested with a view to final modifications for the seventeen Goshawks ordered as F11C-2 fighters. Changes on the production models included a slightly longer chord engine cowl and larger streamlined wheel spats that gave it a similarity to the original XF11C-1.

The first Goshawks were assigned to VF-1 in June 1933, the only squadron to use the F11C-2. No sooner were they settling down in service, than they were caught up in a spate of re-designations. The Bureau were interested in dual purpose fighters, and for dive-bombing the existing fighters were more suited to that role than any other type. The XF11C-1 had already been changed to the XBFC-1; the XF11C-2 was re-assigned as XFBC-2 and the production F11C-2 Goshawks became BFC-2s.

VF-1 was later re-assigned as a light bombing squadron and re-designated VB-2B. Their aircraft underwent some

A re-designed Curtiss Hawk produced the XF11C-1 Goshawk, which was something of a hybrid with wings of the Army's new XP-22, while the fuselage was based on the P-6E Hawks. This is the only known flying shot of this aircraft which became the XBFC-1. It was numbered 9219 and appeared in March 1932. The long chord engine cowling housed a 600 h.p. Wright R-1510-98. Another characteristic of the type was the three-bladed propeller.

modifications to facilitate a bombing role, by eliminating some of the pilots' discomforts. The headrest and windshield were changed, and an additional ten models procured from Curtiss as BFC-2s incorporating these and other changes deemed necessary after service use. For the most part, they were like the F11C-2 with the added turtle back resulting from a new headrest which contained a rubber life raft in a redesigned cockpit. These aircraft were still being used in 1937.

Performance of the later BFC-2 was similar to the F11C-2, but the service ceiling dropped slightly to 24,300 feet. The top speed of 205 m.p.h. in clean fighter configuration, dropped to 195 m.p.h. as a bomber. An internal fuel capacity of 94 to 150 gallons was standard, but an auxiliary fuel tank, of 50 gallons which was very large for that time and raised the gross weight to 4,495 lb., could be slung from the fuselage to extend the range to over 650 miles.

Before the BFC-2 had evolved, an F11C-2 of the original production batch had been held back at the Curtiss plant for the undercarriage to be made retractable. Standard in other respects it was delivered to the Navy at Anacostia in May 1933 as the XF11C-3. This presented the Navy with its second successful retractable gear in a fighter. The wheels and struts folded up into the fuselage sides just forward of the lower wing which necessitated a rather unsightly bulge, but it had the effect of boosting the top speed to 216 m.p.h. at 8,000 feet and raising the service ceiling to 26,000 feet, while in ten minutes 11,500 feet was reached.

The performance figures justified an order for twenty-seven being placed in February 1934 as BF2C-1s. The production models were structurally strenghtened, had additional cooling vents for the engine and incorporated a large turtle back head rest and a semi-enclosed sliding cockpit canopy. Standard armament was used and four 116-lb. bombs could be fitted on wing racks or one 450-lb. bomb could be slung under the fuselage in place of a jettisonable auxiliary fuel tank. Later the XF11C-3 was brought into line with the BF2C-1 standard.

The first of the new series was delivered on October 7th, 1934, to VB-5B who embarked with them the next month on the new carrier, U.S.S. *Ranger* (CV-4). They were the first single-seat craft to serve on the *Ranger* and VB-5B used them until November 1935. Their end was rather abrupt. The last BF2C-1 was delivered in November 1934 by which time the name Goshawk had lapsed and they were known, generally and simply, as Hawks. Those serving with VB-5B in 1937 were getting 'service-weary' early and during manoeuvres wing flutter was encountered. After No. 13 of the Squadron (5-B-13) out on a routine practice bombing run from San Diego, had its wing buckle and fold, the BF2C-1s were grounded; later all were condemned. Those at San Diego had engine, instruments and sundry items removed, holes drilled throughout the structure, and were dumped from a barge into San Diego bay. The BFC-2s with VB-2B (High Hat Squadron) continued to serve satisfactorily into 1937 but after the incident related the Curtiss Hawk basic design as far as the Navy was concerned was finished and Boeing F4B-4s were brought out of storage as replacements.

Naval requirements for fighter-bombers prompted Boeing to submit their Model 236 which became the final variant of the Boeing ' bipes ' as the XF6B-1. Built to utilise

Second of the Goshawk series, the Curtiss XF11C-2 which arrived at Anacostia on May 2nd, 1932. An excellent machine to fly, it suffered several mishaps during tests. It was force-landed twice in May and was severely damaged on August 1st, 1932, during arresting tests.

the new 625 h.p. Twin-Wasp, it appeared externally to be much like the F4B Boeings with a revised, heavier landing gear and a long chord engine cowling. It had been ordered on June 30th, 1931, and arrived at Anacostia for trials just under two years later.

The new Boeing was variously modified and the undercarriage received particular attention, while various engine cowlings were tried, but its chief drawback was its gross weight, being 3,704 lb. as a fighter and 4,282 lb. as a bomber. Well stressed to a safe pull-out from a dive at 9 'G' it appeared most suitable for dive-bombing, and was redesignated XBFB-1 mid-1934. However, too heavy and slow for the 'F' part of its designation, it did not warrant production and the single example was finally expended in barrier tests at Norfolk in 1936.

The Bureau Design No. 113 brought the first naval fighter venture from Douglas Aircraft and the third attempt by Vought. The specification called for a conventional two-seat biplane fighter powered by either a geared and supercharged R-1535 Pratt & Whitney or R-1510 Wright engine, of approximately 700 h.p. By April 1932, seven manufacturers had sent in design proposals, but the Douglas and Vought bids were deemed to be the most satisfactory and contracts were signed on June 30th for one Douglas, designated XFD-1 (BuAer No. 9223), and one Vought XF3U-1 (BuAer No. 9222).

While the Wright engine was suitable, both Douglas and Vought chose the Pratt & Whitney Twin-Wasp R-1535-64 of 700 h.p. which permitted a long chord, small diameter cowling and thus presented less frontal area. Both aircraft were very similar in appearance as well as in their basic construction of all-metal, fabric covered. The Vought XF3U-1 made its maiden flight from the Vought factory in May 1933 and went to Anacostia for flight tests the following month. The first use of landing flaps on a VF type aircraft appeared on the XFD and XF3U models, as the Bureau design had specified that they should be fitted at the trailing edge of the upper wings.

The Vought XF3U-1 attained a top speed of 214 m.p.h. at 8,000 feet and 200·6 m.p.h. at sea level for a landing speed of 67·5 m.p.h., and a climb to 12,700 feet in ten minutes. The Douglas XFD-1 which arrived at Anacostia a few days before the Vought, on June 18th, attained a top speed of 204 m.p.h. at 8,000 feet and 194 m.p.h. at sea level for a landing speed of 65 m.p.h., and a climb to 12,000 feet in ten minutes. Trials showed that both the aircraft had excellent flying characteristics. They were manoeuvrable, with a performance that would have favoured many single-seat fighters and they were considered particularly safe aircraft to fly as there were positive stall warnings.

In competition at the same time, and both promising as far as meeting the original specifications went, it would appear that one or the other, depending on an analysis of the tests, would get a production order. However, this was not to be, mainly because interest in the two-seat fighter was waning again and both the XFD and the XF3U were dropped.

Vought was somewhat more fortunate than Douglas. Their XF3U had a slight edge over the XFD in speed, range and climb. To use its potentialities they suggested that it be tested in the scouting role. It was then found that with modifications, including increased fuel capacity, it would exceed by a top speed of 20 m.p.h. and a range of 223 miles the performance of existing standard service scouts. The Navy were interested in a scout bomber, and production of 84 SBU-1s resulted.

While the Bureau had, with deliberation, specified conventional features in their Design No. 113, they had placed, at the very same time, an order with Curtiss for a two-seat high-wing all-metal monoplane fighter.

In the fall of 1933 this parasol monoplane arrived at Anacostia for trials. It had a deep belly which allowed a retractable landing gear to be accommodated in much the same manner as the BF2C and Grumman biplanes. An enclosed sliding canopy was used over the mid-section tandem cockpits. A carrier hook and flotation gear were fitted. A ground adjustable controllable pitch propeller was used with direct drive from the 625 h.p. Wright R-1510-92 twin-row radial engine. It was very early then for the variable pitch propeller, but this was on the way, the

The F11C-2 design adapted to a fighter-bomber resulted in the BFC-2 of which the first of ten is depicted at Floyd Bennett Field. The turtle-back of the fuselage was hinged to allow access to a raft stored within the rear fuselage and a feature of the modified cockpit was the inclusion of a canopy. The standard power unit, as in all the 'Hawk' series, was the Wright R-1820-78 engine of 700 h.p. Delivery of the BFC-2s commenced in February 1933.

Hamilton Standard Propeller Company having received their first naval contract in 1931 and September 31st of that year saw the first naval tests on a Curtiss F6C-4.

To stow aircraft satisfactorily aboard carriers meant providing the required surface area for lift by wings of broad chord, rather than long span. Most VF types were limited to a maximum span of 35-36 ft. A monoplane, however, with the necessary lift to be provided in one wing instead of two, might well be expected to exceed the 36 ft. and the new Curtiss XF12C-1 had a span of 41 ft. 6 in. For this reason, it became the first VF type to have provision for wing folding. The wing fold line was just outboard of the cabane struts, hinging at the rear spar; the wing strut hinged and folded back at the fuselage joint line together with the wing. Thus, the entire wing assembly folded back, and was operated simply by unlocking and pushing the wing assembly by hand. The wings featured Handley-Page full-span leading edge slots and trailing edge long span flaps.

During October and November 1933, the XF12C-1 was tested at Anacostia. In spite of its gross weight of 5,379 lb., it did exceptionally well to reach a maximum of 217·4 m.p.h. with a landing speed, due to slots and flaps, of only 64 m.p.h. A service ceiling of 22,500 feet was recorded but the climb was rather sluggish for a VF type and its manoeuvrability did not compare favourably with service fighters. The

Third of the Curtiss F11C series of which the predominant feature was a retractable undercarriage. Construction was of molybdenum steel tubing with part metal and part fabric covering. This XF11C-3 was later re-designated as XBF2C-1.

internal fuel capacity gave an estimated range of 750 miles, and Curtiss showed that this could be extended to over 1,000 miles with additional fuel tanks.

The XF12C-1 appeared to have the attributes of an observation aircraft and by December 1933, it had been fitted with the large Wright R-1820-80 Cyclone engine in a revised cowling and re-designated the XS4C-1 which, in January 1934, again changed to XSBC-1. During tests as a scout-bomber in June 1934 it crashed and was rebuilt, then, during a test dive in September 1934, the wing folded and it crashed again. Re-built as the XSBC-2 with biplane wings, it led, through subsequent modifications, to the well-known SBC-3 and SBC-4 'Helldivers' of pre-war. Incidentally five of these, being the residue of U.S. Navy deliveries to France in 1940, reached the R.A.F. and were used in Britain for ground training as Curtiss Clevelands.

The passing of the XF12C-1 to sire the famous SBC Helldivers proved to be the last two-seat fighter concept for the U.S. Navy and Marine Corps until the latter part of the war. Meanwhile, two-seaters that had evolved two years previous were, from May to November 1933, being delivered to the Navy. These were the 'Fi-Fis' as the Grumman FF-1s were colloquially known. Their retractable undercarriages were not then an innovation but it can be said that they were the first naval fighters to be so equipped. It was also the first service type fighter to have an enclosed cockpit canopy.

The original XFF-1 was brought up to production standard which entailed a more powerful version of the Wright R-1820 engine with a rather larger propeller. This had the effect of increasing the maximum speed to 206·8 m.p.h. at 4,000 feet at its gross weight of 4,677 lb., and dropping the service ceiling to about 21,200 feet, but increasing the rate of climb to the extent of reaching 10,000 feet in seven minutes. An additional ten gallons was added to the capacity of the main internal fuel tank, to increase the range to around 860 miles at cruising speed.

The first 'Fi-Fi' reached VF-5, the only squadron to operate them, in mid-1933 and were used until November 1935 mainly on the *Lexington*. Developed as a scout, an order was placed for an FF-1 with increased fuel capacity as the XSF-1 (BuAer No. A-8940). The configuration, all dimensions and power plant, was the same as for the FF-1s. The only prominent difference between the 'Scout' model and the 'Fighter' version was the longer chord cowl used on the SF-1 types. An order for thirty-three SF-1s (A-9460-9492) was placed and eventually they equipped VS-3.

After yeoman service the FF-1s were turned over to reserve units, the SF types to training duties. Dual controls were installed, and with this and other modifications they were redesignated FF-2s and remained in use until as late as 1938. Marine Air Reserves also utilised them and in 1941 two were still in use by the Marine Corps, one with Base Air Detachment One, Quantico and the other with Detachment Two at San Diego.

When BuAer Design No. 120 was set out in 1933, it produced the last U.S. naval fighter with fixed landing gear. This called for a single-seat VF class shipboard fighter,

The days of the Boeing naval fighters were over and the firm made several bids, including the XF6B-1 shown, to regain their position. However, this machine, shown after modification, failed to reach production. Once famous for naval fighters, Boeing became noted for Army bombers.

The Navy's specification for a converted two-seat biplane fighter brought a response from seven manufacturers, of which two were asked to produce prototypes—Vought and Douglas. The former (Vought XF3U) is depicted; together with the Douglas, it was the first VF type to feature landing flaps.

built around the 600/625 h.p. Wright XR-1510-26 twin-row radial engine. Contracts were awarded to the Grover Loening Aircraft Company Inc., of Garden City, N.Y., and the Berliner-Joyce Aircraft Corporation of Baltimore, Maryland. Loening was assigned the XFL-1 designation and Berliner-Joyce XF3J-1 by a contract of June 30th, 1932.

The Loening Company were currently engaged on development and research work for the Navy and they were also building a few commercial aircraft. With the XFL-1 proposal, Loening developed and improved upon the basic design and submitted it to the Navy, but as calculations did not show promise of improvement upon existing types, the project was cancelled by mutual agreement before construction commenced.

The Berliner-Joyce Corporation was rather more successful in that the XF3J-1 reached the construction stage but there were delays due to administrative changes in the firm. In 1932 the Corporation became a division of North American Aviation, then a 'holding company', and in May 1933, it consolidated management with the General Aviation Manufacturing Corporation with which it merged the following year. Not until March 1934 was the XF3J-1 completed and then trials were not conducted until the following September.

In spite of a 'racy' appearance, this last ditch attempt to prolong the life of the fixed undercarriage, biplane fighter failed. It had an all-metal framework with a semi-monocoque fuselage and fabric-covered wings, and was equipped with full naval gear. During initial trials its best speed was 209·3 m.p.h. at 6,000 feet, and an internal fuel capacity of 120 gallons allowed a range of 720 miles at a cruising speed. In under-wing bomb racks up to four 116 lb. general purpose bombs could be carried. However, other designs promised to be superior before the XF3J-1 could be put into production or modified for possible service use. Since it was prone to excessive vibration, it was officially abandoned for that reason in September 1935. This was the last aircraft to be produced under the name Berliner-Joyce and it cannot but be remarked that this firm had singularly bad luck with both their Army and Navy contracts.

The conventional biplanes were at last being dismissed and after a thorough evaluation three two-seat fighter designs, the Curtiss XF12C-1, Vought XF3U-1 and Douglas XFD-1, were all discarded. Development and procurement of such types stopped and by 1935 the last of the two-place fighters were being retired from fleet service and the type passed out of operational use.

The monoplane was making its mark and the U.S. Air Corps were up with the times by ordering the Boeing P-26 'Peashooter' in quantity. Further interest in this configuration was shown by the Navy on March 20th, 1933, with an order to the Boeing Company for an experimental model known as the XF7B-1, which paralleled the Army's YP-29A. Delivered to the Navy on November 11th, 1933, it incorporated to a degree the features of the modern all-metal fighter with a controllable pitch propeller, a partially retracting landing gear and an enclosed cockpit. The XF7B-1 was perhaps just a little too far advanced for the

When the Curtiss XF13C-1 was delivered to Anacostia on February 10th, 1934, it had already flown thirty-four hours on demonstrations. Before trials were completed in May, it went to the N.A.C.A. for vision charts to be made and was also flown with an automatic trailing hook.

Navy at this time. Lest this appears to be a re-occurrence of the 'diehard attitude', this statement needs some qualification. While a top speed of 233 m.p.h. was a marked improvement, the design with its various innovations had to be proved safe, practical and serviceable. The Navy had the additional problems of landing these fast aircraft, with higher landing speeds, on small carrier decks. If tests are prolonged, the public, who have perhaps been impressed with a sight of the prototype flashing past at an airshow, agitate for the type to be in service and label the Navy's cautiousness as procrastination. On the other hand, should the Navy act quickly and order in quantity a type that fails in service, they may well be called to account to Congress for the squandering of the taxpayers' money.

With the XF7B-1 there were complaints by the testing officers of too fast a landing speed, poor visibility and a manoeuvrability not comparable to existing biplanes. The machine was returned to Boeing's for the installation of flaps to reduce landing speed, and for the open cockpit to be converted to afford better visibility. Subsequent trials showed a reduction of landing speed from 78 m.p.h. to 66 m.p.h. The service ceiling was 26,900 feet but like other early monoplane fighters, the decreased surface area was a handicap to the climb rate and it took over eight minutes to reach 10,000 feet. Repeated refinements with various cowlings, landing gear fairings and weight distribution barely balanced out in performance the additional weight of new equipment and instruments deemed necessary.

Structurally the XF7B was designed for a load factor of 9 'G' but in March 1935, during terminal velocity dives, the XF7B was inadvertently over-stressed to over 12 'G'. At 415 m.p.h. the windshield collapsed and pieces blew back, cutting the pilot's face, but it was kept under control and landed safely. An examination showed only slight damage and minor strain, but following this mishap the Navy retired the aircraft and put it on the scrap-heap as being beyond economical repair. And so the first of the low-wing monoplane fighters, to be tested by the Navy, was rejected. Not an auspicious start!

Within two months of ordering the Boeing monoplane, the Bureau had contracted with the Northrop Corporation of Inglewood, California, for an experimental fighter, the XFT-1. This was a compressed version of the firm's highly successful commercial 'Gamma' mailplane, of all-metal construction including the covering of stressed aluminium. This short, stubby machine at Anacostia in March the following year, proved the fastest fighter type so far tested by the Navy. It featured trailing edge flaps, and ailerons designed to drop in conjunction with flaps which reduced the landing speed to around 65 m.p.h. Powered by a Wright XR-1510-8 engine of 650 h.p. at 6,000 feet initial tests proved it to be a rather vicious speedster with a maximum of 235 m.p.h. and a service ceiling of 26,500 feet. As with the Boeing, poor visibility and high landing speeds were the major complaints, but unfortunately, it was also a 'spinner' i.e. it had an unpleasant tendency to spin. Its large wheel fairings, much in vogue at the time, were a constant source of maintenance difficulty. Modifications were made continuously and by mid-1935 final recommendations were presented and it was returned to the factory for re-work. Among the recommendations made, was a change in power to a Pratt & Whitney R-1535-72 Twin-Wasp

The twentieth production Grumman FF-1 of Squadron VF-5, the first and only fighter unit to operate the 'Fi-Fi', bearing the blue tail of the U.S.S. Lexington. In outward appearance it differed little from the XFF-1 shown on page 59. This FF-1 was flown by Lt. McDonough.

engine which was in production and thereby easily maintained. The wheel spats were modified so that the outer surface panels gave access for maintenance and the wheels were protruded to facilitate carrier landings. By April 1936 the modified aircraft returned to Anacostia as the XFT-2.

With all the changes made the only apparent outward change was a deep drawn metal engine cowling of reduced diameter. When trials got under way again, there was found to be very little change in performance. The 650 h.p. Pratt & Whitney gave a top speed of 234 m.p.h. which was slightly less than the XFT-1, possibly due to a weight increase. More unfortunately it still had a tendency to spin. Again modifications were made at Anacostia, but it came to a point when the test pilots were chary of flying this little Northrop, not without good reason for it dropped into a spin at the least provocation. Moreover, these spins were difficult to control. The aircraft was grounded until Northrop could effect a re-design.

In early July 1936, a Northrop test pilot was sent to Anacostia to fly the XFT-2 back to California for further company trials and alterations. Official Navy orders were that it was not to be flown back. But officials awoke one morning to find it gone! The Northrop pilot had arrived and taken off, against instructions, before sunrise. The next

The first low-wing monoplane fighter tested by the U.S.N. was the Boeing XF7B-1 of 1933, which was similar to the U.S. Army's YP-29A. It was of all-metal construction, with a retractable undercarriage. Testing was carried out with open cockpits, and enclosed as shown.

word came that afternoon from a farmer in Pennsylvania. The aircraft, flying through turbulent air over the Allegheny mountains, had spun in, injuring the pilot and completely destroying the aircraft. There the project ended.

Curtiss made a second bid to supply the Navy with a single-seat monoplane fighter. Possibly because of the inherent stability of the high wing monoplane with which they already had experience, Curtiss kept to this configuration. Smaller than their other monoplane, it differed externally mainly by an enclosed cabin type cockpit and by wings attached directly to the fuselage instead of parasol fashion. Construction was normal for the period with an all-metal fuselage and tail surfaces, and fabric over the metal framework of the wings and rudder. The retractable landing gear was similar to the BF2Cs and XF12C-1s.

The XF13C-1 incorporated several unorthodox features and during its early development was termed a ' mystery ship ' by the press. The original design and construction was carried out as a company sponsored project, before the Navy awarded a contract. The naval designations tend to confuse the actual sequence of events in its development. It was the Curtiss Model 70 and designed before the XF12C-1 which was Model 73. The 70 was designed as a carrier fighter from its inception, that could be fitted with a set of biplane wings for ' better performance '. Such an arrangement would be of advantage for aircraft alternating between carrier work and land stations. For years the XF13C-1 was marked down simply as another 'plane that failed', but in retrospect it is felt that it contained the key to a proper evaluation of the pros and cons of biplane versus monoplane and that, had it been further developed, might have provided at that time, a fighter with the attributes of either as the situation demanded. This recalls the experiments during the War in Britain with the famous Hawker Hurricane, fitted with a slip-wing, so that in biplane form it had increased lift for take-off with overload.

On November 23rd, 1932, the Curtiss machine was ordered and was ready for trials early 1933. As a convertible wing model, the Navy allotted XF13C-1 in monoplane configuration and XF13C-2 as a biplane in which form it first flew. It was a logical progression to first get the feel of the machine as a biplane, but seemingly illogical was the fact that the airframe was marked XF13C-1.

The aircraft was subsequently tested at the Buffalo plant in both versions. A paradox was the fact that the biplane version was lighter than the monoplane by 87 lb. This was due to the heavy strut bracings and their fittings in the monoplane form. The biplane had the shortest take-off run of 504 feet, compared with 617 feet; it also had a higher ceiling and landed $3\frac{1}{4}$ m.p.h. slower than the monoplane's 61·2 m.p.h. All these were factors in which the Navy had a vital interest, but inevitably the drag of the lower wing resulted in a reduced top speed which was only 205 m.p.h., whereas the monoplane reached 223·9 m.p.h.

The last naval Berliner-Joyce and the last fighter project for the Navy with a fixed undercarriage. Excessive vibration and a mediocre performance led to its rejection. The planned armament, included provision for carriage of four 116 lb. bombs.

The Northrop Corporation was controlled by the Douglas Aircraft Company which was, in turn, a subsidiary of Northwest Aviation. An adaption of the Northrop Gamma commercial aircraft design brought about the XFT-1, shown here at Anacostia on April 18th, 1934.

Following factory tests, the aircraft was flight delivered to Anacostia in monoplane form on February 10th, 1934, followed a few days later by a second set of biplane wings. The monoplane wings were equipped with leading edge, full span slats and trailing edge, long span flaps; the new biplane set were of narrower chord and the ailerons were full span and on the upper wing only.

After satisfactory flight trials in both forms it was sent to Dahlgren for armament tests where recommendations included replacing the two ·300 machine-guns with the usual one ·300 and one ·50 machine-gun to maintain the current standard; a matter easily rectified. Later it went to the U.S.S. *Saratoga* for carrier trials which was normally the final stage before approval or rejection. From August to October of 1934 evaluation continued. While generally satisfactory to a degree that a production order was anticipated, some changes were thought advisable. The cabin position under the wing provided poor 'combat vision' for the pilot compared with the conventional open cockpits situated behind the wings. There were also some stowage problems aboard the carrier due to the high fin and rudder.

Next month, the XF13C was loaned to the Army for tests at Wright Field. Although its flight characteristics received favourable report it was considered unsuitable as a military aircraft, let alone a fighter. It was returned to Anacostia and thence to the Curtiss works in February 1935. Practically a new concept was outlined from the many alterations and modifications recommended. With nothing more revolutionary than the vertical tail surfaces reduced in height and rounded, the XF13C re-appeared for stability tests resulting from this relatively minor modification. As that proved satisfactory it then returned to Buffalo for the rest of the modifications.

In its final form, tested as XF13C-3, this convertible Curtiss appeared in May 1935 as a monoplane with increased fuel capacity and leading edge full length slots and flaps. The power plant was the new XR-1510-12 Wright engine of 700 h.p. at 7,000 feet. Its weight had risen to 4,721 lb. but the performance was up also and the top speed was boosted to 232·8 m.p.h. at 7,000 feet with a climb to 5,000 feet in 2·5 minutes. Unfortunately landing speeds were increased too, to 64·9 m.p.h. with slots and flaps employed, and take-off runs and landing distances were greater. Plagued constantly with problems in slat mechanism and engine maintenance that almost daily interrupted trials, it was finally decided to terminate all tests in October. The aircraft, together with its experimental engine, was handed over to the N.A.C.A. at Langley Field. There, it was used for several years as a test bed for constant speed and adjustable pitch propellers, cowlings, slots and flaps. In November 1937 it was turned over to the Marine Corps who used it as a utility with VMJ-1 at Quantico. A year later, it was returned to the N.A.C.A. who dismantled and stored it in their hangar. Its final disposal cannot be ascertained.

If the Bureau were trying to press the biplane to its ultimate, at least the merits of the monoplane were being assessed. Various refinements that increased performance were being introduced on biplanes that could be applied equally to monoplanes. The retractable undercarriage had been introduced on the Grumman FF-1. Variable pitch propellers used on six F4B-4s of VF-3 aboard the U.S.S. *Langley* and one F4B-4 of VF-4 aboard the U.S.S. *Saratoga* in August 1932, marked the initial service acceptance of this type of propeller. Later that same year, contracts were placed for the production of 125 GF radios by the Aviation Radio Corporation, for installation in VF type aircraft, and so standardise radio telephony, for all fighters. These were all improvements that would enhance speed, climb or tactics in the transition that was taking place from the biplane to the monoplane.

A refined version of the airframe shown above, with a Pratt & Whitney R-1537 engine in place of the Wright XR-1510 originally installed. It was calculated that the 120 gallons of fuel carried would give it a 900-mile range with two 116 lb. bombs.

CHAPTER NINE

Grummans Forge Ahead

A short stubby Grumman based on the FF-1 was the XF2F-1 which had a Pratt & Whitney engine in place of the Wright engine of the earlier version. It was a tricky machine to fly, but it proved to be very robust.

While many aircraft types had been tested, few fighters had been awarded large scale production orders. The immediate need was a fighter to replace the Boeing F4Bs and the Curtiss Hawks that formed the bulk of the Navy and Marine Corps equipment in VF formations. Procurement, then and now, has been subject to funds, which in turn is conditioned by the political situation. The 'Wall Street crash' in 1929 brought several years of depression with repercussions on both sides of the Atlantic. Such limited funds as were available for naval aviation, were spent on experimentation rather than production. However, under the National Industrial Recovery Act of June 16th, 1933, procurement of new aircraft and equipment was further assured with $7,500,000 allotted to the naval aviation. The former 1,000 plane programme was to be maintained and a new build-up got under way, together with funds for developing navigational and other instruments designed to give operational efficiency. In 1934 under the Vinson-Trammell Act two new carriers, the *Yorktown* and *Enterprise* had their keels laid.

With new funds the need for the replacement of standard fighters and the dependability of the Grumman FF-1 led to a new single-seat Grumman fighter, that was to take the lead from Boeing. A contract placed on November 2nd, 1932, for a single prototype XF2F-1, was met promptly so that less than a year later this squat, pot-bellied, little biplane was ready for test. With a span of only 28½ ft. it presented no storage problem aboard carriers. Due to its small size some stability problems were encountered and it proved tricky to fly, albeit extremely strong. Although like the ill-fated Northrop, it had a tendency to spin, it could, as a biplane, be more easily controlled.

Trials were run at Anacostia, Dahlgren and Hampton Roads. Minor changes and modifications were recommended for incorporation into a production version, but relatively few compared to some of the earlier VF designs. Powered with a Pratt & Whitney XR-1535-44 engine rated at 625 h.p. at 8,500 feet, it achieved 229 m.p.h. with cruising speed of 150 m.p.h. and a range of 540 miles, while the landing speed was about 65 m.p.h. The service ceiling was 29,000 feet and it climbed to 10,000 feet in 4·7 minutes. All performance tests were made with normal combat gross weight of 3,490 lb. and were sufficiently impressive to put it into production.

On March 17th, 1934, an order was placed for fifty-four F2F-1s. Of conventional construction they had an all-metal stressed skin monocoque fuselage, which was longer by a few inches than the prototype to give better stability; the wings and control surfaces were of metal framework, fabric covered. The cockpit enclosure on production models was

Production version of the model above, the Grumman F2F-1. Squadron VF-2B received this type in 1935 and were still using some as late as 1940. The bulk were delivered to the U.S. Navy but the U.S.M.C. did acquire two in 1938. The aircraft illustrated served with Squadron VF-2 and was from the first production batch.

enlarged and modified and the engine cowling completely re-designed to provide efficient cooling and streamlining for the Pratt & Whitney R-1535-72 engine. Unlike the single exhausts of the prototype, the new version had a collective exhaust pipe. The retractable undercarriage folded into the lower fuselage recesses by means of a hand-cranked gear and chain.

Early in 1935 the first F2F-1s appeared and went to VF-2B who were still operating the type as late as June 1940. VF-3 received a few, supplementary to their FF-1s during November 1935 as an interim measure. VF-7 formed in late 1937 with F2F-1s on a make-shift basis until June 1940. The U.S.M.C. received only two, and then not until 1938; they were assigned to VMF-2.

Although the F2F-1s served faithfully for many years, pilots complained of a tendency to spin and directional instability; undesirable traits that were of much concern to Grumman and the Navy. The problem this presented was taken in hand by ordering an additional F2F-1, BuAer No. 9997, in mid-1935, for study by both the Navy and Grumman to compare matters for a new model under construction. Strangely, in view of its tubby appearance, pilots also complained of cramped cockpits.

In squadron service the F2F-1 could show a top speed of 231 m.p.h. at 7,500 feet and 205 m.p.h. at sea level. At its gross weight of 3,847 lb., the service ceiling was 27,000 feet and it climbed to 10,000 feet in five minutes. On internal fuel of 110 gallons the range was 700 to 750 miles at a cruising speed of 210 m.p.h., but 900 miles could be made at one quarter throttle. The aircraft were easily serviced, and the most violent manoeuvres did not appear to cause strain and in this respect dive bombing presented no hazard to these sturdy aircraft. On August 2nd, 1935, the last F2F-1 was delivered, bringing the total, including the prototype, to fifty-six delivered over eight months.

From 1935 Grumman monopolised the fighter scene until the War. When the American Airplane and Engine Corporation, producers of the American Pilgrim 100-A, ten-seat cabin transport, went out of business in 1932, Grumman moved into their plant at Farmingdale, Long Island.

While the first Grumman fighters, the two-seat FF-1s, were being withdrawn from front-line service, and the F2F-1s were entering service, Grumman already had the XF3F-1 built. Based on its predecessor, it had the same basic construction and even the same power unit, yet it was to cause much trouble and expenditure before it was acceptable to the Navy.

On the morning of March 22nd, 1935, after careful ground checks, Grumman's chief test pilot, Jimmy Collins, who was well known in aviation circles, took the XF3F-1 up on initial company trials. It seemed to be handling beautifully and performing well. Collins landed and made his recommendations for some handling and stability improvements. With these modifications completed the XF3F-1 was ready for official demonstrations before the Navy Board, and for this Jimmy Collins was again at the controls. Once again the Grumman performed with ease. As portrayed so many times in films, the terminal velocity dive marked the climax to testing. From 15,000 feet the F3F-1 was nosed over and put straight down, on full throttle into a terminal velocity dive. At 5,000 feet it was planned to pull out and the terrific amount of 'G' pressure would show any structural defects in the machine. The Navy required several such dives to be made. The last dive in this case took place under many eyes and those watching grew apprehensive as the machine appeared larger, and the engine roar became louder, without sign of a pull-out. Then, with a great thud it crumpled into a mass of metal as it hit the ground

An improved larger version of the Grumman F2F, opposite, was the XF3F-1 with which much trouble was experienced on test and two of these aircraft crashed. The XF3F-1 is shown here during Grumman's trials with Lee Gehlback at the controls. The engine was a Pratt & Whitney R-1535-72.

in a small cemetery across the road from the Grumman plant.

A great test pilot, and a promising design had been lost. The failure was not explained, but there was confidence in the design and another experimental XF3F-1 was ready for trials in May and delivered to the Navy. Again it was proved to be highly manoeuvrable. Then came the final official tests at Anacostia with spins and terminal velocity dives. Lee Gehlback, a Grumman test pilot, was at the controls as the second XF3F-1 started its stringent tests on the morning of May 17th, 1935. During its second dive the little Grumman refused to pull-out, and went into a spin. Again the horrified spectators watched as the machine went down out of control. Gehlback fought for control but realising it was futile baled out at the last minute. Another heavy crash and a second XF3F-1 was washed out, but fortunately this time the pilot was alive. Gehlback reported that the design was directionally unstable at high speeds and the machine fell into uncontrollable spins. Another XF3F-1 was built, and this, the third, came through all tests with flying colours.

69

This Grumman F3F-1 production version of the model shown on the preceding page, is numbered in the first fighter batch of the 10,000 series (10220 of the batch 10211-10264) although the initial digit is not marked. This change is evident from the photographs in this chapter.

During trials top speeds of 226 m.p.h. were attained and a service ceiling of 29,500 feet was estimated. A climb to 10,000 feet was made in 5·5 minutes at a gross weight of 4,094 lb. Production of fifty-four aircraft as F3F-1s ensued. Directional instability was corrected on these by a slight change in wing arrangement and by an additional three inches added to the fuselage length. Like the XF2F prototype, the 'bugs' in the XF3F were worked out through recommended modifications, including more horsepower, incidence changes, added control areas, until the undesirable flying characteristics of the experimental models had been eliminated. In service, there was no criticism on that score. On April 19th, 1936, the first production models were received by VF-5 as replacements for their FF-1s: re-designated VF-4 they later took their F3F-1s aboard the *Ranger* and used them until June 1940. In June 1936, VF-3 received F3F-1s and VF-6 was fully complemented with them by November, by which time the last had been delivered.

Production F3F-1s were powered by a Pratt & Whitney R-1535-84 Twin-Wasp Junior which gave a top speed of 231 m.p.h. at 7,500 feet and a crusing speed of 215 m.p.h. with a cruising range of 720 miles. Their climb was far superior to their contemporary monoplanes at an initial 2,700 feet per minute. The armament remained the conventional one ·300 and one ·50 Browning machine-gun located in the fuselage to fire through the propeller arc.

The fifty-four F3F-1s had received BuAer Nos. 0211-64. By this time the 10,000th number had been allotted, and to maintain a four-digit presentation of the number, the '1' indicative of 10,000 was dropped, e.g. the F3F-1 No. 10211, appeared with only 0211 on the vertical fin. The number 10211, however, appeared on the official Navy records. This applied to subsequent assignments until the War when numbers climbed to six, and later, even seven digit serials, at which time only the last four or five figures were applied, the first digits being dropped.

While Europe was re-arming to contain a resurgence of power in Germany, the expansion in American arms resulted more from the National Recovery Act than participation in an arms race. American aircraft were used in the Spanish Civil War, but the United States joined with other nations in placing an embargo on arms and officially remained strictly neutral. The Japanese were regarded as a potential enemy but their strength, and particularly the quality of their equipment, was seriously under-estimated. The bombing of the American gun-boat *Panay*, sunk by Japanese bombers during the Sino-Japanese War, did bring the Far East into calculations. That the Navy took things seriously is evident from the movements of the U.S.S. *Ranger* in January 1936 when it arrived at Cook Inlet, Alaska, to commence three weeks tests under extreme cold weather conditions. This was the first serious study of the effects of low temperatures on the operational efficiency of carrier-borne aircraft.

Both the Navy and Marine Corps demanded the best of equipment and they pointed at the unrest in Asia and Europe to emphasise their case. From decisions taken some years previous, the U.S.S. *Yorktown* (CV-5) was placed in commission at Norfolk on September 30th, 1937, and on May 12th, 1938, the 'Big E', the U.S.S. *Enterprise* (CV-6) was commissioned at Newport. The new Grummans that were replacing the Boeings were merely a case of biplane replacing biplane. It was evident from trends in Europe and, indeed, in the U.S. Army, that the future fighter would be a monoplane and design and study was underway. Meanwhile, to not merely replace, but to expand to meet the new establishment for the two new carriers, action was taken from mid-1936 to continue with biplanes as an interim measure, pending the acceptance of a service monoplane. This led to improved models based on the F3F design.

Fitting one of the large, 850 h.p. nine-cylinder Wright R-1820-22 Cyclone engines to the latest biplane was suggested in mid-1936. This entailed the use of a much larger cowling, but of shorter chord. The speed was boosted to 260 m.p.h. at 7,500 feet, with the cruising speed as high as 241 m.p.h. A record service ceiling, at least for a fighter, was the 32,000 feet achieved by this model, the XF3F-2. In spite of this performance the landing speed was still only 68 m.p.h. The range on a 130-gallon internal tank was 975 miles at 125 m.p.h. Having proved itself early in 1937 a production order was placed on March 23rd for eighty-one, which constituted the largest single order placed to date for a VF aircraft. By December the first F3F-2 production models were ready for delivery to VF and VMF squadrons, the first being VF-6 which went aboard the new U.S.S. *Enterprise*.

Marine squadrons received their first F3F-2s in June 1938 when eighteen were assigned to VMF-1 at Quantico and seventeen to VMF-2 at San Diego. In fact, apart from the first eighteen to bring VF-6 up to a full complement, practically all the F3F-2s went to the Marines. This was an encouraging sign for the Corps. It could not be said that they had always to put up with handed down equipment; no doubt they had the best equipment that the budget would allow, but in a number of cases their equipment had been on a hand-down basis from the Navy. Their first F3F-1s, six for VF-4M (re-designated VMF-2, July 1937) at San Diego, were received after they had been a year in Navy service. However, with the F3F-2, they were receiving direct deliveries of the latest fighter available.

With minor refinements, that achieved little more than an additional 4 m.p.h., a basic F3F-2 was produced as the XF3F-3 and became the prototype of twenty-seven similar production models. The first eighteen of these went to VF-5 aboard the *Yorktown* to re-place the F2F-1s. Two others went to bring VMF-2 up to strength and the remaining seven were stored for replacements as required. Thus, the last of the biplane fighters for the Navy and Marine Corps went into service. The last F3F-3 came off the assembly line on May 10th, 1939, and later that year, their special water injection systems were removed and they were re-designated as standard F3F-2 models.

In service use the human element was the most difficult to overcome as far as retractable undercarriages were concerned. Many wheels-up landings were made unwittingly. One incident was not without its humorous side. On an occasion at San Diego, a complete squadron of F3F-1s were returning from a training flight. One pilot, failing to lower his gear, hit the ground causing the machine to flip over on to its back. Apart from a bent propeller and dented cowling it was undamaged. As an ambulance raced out to the scene of the accident, the wheels, to the astonishment of onlookers, slowly came out of their wells and went down—or rather up—into the extended position. When the crash crew reached the Grumman, they found the pilot hanging upside down by his safety belt, cursing and swearing, cranking the wheels down as if for dear life, in an apparent endeavour to put things right!

When the Grumman biplanes were retired from active service they were relegated to training use as advanced fighter trainers. In the hands of over zealous students, many incidents occurred including 'wheels-up' landings due, mainly, to the 'trying too hard' attitude of pilots. Generally this resulted in a bent propeller and a scratched or dented fuselage. It came to a point where mechanics were tired of replacing the aluminium outer skin, and decided that it was easier to repair the damaged portion with fabric. That the structure itself stood this treatment without buckling is a tribute to this rugged biplane.

While the interim biplanes were maintaining squadron strength, the Bureau were weighing the pros and cons of which fighter to next place into service. From a design competition let to the industry in 1935, Grumman and Brewster designs were chosen as the most promising. The Bureau, on November 15th, 1935, approved the two designs and initiated developments; Grumman had a biplane—to be designated XF4F-1 and Brewster a monoplane—XF2A-1. (There was no XFA by Brewster as the 'A' designation had originally been assigned to the Fokker/Atlantic/General Aviation Company fighter of 1932; it should also be appreciated that for first designs XF1A for example is abbreviated to just XFA.)

While these developments were under way another monoplane was tested by the Navy, the Curtiss 75 Hawk which had been tried by the Army Air Corps. Through successive modifications it became the firm's P-36 and later P-40. It was demonstrated by Curtiss at Anacostia for naval officials. Thus the Navy had the opportunity to evaluate a truly modern monoplane fighter. Their views, however, were not favourable. It was not considered sufficiently manoeuvrable and violent vibration was said to be encountered in a tight turn or snap roll. The Hawk 75 was returned to the Curtiss Company. If it had not evoked sufficient interest as a type, at least it had shown the Bureau the high speeds the monoplane could attain and the inadequacy of their biplanes in this respect.

Seversky (now Republic Aviation) had produced a navalised version of the U.S. Air Corps P-35. The Army had ordered the P-35 in quantity. This basic design went on to become the famed P-47 Thunderbolt. Study by the Navy of this design, with its many advanced features, prompted Seversky to offer a modified version, more suited to naval use. As a company owned and sponsored project, this aircraft was sent to Anacostia on September 24th, 1937. Registered as a civil aircraft, NX1254, it was designated XNF-1 by the Navy for record purposes. Powered by a Wright R-1820-22 engine of 950 h.p., the XNF-1 had a top speed of 267 m.p.h. at 15,200 feet. Later, official naval trials reported the maximum speed as 250 m.p.h. Tests showed a service ceiling of 30,700 feet and a climb to 10,000 feet in 4·2 minutes. The landing speed, with flaps, was

The original Buffalo, in its initial form when it was known officially only as the Brewster XF2A-1. This was the first full-scale airframe to be wind-tunnel tested in the U.S.A., which was performed in the Langley Memorial Aeronautical Laboratory.

First of the Wildcats built, BuAer airframe No. 0383. The original F4F-1 biplane project was not built and in its place came the XF4F-2 monoplane, which, enlarged and re-engined, flew as the XF4F-3. It is seen in flight on February 12th, 1939.

69 m.p.h. It was not, however, considered to offer sufficient improvement over the Grumman biplanes to warrant production. In one respect it proved most useful, in that the flight test data provided an excellent yardstick on which to gauge the efficiency of the designs from Grumman and Brewster. Later, it was tested on a more or less parallel programme with these two, and rejected.

The Navy were working with the times in plans, if not in production. On July 10th, 1936, they cancelled the Grumman XF4F-1 design as it was evident that the biplane had reached its ultimate. Construction was to have been the same as the previous biplanes with either Wright XR-1670 or Pratt & Whitney XR-1530 radials of 800/875 h.p. Looking more like the F3F-1, than the F3F-2 model, drawings show that it would have had equal span wings of 27 ft. and a fuselage length just over 23 ft. A new contract was then issued to the firm for midwing monoplane, the XF4F-2, with a 950/1,000 h.p. Pratt & Whitney engine—in the event the Pratt & Whitney R-1830.

The Brewster design, was also modified in 1936. A midwing monoplane from inception, it was now to incorporate the 950/1,000 h.p. Wright Cyclone G. Since there was no major change in configuration the designation remained XF2A-1. By the installation of these engines the Navy hoped for a top speed bordering on 300 m.p.h.

The Brewster made its maiden flight in December 1937 and commenced trials at the Brewster Aeronautical Corporation's plant at Long Island, New York. Designed by Dayton T. Brown and R. D. MacCast, this stubby little monoplane was of all-metal construction and flush riveted, stressed metal skin throughout, except for fabric over metal framework on control surfaces. It had many new features, including split flaps and hydraulically operated retractable undercarriage. Power was by a Wright R-1820-22 Cyclone 850 h.p., driving a three-bladed Hamilton Standard hydromatic propeller. The armament comprised the usual ·300 and ·50 forward firing machine-guns located in the top cowling, but there was additionally provision for a ·50 machine-gun in each wing, to fire outside the propeller arc.

Navy trials followed early in 1938. The Wright engine put out 750 h.p. at 15,200 feet at which altitude 277·5 m.p.h. was attained. The service ceiling was 30,900 feet with a credited 2,750 feet per minute initial climb. Its gross weight of 4,830 lb. and landing speed of 67 m.p.h. was well within the limits allowed for shipboard use, but its over-all wingspan of 35 ft. caused a stowage problem aboard ship until a suitable folding wing mechanism was perfected. While these figures were impressive, all the requirements were not met, particularly the top speed. On April 21st, 1938, this prototype XF2A-1 was delivered to the Langley Memorial Aeronautical Laboratory of the N.A.C.A. This marked a new era of testing aircraft. This Brewster was the first full-scale aircraft to be wind tunnel tested in the U.S.A. to determine means of decreasing drag, and improving streamlining to increase the speed.

Results indicated that by improvements its top speed could be increased by 31 m.p.h.—and so reach the de-

First of the monoplane fighters actually in service was the Brewster F2A-1 Buffalo of which the third production model is shown with VF-3. Only sixteen of the fifty-four ordered were delivered to this standard, the remainder being F2A-2s as shown opposite.

sired 300 m.p.h. Modifications, apart from streamlining, such as revising the cowling, included raising the cockpit canopy to allow more head room and a larger fin and rudder area to improve stability. The revised version was accepted by the Navy after trial and on June 11th, 1938, an order was placed for fifty-four production models as the F2A-1. With the same type of power unit, a top speed of 304 m.p.h. at 17,000 feet was attained with 271 m.p.h. at sea level. The climb rate was improved to an initial 3,060 feet per minute. The maximum range at cruising speed was 1,095 miles on the 160 gallon internal fuel capacity. Further tests continued on the XF2A-1, concurrently with production models coming off the assembly line. Meanwhile, the Grumman was reaching the production stage.

The XF4F-2 had flown before the XF2A-1, on September 2nd, 1937, in the hands of Robert L. Hall. The flight was a great success and it appeared to be a ' winner ' there and then, but it had its vicissitudes before it became the famed Wildcat. To the Navy at Anacostia where it arrived

While the XF4F-2 was being repaired the Bureau were evaluating results. The Brewster was considered to have the most possibilities and thereby won the initial contract for production. The Seversky was out—it may have been touch and go with the Grumman, too. The XF4F-2 design undoubtedly had its merits and was regarded a close second to the Brewster. Negotiations between the Navy and Grumman led to a decision to incorporate several modifications on the original design and to use a new engine. This resulted in another contract to Grumman for a redesigned model as the XF4F-3 in October 1938.

The re-designed fuselage, bearing the same BuAer airframe number (0383), was mated to the new Pratt & Whitney XR-1830-76 two-stage, two-speed supercharged Twin-Wasp which gave 1,050 h.p. at 11,000 feet. On the first day in February taxying tests were run and the engine performance checked, which brought out a problem that plagued the machine for some time, adequate cooling. The cowling was modified several times.

Improved model of the F2A-1, the Brewster F2A-2 Buffalo. In addition to two ·50 machine guns in the cowling, some F2A-2s had similar guns mounted in the wings making the first substantial increase in the armament of single-seat fighters for twenty years.

on December 23rd it appeared small and frail. Grumman's estimate of 290 m.p.h. top speed held true, for at 10,000 feet the prototype reached just that, but the service ceiling was below estimates, at 27,400 feet. Throughout January 1938 the test pilots, Robert L. Hall and Selden Converse, put it through trials, but engine trouble continuously delayed the testing of this promising type.

The XF4F-2 was never delivered to the Navy officially. On February 14th it was returned to the Grumman plant for minor changes and additional work on the engine. By the 16th, Hall had it at Dahlgren Naval Proving Grounds for final demonstration trials. Here spins, dives and armament tests were run. Dives were made in clean configuration and also with underslung bombs. Pull-outs up to 8·5 ' G ' were accomplished and a good recovery from spins was reported; this must have brought sighs of relief all round—not least from the test pilots. On March 1st at Anacostia direct competition tests with the Seversky XNF-1 and the Brewster XF2A-1 were satisfactorily run and April brought the deck-landing trials. During these, the engine failed and the machine was brought down to a forced landing, whereupon it overturned and was badly damaged. Lt. Gurney emerged unhurt, but the aircraft had to be returned to Bethpage for extensive repairs.

With Bob Hall at the controls, the XF4F-3 made its first flight on February 12th, 1939, at the Grumman Plant. Later, it was flown to Anacostia for further testing and May found it back at Bethpage for alterations. The wing dihedral was increased, aileron area reduced, and the vertical tail surfaces were enlarged. The XF4F-3 now differed greatly from is first form; the wing span was increased to 38 ft. to give an area of 260 square ft., and, in fact, about the only part of the XF4F-2 model that remained on the XF4F-3 was the undercarriage and part of the fuselage. The gross weight had gone up to 6,103 lb.

Trials at Anacostia now showed a top speed of 334 m.p.h. at 20,500 feet and an initial climb of 2,800 feet per minute. The service ceiling was 33,500 feet. Like the Brewster, in addition to standard armament in the fuselage there was provision for two ·50 machine guns in the wings. The XF4F-3 was flown to Philadelphia for carrier evaluation trials and night flying tests in May during which further troubles were encountered with engine cooling. Finally, towards the end of the year, it was turned over, like the Brewster, to the N.A.C.A. for full-scale wind tunnel tests at Langley Field.

War came to Europe in September 1939 with the German invasion of Poland. On September 5th the United

States President, Franklin D. Roosevelt, proclaimed the neutrality of the United States in what was then a European War and directed that the U.S. Navy form a Neutrality Patrol. From the lessons of actual combat it was evident that weights had to go up. Heavier armament and armour plating were indicated. Battles were taking place at great heights and speeds well in excess of 300 m.p.h. were being attained by standard German and British fighters. However, whereas the U.S. Army Air Corps were learning much from the activities of the Royal Air Force, there was little the Bureau of Aeronautics could learn from the Fleet Air Arm of the British Navy which was operating biplane fighters, obsolete by R.A.F. standards. Not until 1940 were they to get monoplane fighters. This was the result of a national policy of creating the Royal Air Force as a separate service that embraced all service aviation—to the neglect of naval aviation. The Royal Navy did gain operational control shortly before the war, but they had much leeway to make up and research and development was not in their hands.

As 1939 came to a close, the U.S. Navy received its first service monoplane fighters, but the Bureau was not altogether satisfied with their performance in relation to contemporary fighters. A Wright R-1820-40 Cyclone of 1,200 h.p. had been tried in the XF2A-1 prototype (BuAer No. 0451) which thereupon became the XF2A-2. This increased the top speed to 325 m.p.h. at 16,000 feet with a range of 1,015 miles on its 164 gallon internal fuel capacity at a cruise setting of 144 m.p.h., and service ceiling improved to 35,000 feet. It was decided to change the contract from F2A-1s to F2A-2s immediately. The twelfth and subsequent F2A on production was brought up to this new standard. Meanwhile the first of the production models were going to VF-3 aboard the *Saratoga*; they received nine on December 8th, 1939. Until November of the following year they operated the new monoplanes together with F3F-1 biplanes.

The first few Brewster monoplanes were kept aside for further valuation tests and BuAer No. 1386 was retained at Anacostia in connection with engine cooling. Modifications were continuously made, many effective retrospectively on Brewsters in service, including a new variable pitch propeller, increased fuel capacity and improved flotation gear. The 30th production F2A-2 (BuAer No. 1426) was 'pulled out' from the line for additional trials as the Navy's first 'combat modification' aircraft. Modifications were based on reports from the war in Europe. For the first time, since the first Navy single-seat fighters of 1922, an increase in armament was introduced in service; to the two forward-firing ·50 machine guns in the cowling, two additional similar guns were added, one in each wing. Armour protection was added for the pilot and similar protection was fitted around the fuel tanks. Inevitable, with this further increase of weight, the performance, already down with increased equipment, fell further. The service ceiling dropped by 1,000 feet and the maximum speed fell 5 m.p.h with the additional weight that totalled some 900 lb.

The delay between placing orders and receiving deliveries was greatly facilitated by foreign purchases boosting the industry. An important factor in the considerable expansion of the American aircraft industry in 1938-41, that geared production so well for the emergency of late 1941, was the placing of substantial orders by European countries. A British commission in America had ordered 120 Brewster Model 339 aircraft, the equivalent of the U.S. Navy's F2A-2s, in late 1939. Paradoxically, these U.S. Naval fighters were not for the Royal Navy, but for the Royal Air Force as a normal land-based fighter! The first tested under the civil registration NX149B became the R.A.F.'s prototype, W8131, and they named the type Buffalo; this seemed particularly apt, for it was a thick-set machine, originated from America, and the name was alliterative with Brewster. The bulk went to the Far East where they were used jointly by the Royal Air Force and Royal Australian Air Force at Singapore.

The Buffalo was called several unofficial names such as

A new concept and configuration, the Grumman XF5F-1 Skyrocket. The poor visibility afforded the pilot because of the large engine nacelles forward of the cockpit was most unsuitable for carrier landings. Its performance of 358 m.p.h. at 17,500 feet was unprecedented.

Based on the Army's P-39 Airacobra was this Bell XFL-1 Airabonita with a conventional undercarriage, in place of the tricycle arrangement of the former, to facilitate the functioning of the arrester gear. As in the P-39, the engine was mounted behind the pilot.

'Peanut Special' and 'Flying Barrel', but Buffalo stuck and it also came into general U.S. Navy use, a service where aircraft names were the exception, not the rule. Earlier names such as Hawk, Helldiver and Comanche were company trade names, not official names. From the adoption of the name Buffalo onwards, it became usual to accept a name as well as a designation. It appeared to give an aircraft a certain personality and it was operationally expedient, since it was easier to memorise and say 'Buffalo' than 'F2A' particularly when the latter was but one designation in a system of letter/figure combinations for many different types of aircraft.

Brewster had substantial export orders for the Buffalo; Britain, Belgium and the Netherland East Indies were all interested and in all 340 were shipped abroad. The U.S. Navy even forewent the bulk of their original contract for the type. Their F2A-2s entering service were withdrawn and those on the assembly lines, which were completed by February 1940, were sent to Finland to assist in that country's magnificent struggle against Russian invasion. In all forty-four were despatched, which is believed to be all the F2A-2s and the last F2A-1, but it is possible that some of the former were F2A-1s re-engined before despatch to conserve the latest Wright engines. This left ten F2A-1s in U.S. service, nine with VF-3 and one with VMF-221; the latter had also received two F2A-2s but they were withdrawn for Finland at a later date. Although outclassed by the fighters opposing it in all theatres of war, the Buffalo saw service on a surprisingly widespread scale from the far north of Finland to the Dutch East Indies, apart from its limited operational actions with the U.S. forces in the Pacific. Less known, perhaps, are its brief appearances with the Fleet Air Arm in the Crete campaign and, after the fall of Malaya, in Burma where surviving R.A.F. Buffalos co-operated with P-40s of the American Volunteer Group.

Apart from the promising F4F design a completely new design had been mooted by Grumman in 1938. Preliminary proposals were for a twin-engined, single-seat fighter. In the spring of 1939 the project was presented to the War Department, which granted it a patent, embodying new and radical features. From there it went to the engineers at the Naval Aircraft Factory for re-design and study of the constructional principles. A contract was then awarded to Grumman by both the Army and Navy. Further engineering changes were made by Grumman and on April 1st, 1940, the single prototype, BuAer No. 1442, made its first flight. Of all-metal construction, it had an abbreviated fuselage with the nose section ending at the main spar of the wing. A conventional undercarriage was used and it had complete carrier equipment and folding wings. This was the XF5F-1 and a similar aircraft, built for the Army with a tricycle undercarriage, was the XP-50.

The XF5F-1 suffered many delays and there were re-occurring engine cooling difficulties with the 1,200 h.p. Wright radials, so that not until February 1941 did it reach the Navy for official trials. A top speed of 358 m.p.h. at 17,500 feet and 312 m.p.h. at sea level was unprecedented for a fighter and its initial rate of climb of 3,000 feet per minute, with 10,000 feet gained in four minutes, earned the XF5F-1 the nickname of 'Skyrocket'. At the end of

Third version of the Buffalo, the Brewster F2A-3 of which the example shown was the first of the 108 produced. This fighter had the heaviest wing loading of any VF type up to 1941. Within a year most had been relegated as fighter trainers.

July 1941 it returned to Grumman for extensive modifications resulting from these trials.

Originally two 23 mm. Madsen cannon were to be employed in the fuselage nose but Denmark had since then been overrun by the Germans and four of the standard ·50 machine-guns were substituted. To improve stability, the nose had been elongated to protrude ahead of the wing. Further evaluation trials were run but after some 211 flights the project was dropped in favour of a completely redesigned model. Only one Skyrocket was built.

In its existing form the Skyrocket was rejected for many reasons, not least of which was poor visibility, due to the wing-mounted engines obstructing the view of signal officers aboard carriers when landing. The Skyrocket at least proved the feasibility of a twin-engined naval fighter and the Navy requested further study and evaluation which led eventually to the F7F. Due to pressure of work in the Grumman Company, this was considerably delayed and the XF5F-1 Skyrocket itself was not officially delivered and accepted by the Navy until much later, by which time the design had been superseded.

During the time the Skyrocket was under trial the Bell Aircraft Corporation of Buffalo, New York, made their only venture into naval aircraft. The Bell P-39 Airacobra of the Army Air Corps had several features that might prove of interest. In mid-1939 one experimental model was ordered as the XFL-1 and on May 13th, 1940, the first flight was made. From general appearance it resembled the P-39 Airacobra but the planform and profile were extensively altered. It had a greater wing span of larger chord and shorter fuselage than the P-39.

The Bell XFL-1 had a conventional landing gear together with a carrier hook, flotation gear and sundry naval equipment. Power was supplied by an Allison XV-1710-6 in-line engine, located behind the pilot, who straddled the long shaft that ran from the engine to the reduction gear-box in the nose. The pilot's seat was rather higher in the Airabonita, as it was called, than in the Airacobra and the canopy was correspondingly higher. The armament consisted of two ·300 Colt-Browning machine-guns in the upper nose cowling and one ·50 M-2 Browning machine-gun firing through the propeller shaft. Bomb containers could be mounted under each wing.

Under official naval evaluation from July 1940, the Airabonita was regarded with some disfavour. From the outset the liquid-cooled engine was against the present policy. Longitudinal stability was found to be poor and following a wind tunnel test report on a model, in September, 1940, the vertical tail surfaces were enlarged. The emergency exit of a hatch in the rigid canopy was condemned by pilots who preferred a jettisonable canopy. Vibration, caused by the long shafting, was thought to limit its useful life. Such considerations as these are equally as important as speeds which were: a maximum of 335 m.p.h. at 11,000 feet and 305 m.p.h. at sea level, with a landing speed of 72 m.p.h. The reported service ceiling of 30,000, fell short of that under test. All things considered, the Airabonita was not suitable as a naval fighter and so the project was terminated at this stage.

Funds were available for large-scale orders once it could be ascertained that the initial production versions would be serviceable. Congress on May 14th, 1940, outlined a naval build-up which included provision for 4,500 aircraft, and this was raised the very next day to 10,000 aircraft.

The Buffalo was the only monoplane fighter sufficiently developed to hold any promise as a front line naval fighter in the event of America becoming embroiled in the war. Hopes were pinned on the Grumman XF4F-2 which was undergoing tests, but it was suffering many teething troubles

The Brewster F2A-1, variously known as the Buffalo, Peanut Special or Flying Barrel. This example at Roosevelt Field, Mineola, N.Y., on April 10th, 1940, carries both U.S. civil registration and Belgian roundels, being one of a batch initially ordered for that country, but due to events in Europe never delivered.

and the earliest estimate for production deliveries was late 1940/early 1941. A lot devolved on the development of the Brewster and Grumman, not only for the American Services, but as military assistance to Greece, Finland, Denmark, Holland, and France. Countries which were all seeking arms in, what was becoming, the arsenal of democracy for the nations at war.

An improved version of the Buffalo had been projected incorporating increased armour protection for the pilot, newly developed self-sealing fuel tanks, a bullet proof windshield, more modern radio installation and greater power through the use of a Wright R-1820-40 (G105A) Cyclone. The armament would be four ·50 Colt-Browning machine-guns, two in the nose and two in the wings. A production order for 108 of these as F2A-3 Buffalos was placed in the first month of 1940 and deliveries followed from

A Martlet II seen in service with the Royal Navy. This is one of a batch of fifty-four aircraft serialled AJ100-AJ153. This model had folding wings and was armed with six machine guns. Total British purchase of Martlets was about 200 aircraft, some of which were delivered direct to the Far East; a few of these were lost at sea en route.

July of 1941. The last was delivered to the Navy in the fateful December of 1941 but the type remained in production until April 1942, to fulfil an order for the Dutch East Indies. This terminated the entire Buffalo series. VF-2 was given additional Buffaloes of the final version to supplement their F2A-2s. VF-3 received deliveries and VMF-221, by October 1941, was over-complemented.

Before production had terminated one final experiment was conducted with the first of the F2A-3s, BuAer No. 1516, which had been held back for general testing. This was fitted with a pressurised cabin and re-designated XF2A-4. The performance of this version was too poor to warrant further development.

Fortunately, indeed most fortunately, the Grumman was showing promise. The N.A.C.A. had made its recommendations for improvement and Grumman had tooled up for production. This was confirmed on August 8th, 1939, by an order for fifty-one, and in February 1940 the first of the Grumman F4F-3s appeared. Production was slow at first to iron out kinks, and the first two, BuAer Nos. 1844-5, were in the nature of pre-production models. As with the Brewster, so with the Grumman, a series of modifications were incorporated; a pressurised fuel system was installed in all the aircraft and from the third F4F-3, the first true production model, a stronger undercarriage was fitted and armament was brought up to four ·50 machine guns in the wings and firing outside the propeller.

Tests showed that this monoplane now had a top speed of 331 m.p.h. at 21,300 feet, with a service ceiling of 37,000 feet. The Navy stated that although the F4F-3 did not fulfil all guarantees as stated in the contract, it was acceptable as a service type. BuAer No. 1845 was turned over to the N.A.C.A. at Langley Field for further tests, involving engine cooling problems. There, various cowlings and cowl flaps were tried, propellers with and without spinners, and propeller cuffs. The findings were eventually incorporated on subsequent production aircraft. Tests continued with the prototype XF4F-3 until it crashed at Norfolk, Virginia, in December 1940.

Like the Buffalo, there was considerable European interest in this design and the French ordered 100 F4Fs on a contract signed in the autumn of 1939. However, suitable engines were at a premium as manufacturers were taxed to the limit of production, and the commercial 1,200 h.p. Wright R-1820-G205A Cyclone was the only engine available. Although this engine had not then been tried in the Wildcat, the French ordered and tested it. Their first model emerged for tests in May 1940 just as France was invaded, and deliveries were commencing as France fell in June 1940. Britain took over deliveries and on July 24th, 1940, they received the first. Since the first four on U.S. Navy contract had been retained for experimental work, the first production delivery of an F4F type was to Britain. By October, eighty-one had been delivered to the Royal

The first production version of the Wildcat, the U.S. Navy's only efficient VF type in 1941, was the F4F-3. Used by Marine Corps Squadron VMF-121 the example shown at Louisiana bears the white cross marking of 'White Force' during air exercises.

Ninety-five of the 280 F4F-3 Wildcats were fitted with Pratt & Whitney R-1830-90 single-stage supercharged engines and as such were known as F4F-3As. The first of these, BuAer No. 3905, is depicted.

Navy, who thereby received their first monoplane fighter, which they named the Martlet.

When France had accepted the Wright commercial engines, the U.S. Navy decided to do an evaluation and a modification contract was ordered for the third and fourth F4F-3 production machines, BuAer Nos. 1846-47, to be fitted with Wright R-1820-40 Cyclone engines with a single-stage two-speed supercharger. They were first flown in June 1940, and delivered to Anacostia in July as XF4F-5s. During trials, the top speed was only 306 m.p.h. at 15,000 feet which was the same as the British found with their Martlets. The service ceiling however, showed a slight improvement over the standard model. The two XF4F-5s were tested throughout 1940 and many comparative trials were run with them and also on the first production F4F-3 and F4F-3A.

The main difference between the F4F-3 and F4F-3A was the engine installation. Both had Pratt & Whitney engines rated at 1,200 h.p. at take-off, but whereas the '-3' had the R-1830-76 or R-1830-86 engines with two-stage, two-speed supercharger which gave a top speed of 331 m.p.h. at 21,300 feet, the 3A had the R-1830-90 with a single stage two-speed supercharger which gave a speed of 312 m.p.h. at 16,000 feet. The latter had a better initial rate of climb, but suffered a loss of some 3,300 feet on the 37,000 feet service ceiling of the standard F4F-3. This was a compromise forced on the Navy to facilitate production in view of an engine shortage. Initially the F4F-3A had been designated F4F-6.

While U.S. Navy and Marine Corps squadrons prepared to put the F4F in service, the British already had the Martlet in action. On Christmas Day, 1940, a Junkers Ju88 attacking the naval base in the Orkney Isles, was shot down by defending Martlets. In this encounter the F4F became the first American fighter with the British to shoot down an enemy aircraft.

Of the U.S. Navy and Marine Corps fighters of the Second World War the best known is probably the Corsair, which was conceived as early as February 1936 and of which preliminary specifications and drawings reached the Navy in 1937. It was born under the Chance Vought Division of the United Aircraft Corporation, but due to re-organisation was produced by the Vought-Sikorsky Division at Stratford. The Pratt & Whitney Aircraft Corporation were a division of United Aircraft and this new fighter project was to receive the new eighteen cylinder double-row Wasp that the firm was developing. A contract was awarded for one experimental aircraft, the XF4U-1, after the design had been examined by the N.A.F.

Some two years elapsed before the prototype flew. Many problems were involved with the engine envisaged, which at 2,000 h.p. was of almost unprecedented power. It was even looked upon by the Navy as unsuitable for carrier use, but as a land-based fighter it appeared unequalled. With the power available, a special, highly efficient airscrew was designed—the three-bladed Hamilton-Standard hydromatic, constant speed, fully feathering propeller. This large propeller necessitated either long undercarriage strutting or, as the design proved, an inverted gull wing. This wing not only solved the undercarriage problem but gave less drag than the conventional form.

This ingenious design evolved under an engineering staff headed by Rex. B. Beisel of the Vought concern. In February 1939 a model was wind tunnel tested and the full size mock-up was reviewed by naval officials. Shortly thereafter, the go-ahead was received to proceed with construction as a fully navalised aircraft. On May 29th, 1940, the XF4U-1 made its maiden flight at Stratford, Connecticut with Lyman A. Bullard, the firm's chief test pilot, at the controls. Spot-welding, developed by the N.A.F. was extensively used on the Corsair. The all-metal fuselage was of full monocoque construction and the cantilever wing utilised N.A.C.A. 23015 aerofoil. Leading edge areas of the wing and all inboard wing panels were metal covered, but the remainder of the outer panels were fabric covered on all early models. Control surfaces were of metal framework, fabric covered. The landing gear was hydraulically retracted to be completely sealed, with doors fitting the contour of the wing undersurfaces. The front of the wheel struts had large flat plates which were used as air brakes when the wheels were lowered. The tail wheel and later the carrier hook were retractable by hydraulic control.

The performance of the XF4U-1 was actually above the preliminary estimate. Using an early version of the R-2800 series of engines, the XR-2800-2 of 1,850 h.p. at take-off, maximum speeds nearing 400 m.p.h. were obtained for the acceptable landing speed of 75 m.p.h. The service ceiling was over the 35,000 feet mark and the initial climb rate was 2,650 feet per minute. The armament consisted of the existing standard of two synchronised ·300 Browning machine-guns in the nose cowling; additionally, provision was made for two ·50 machine-guns in the wings, firing outside the propeller arc.

On October 1st, 1940, the XF4U-1 was flown to Hartford for a last minute check of the engine before delivery to the Navy. During this timed flight the prototype reached 405 m.p.h. in level flight. The XF4U-1 passed performance tests satisfactorily but, as with most new types, modifications were deemed necessary. These ranged from poor canopy visibility, which was a simple matter of redesign, to instability for which it was turned over to N.A.C.A. The engine brought its own problems particularly with the supercharger and ducting systems. Rectification of these shortcomings took time but there was promise of production by a 'letter of intent' from the Navy in March 1941 followed by a large production contract awarded in July. If Grummans were to the fore at the moment, there was certainly a promise that this might be changed in the near future.

While hopes were pinned on the XF4U for the future, the F4F was the fighter of the present. The Wright-powered Martlet Mk.1 was reasonably satisfactory. There were several snags with the two-stage engines which were being further developed to meet new contracts, so that one hundred models with the Pratt & Whitney R-1830-90 single-stage supercharged engines were ordered as the Martlet II. Performance in consequence suffered somewhat, but the Royal Navy could not afford to choose. By March 1941 the first Marlet IIs were ready for shipment. After British representation, studies had been made as to the feasibility of folding wings, but delays occurred due to the engineering problems involved. The first few Martlet IIs had fixed wings, but all subsequent featured folding wings.

Further Martlets found their way into British hands. In late 1940 Greece, engaged in an heroic defence against Italian troops operating from Albania, requested American aircraft. The U.S. Navy agreed to divert their first F4F-3As on order and on March 18th, 1941, thirty were shipped to Greece. But on April 6th, German troops threw their weight into the battle and the country was overrun. The F4F-3As en route were diverted to Gibraltar where the Royal Navy took them over. Since these were intended as landplane fighters, the Greeks did not require folding wings and they went on British records as the Martlet III.

Following the British lead in ordering the Martlet to incorporate folding wings, the Navy ordered a test article to utilise this feature. One prototype was built incorporating a system similar to the hydraulically operated mechanism of the Martlet II. However, it was not until April 14th, 1941, that this aircraft, the XF4F-4, made its maiden flight at Bethpage. In May it was turned over to VF-42 for carrier trials aboard the U.S.S. *Yorktown*. Everything went according to schedule, but the added weight and complexity of the hydraulic system was deemed unnecessary. Tests showed that the wings could be folded or unfolded manually by the deck crews equally as fast as with the hydraulic system. The saving in weight would allow wing folding with performance to equal that of the fixed wing F4F-3 model. Folding wings were becoming an operational necessity with the new class of small escort carriers being planned. An order was placed for production models as the F4F-4.

The U.S. Navy was getting rather perturbed about the late deliveries of their F4F aircraft which they were beginning to call Wildcats, in spite of the R.A.F. having chosen Martlet—a not very inspiring name it must be admitted. The diversions to other governments resulted from decisions taken at Secretary of State level, and the Secretary of Navy had to acquiesce. Further substantial orders were placed with Grumman, but by December 1940 only twenty-two F4F-3s had been delivered for assignment to service squadrons, and those went to VF-4 at Norfolk. Later that month, this squadron became VF-41 aboard the U.S.S. *Ranger* on Atlantic duty. As the year turned, VF-7 (redesignated VF-72) replaced their F3F-1 biplanes with the new monoplanes and went on Pacific duty. As production got into stride, so re-equipment took place. VF-71 (ex VB-7), flying Vought SBU and SB2U scout bombers, reorganised as a fighter unit in March with eighteen Wildcats. The Marine Corps started receiving theirs during 1941; at Quantico sixteen F4F-3As were assigned to VMF-111, eighteen F4F-3s to VMF-121 and seven to BAD-1. In Hawaii, VMF-211 received fifteen and VMF-221 a single Wildcat. So the build-up went on.

When the Japanese struck at Pearl Harbor the total fighter strength for the U.S. Naval Services comprised 183 Grumman F4F-3s, sixty-five Grumman F4F-3As and approximately ninety F2A-3 Brewster Buffalos. Of the actual units with which they served, a quarter were not fully organised, and others were not up to strength, while pilots had not received sufficient time to get used to their new machines before facing the supreme test of war. From all the experimentation and development that had taken place in peacetime, and of all the fighter types that had been reviewed, only two were classed as combat-worthy and the test of war was to eliminate one of these, leaving Grumman's well to the forefront to meet the test of war.

The characteristic inverted gull wing, shown here to good effect, reveals this as a Corsair. This is in fact the prototype of the series which first flew on May 29th, 1940. This was the first VF type to top 400 m.p.h.

CHAPTER TEN

Fighters of Fame

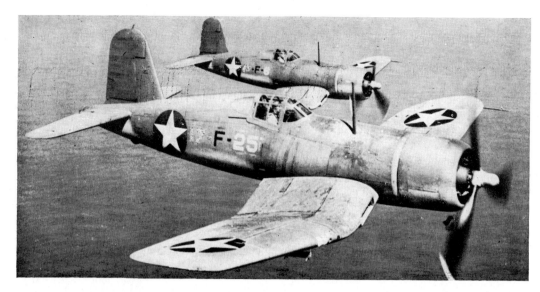

Early service Corsairs with VF-17. Note the early type of cockpit canopy. This was replaced by a semi-bubble type on later aircraft. From the time VMF-124 first took the F4U into action in February 1943 until the Japanese surrender in 1945, Corsairs flew 64,051 sorties, including one by that famous pre-war airman, Charles A. Lindbergh.

At 07.58 hours on Sunday, December 7th, 1941, Rear Admiral Bellinger signalled from his Headquarters at Ford Island in the Hawaiian Group—'AIR RAID — PEARL HARBOR — THIS IS NO DRILL'. This was three minutes after the first bombs had dropped on the Island. Although the primary targets were the battleships in Pearl Harbor, Japanese fighters and dive bombers were first directed to the air bases to prevent interference by intercepting aircraft, and some thirty-three aircraft, about half the total on Ford Island, were destroyed. At Ewa, the Marine Corps group air station, forty-eight aircraft were parked, in some cases wing-tip to wing-tip; they were destroyed with such ease by a formation of Japanese 'Zekes', that the next attacking formation concentrated on buildings and installations. All ten Wildcats there were destroyed and in fact, the only Marine aircraft to escape damage was an R3D (Douglas DC-2) under repair at Ford Island.

Kaneoke, the newly established seaplane base, across the Pali from Honolulu was attacked by 'Vals'. Of the thirty-three PBYs (Calatinas) that were on the station at the time, twenty-seven were destroyed and six damaged; while on the Army fields of Hickham and Wheeler, sixty-five aircraft, many of them B-17 Flying Fortresses, were left a tangled mass of metal. In all, 188 Army and Navy aircraft were destroyed in two hours. The Japanese declaration of war on the United States and the British Empire, received later, came after the holocaust of Pearl Harbor, as something of an anti-climax.

No Navy or Marine aircraft had been able to take-off from Ewa, where the destruction included nine Wildcats. Two days after the fateful 7th, Lt. D. D. Kliewer and Sgt. W. J. Hamilton, intercepting a Wake Island raid, shot down a Japanese bomber; this was the first enemy aircraft to fall to the guns of the Wildcat. Little more could be hoped for with the depleted strength, except from ground defences, until more aircraft arrived from the States.

That day, the *Saratoga* with eighteen Buffalos of VMF-221 aboard, was docked at San Diego. Within twenty-four hours the carrier sailed to relieve the beleaguered Wake Island. They arrived too late. Wake Island fell on December 23rd, 1941. Regrouped at Ewa, VMF-221 was put aboard the *Saratoga* again and sent to defend Midway, the next stepping stone of the Japanese in their eastward advance across the Pacific. It was several months before they were to have an encounter with the Japs but when it did come, it sealed the fate of the Buffalo as an operational aircraft. In May the original fourteen Buffalos were augmented with seven more and also seven F4F-3 Wildcats. They were alerted on June 3rd when a Catalina reported a Japanese force at sea, and brought to readiness next day when another Catalina radioed that an enemy fleet was heading for Midway. VMF-221 sent up nineteen Buffalos and sixteen Wildcats. They ran into a formation of 108 aircraft of the Imperial Japanese Navy. After a short engagement the score was six Japanese aircraft destroyed, for a loss of thirteen Buffalos and two Wildcats missing. Of the few fighters left on the ground, only two Wildcats were in flying condition after the bombing attack. In one brief encounter, VMF-221 was wiped out on Midway.

The F2A Buffalo had proved a dismal failure in actual combat. Captain Philip White, one of the survivors of the fight at Midway, boldly stated that any commander who ordered a pilot up in combat in an F2A should consider the pilot lost before leaving the ground. Even the Wildcat was sluggish in comparison to the Zero. They did far better than the Buffalos, but as Captain Marion Carl, one of VMF-221's survivors, stated, the 'Zero's' speed and versatility was amazing. Embodied in reports, this news caused the withdrawal of the Buffalo from active service. The Navy and Marines would have to do battle with the Wildcat or nothing at all and more orders were placed for F4F-4s. The design of the Buffalo was basically sound, but with all the modifications that had ensued since its first appearance, the wing loading was so high as to render it unstable, quite apart from seriously affecting performance. One report stated that it took over thirty minutes to reach

The XF6F-3, the prototype of the Hellcat series, which first flew on June 20th, 1942 is shown here in a three-tone paint finish. This type, with a wing area of 334 square feet had the lowest wing loading of any U.S. Naval wartime single-seat fighter.

21,000 feet! During 1942, all remaining Buffalos were used as fighter trainers. The Buffalo had the distinction of being the first monoplane fighter to be put both into production and service with the U.S. Naval Services, although it was by no means the first monoplane in service; the Douglas TBD Devastator, Vought SBU Vindicator and the famed PBY Catalina were all in use before the Buffalo introduced the monoplane configuration to fighters—which is a point not without significance. The success of the above types was yet to come to the monoplane fighter.

Using a Buffalo on May 26th, 1942, Commander C. Finke Fischer, successfully demonstrated the feasibility of jet assisted take-off (JATO) at Anacostia. During this test flight, Fischer used five British solid propellent rocket motors attached to his F2A-3 which showed a 49 per cent. reduction in take-off distance, but apart from this minor distinction, the operational history of the Buffalo is a tale of woe. In U.S. service in the Pacific it was a dismal failure and it was shot out of the sky by the Japanese Zeros as were the Buffalos in R.A.F. and R.A.A.F. service at Singapore. The few in Britain were relegated to training, those in Crete at the time of the German invasion were reported to have been sabotaged, and those sent to Finland arrived too late for the Winter War. In fact the Finnish Buffalos were used by No. 24 Squadron of that country's air force when they became a German ally.

The particular thorn in the U.S. Navy's Pacific side was the Japanese A6M 'Zeke' or 'Zero' fighter, which had first flown on April 1st, 1939, a couple of months after the first Wildcat. Perhaps the most astounding thing is that it came as a surprise, since this aircraft was in operation in China eighteen months before the Pearl Harbor attack. Either the United States intelligence services were inefficient in this respect or their reports were ignored because there was a popular belief, soon to be proved unfounded, that the Japanese were ineffectual fighters and that their equipment was poor.

Fortunately for Britain, America agreed to throw her main weight into the European theatre first, leaving the Navy to contain the Japanese in the Pacific. The Marines, smarting under their initial losses, tried to obtain fighters from the Army, but none could be spared. Quite apart from the American Services need, under the Lend-Lease Act of 1941 the United States provided defence material without cash payment to any country 'essential to the defence of the United States' and quantities of aircraft were being shipped to Britain and Russia in particular. Time was on America's side and time meant production and development.

Japanese seaplane fighters, mainly single-float 'Rufes' had been evident during the Guadalcanal campaign, and the U.S.N. were conscious of the fact that this was a development they had neglected since the 'twenties. As an experiment an F4F-3 Wildcat, BuAer No. 04038, was fitted with twin floats designed, built and fitted by the Edo Corporation of College Point, Long Island. These were tested on February 28th, 1943, by the Grumman test pilot F. T. Kurt. Some stability problems were encountered and additional small, movable fins were applied to the hori-

The first production version of the Hellcat, the F6F-3. This type first entered combat on August 31st, 1943 in the hands of VF-9 during operations off the Marcus Islands. It was armed with six ·50 machine-guns mounted in the wings.

BuAer No. 16373, one of the 4,777 FM-1 Wildcats built by Eastern (General Motors Corporation) photographed after test in overload condition on February 28th, 1944. This type with a wing area of 260 square feet had a much higher wing-loading than the Hellcat.

zontal stabiliser. Although successful, performance naturally suffered and since there were few occasions on which fighter seaplanes would be required operationally, the project was not taken further except to try out later the suitability of the F6F-3 Hellcat and SB2C-4 Helldiver as floatplanes.

Late in 1941, the Navy felt the need for a long range photo-reconnaissance aircraft; a standard F4F-4 Wildcat adapted for this became the F4F-7. On the penultimate day of 1941 the first '7' flew and with the urgency of war tests were successfully completed at Anacostia within a few weeks and were followed up by an order for 100. No armament was fitted and the emergency fuel tank behind the pilot was replaced by a camera. The wings were filled with 555 gallons of fuel and to facilitate this were non-folding. Together with the normal internal fuel tankage, a total of 685 gallons was carried which gave the astounding range of nearly 3,700 miles. Fitted with an automatic pilot, an endurance of twenty-four hours was claimed possible, although it is doubtful any pilot could safely endure such a flight in the Wildcat. Emergency fuel jettisoning pipes were extended aft of the tail section just below the rudder. The gross weight went up to 10,328 lb. which limited it to operation from a land base when fully loaded. However, the task of photo-reconnaissance was being served efficiently by other types more adaptable and the original order was cut-back to twenty-one and subsequently a number of these were converted to standard F4F-4 models.

The last of the Grumman-built F4F series came in late 1942. The new escort carriers with their shorter flight decks necessitated more power for take-off and, if possible, a slower landing speed. Two were ordered as XF4F-8s powered with a single-stage, two-speed, supercharged Wright XR-1820-56 which gave 1,350 h.p. for take-off. The first model, BuAer No. 12228, flew on November 8th, 1942, and went to Ancostia the following month, while the second, BuAer No. 12229, was retained by Grumman for evaluation trials. Both were at Anacostia in early 1943 and comparison tests were made with slotted wing flaps and split wing flaps respectively, of which the latter was adopted. To cope with the torque of the powerful Wright engine, the vertical tail surfaces were increased in height and area. The performance was about the same as the production F4F-4 but a slight decrease in gross weight and the extra power achieved a shorter take-off, a better rate of climb and a slightly better ceiling of 36,400 feet.

Grumman with the XF5F-1 under test, the F6F Hellcat under development, the JRF Goose, J2F Duck and J4F Widgeon all under production along with the first TBF Avenger models, was over-burdened. To meet the demand for Wildcats, Eastern Aircraft was established in the first month of 1942, composed of five East Coast plants, all sub-assembly manufacturers of the General Motors Corporation. The Linden, New Jersey plant took on the assembly of the Wildcat under licence agreements and the XF4F-8s were used as patterns. By April when tooling up was underway, Eastern was awarded a contract for over a thousand F4F-4 models which were designated FM-1. The first, took the air at Linden late the following August.

The Wildcat was producing aces. Lt. Edward Henry 'Butch' O'Hare became the Navy's first ace of the war and the second in U.S. naval history on February 20th, 1942, by shooting down five Japanese 'Kate' bombers in seven minutes during an attack on the *Lexington* off Bougainville. During this battle the Japs lost eighteen bombers to the Navy's two Wildcats. Aces such as Capt. Marion Carl, U.S.M.C. (18½), Capt. James E. Sweet, U.S.M.C. (16½), Warrant Officer Henry B. Hamilton, U.S.M.C. (7) made their first victories flying Wildcats, and Capt. Joseph Foss, U.S.M.C. achieved all his 27 victories by February 1943, in a Wildcat. Not until early 1943 did reinforcements for the F4F arrive in the form of the F4U Corsair. Until that time, the Wildcat held the line in the Pacific.

Probably no other U.S.N. aircraft has ever undergone as many modifications and such varying operational changes, as did the Corsair. In spite of over 500 major engineering alterations and 2,500 minor changes, the Corsair became the Navy's premier fighter. The first of the initial production batch of F4U-1 Corsairs flew on June 25th, 1942. These models differed considerably from the preceding 'X' model. The fuselage was lengthened by over a foot and it contained additional fuel tanks to make up a total of 250 gallons in self-sealing tanks which necessitated moving the cockpit aft, to a position just behind the trailing edge of the wing. The extra fuel, 155 lb. of armour plating and six ·50 machine-guns with 2,350 rounds of ammunition put the weights up considerably to 8,982 lb. empty, 12,050 lb. gross and a maximum allowable of 14,000 lb. In fact it weighed almost as much empty as the XF4U-1 at full weight. Due to a 2,000 h.p. Pratt & Whitney R-2800-8 engine the performance figures went up also, to give a top speed of 417 m.p.h. at 19,900 feet, 359 m.p.h. at sea-level, or 182 m.p.h. cruising. The service ceiling was 36,900 feet.

On October 3rd, 1942, the initial ten F4U-1 Corsairs were delivered to VF-12, then temporarily assigned to the *Saratoga*, for service test evaluation—and were reported on unfavourably! They were found to be difficult to land, with visibility poor. In the final approach to a carrier the view of the Landing Deck Signal Officer was obscured at the crucial moment. As a result the Corsairs were assigned to the Marine Corps, since their operations were restricted to land bases. Meanwhile VF-12 returned to San Diego where, in January 1943, six of their pilots conducted tests on the anti-blackout suits. Utilising the Corsairs, they reported more favourably. By the end of the month, they turned their Corsairs over to the Marine Corps and re-equipped with Hellcats—the Grumman F6F.

The Hellcat had been quickly developed. On the same day that the F4U-1 Corsair had been ordered into production, June 30th 1941, a contract had been signed for the Grumman XF6F-1 as a replacement for the Wildcat. It was developed to do battle at high altitudes as a result of combat reports from the European theatre, but modified subsequently for the requirements of the Pacific theatre. The initial design was approved after the results of a wind-tunnel test at Washington, D.C., of a one-sixteenth scale model, had been evaluated.

Previously, only single prototypes had been ordered for new type designs; now two became the general rule to permit concurrent comparative tests with different models under development. Two XF6Fs (BuAer Nos. 02981–2) were therefore ordered. Basically similar to the Corsair, they were of larger proportions and of low-wing monoplane configuration, which permitted the retraction of the wheels into the wing centre section instead of the fuselage as in the Wildcat. Unlike the Corsair, they were at this stage committed as carrier-borne aircraft and the Navy was prepared to sacrifice top speed and ceiling for manoeuvrability, visibility, rate of climb and a low landing speed. This proved a wise decision and met the operational requirement, at least in the Pacific—and with more powerful engines available, the speed did go up considerably.

The XF6F-2 was a project to employ a turbo-supercharged engine, but due to engine delays, this was not effected until much later. However, the basic airframes were being constructed and the second, BuAer No. 02982, had the successful Pratt & Whitney R-2800-18W engine installed for which the designation XF6F-3 was applied. This combination proved so successful that it was put into production before full evaluation tests were completed. Meanwhile the original airframe, BuAer No. 02981, planned as the XF6F-1, was being variously modified, but not until 1943 did it emerge with an R-2800-27 engine and the designation XF6F-4.

Fortunately trials with the XF6F-3 went smoothly and facilitated its quick introduction. It might at an early stage have been considered as a Corsair replacement, but in effect the production models supplemented the later Corsairs which had been developing concurrently. Production was well in hand during 1942 and early in 1943, VF-9 became the first of the squadrons to be fully equipped with the F6F-3 which they took into action in late August. As F6F-3 was the first production version and F6F-1 existed only on paper, it might be thought logical to designate this F6F-1 and 'forget' the original paper project, but apparently once recorded, an aircraft designation is, to the Bureau of Aeronautics, as inexorable as the moving finger in the *Rubáiyát* of Omar Khayyám.

Apart from minor revisions to the fin and rudder, engine cowling, undercarriage and general fairing, the F6F-3s did not differ much from their prototype. Perhaps the most significant change was the substitution of a Hamilton Standard hydromatic propeller, as used on the Corsair, for the original Curtiss electric fully feathering model. Modifications in service followed much the same pattern as with the Corsair. Its large wing area of 334 square feet gave the Hellcat the lowest wing loading of any single-seat fighter then in American service, which accounted for its ease of handling, particularly for landing, on carriers. For stowage, the wings folded rearwards. That it was well received is evinced by orders, initially for 184, then 250—and then for 3,139 of the first production model, the F6F-3.

Powered by the Pratt & Whitney R-2800-10W engine, with water injection, which developed 2,000 h.p. at sea level and 1,975 h.p. at 16,000 feet, the F6F-3 had a top speed of 375 m.p.h. at 17,300 feet and 335 m.p.h. at sea level. Its normal range on the 250 gallons internal fuel capacity at a cruising speed of 160 m.p.h. was 1,090 miles; with an auxiliary belly tank, this range was extended by 600 miles. On some later models, this performance fell slightly. The Hellcat did not have the speed or climbing ability of the Corsair, but more than made up for this, through lighter

A Corsair built by Goodyear, an FG-1D. The 'D' version was equipped with pylons for the carriage of bombs and these fittings are clearly visible here. The machine depicted, BuAer No. 92215, is shown in post-war Reserve service at New York N.A.S.

The F4U-1C version of the Corsair in which the standard armament of six ·50 machine-guns was replaced by four 20 mm. cannon. The total number of all versions was 12,571 — more than any other type ever produced for the U.S. Navy. Initially the Corsair proved to be a difficult aircraft to operate from the decks of carriers.

weight, by being much more manoeuvrable and, above all, easy to land on a carrier. The load capacity was nevertheless good; two 1,000 lb. bombs could be carried in addition to the fixed armament of six .50 machine-guns mounted in the wings with a total of 2,400 rounds of ammunition. The gross weight of 12,400 to 13,025 lb. included 200 lb. of armour plate.

While in Britain the R.A.F. and Royal Navy had to place their orders through a Ministry acting as an agency— the Ministry of Aircraft Production which controlled all aircraft research, development, production and inspection— the two American Services each placed their own orders and had their own establishments for development. But even so, the aircraft ordered and even held by these two U.S. Services could not be called their own. A Joint Aircraft Committee under the Lend-Lease administration decided priorities and allocations. In some cases orders were placed direct with American firms and in some cases U.S. orders were diverted, so that some were numbered as U.S. naval aircraft and others were not, an issue which confuses service deliveries, e.g. the BuAer Nos. 55784–56483 allotted to 700 Corsairs were re-allotted JT100–799 by the British, but even then there were various diversions to a number of other countries.

The Corsair was reaching squadrons early in 1943. VMF-124 first took it into combat on February 2nd, 1943, when they flew escort for the PB4Y Liberators of VP-51 on a daylight attack against enemy shipping around the Kalili area of Bougainville.

Operating Corsairs from carriers initially brought trouble when VMF-124 and VMF-213, the first Marine squadrons aboard carriers, took off from the *Essex* on one occasion; the first one spun in, then another, and a third dropped into the sea. As strikes against Okinawa started, four more were lost in water landings, spins during final approach or deck crashes. In contrast, during the Hellcat's initiation nothing untoward occurred. It was some time before the Corsair became a standard carrier fighter.

From the end of 1941 F4F-4 Wildcats were being delivered. A proposal for a change to the Pratt & Whitney R-1830-90 engine was mooted and registered as the F4F-4A, but this was cancelled. The F4F-4B was a special version under Lend-Lease for Britain's Fleet Air Arm where they became the Martlet IV. Shortly afterwards the British adopted the American name Wildcat too; their Mks. I to III were often still called Martlets, but Mk. IV and above became Wildcats; officially the name Wildcat applied retrospectively to all marks. The F4F-4 became the most widely used of the Wildcat series and during the first year of war they provided the bulk of the fighter strength in both U.S. Navy and Marine squadrons.

Of the many special projects tried with the Wildcat

Four-cannon armament was also a feature of the Curtiss XF14C-2, a fighter that failed to meet naval requirements. The XF14C-1, planned for an in-line engine, was cancelled in favour of this experimental version with a turbo-supercharger and contra-rotating propellers.

was the possibility of increasing the range of single-engined aircraft for special long-range operations. In this connection Lt. Cdrs. W. H. McClure and R. W. Deubo had their Wildcats hooked up by tow lines to a Douglas Boston. After cutting their engines, they were successfully towed as gliders at about 7,000 feet for an hour, averaging 150 m.p.h.

F4F-4 Wildcats, with detail changes were built by the Eastern Aircraft Division of the General Motors Corporation as the FM-1. They were immediately followed on the line in the fall of 1943 by the FM-2 and production continued well into 1945. Their performance was similar to the XF4F-8s which had the same engine, the Wright R-1820-56. Speeds were about the same as the F4F-4's but the service ceiling of 35,600 feet was an improvement, and their rate of climb was particularly good. Compared with the FM-1 they had a taller fin. Both FM-1s and FM-2s were supplied to the Royal Navy as their Wildcat Mks. V and VI respectively. Altogether 1,082 Wildcats were delivered to meet British requirements in various theatres of war.

was the fixed wings of the FG-1. Combat experience reports coming back from the Pacific area with notes pertaining to in-the-field modifications to suit specific requirements, led to production modifications. The F4U-1 series emerged under new contracts as the F4U-1C and -1D. A contract was awarded to Vought for 190 F4U-1C Corsairs. The 'C' denoted an armament change to four 20 mm. cannon in the wings, taking the place of the ·50 machine-guns. Other changes were made, notably in the cockpit and canopy. The pilot's seat was raised seven inches and a clear unobstructed canopy of corresponding height was employed, to improve the all-round vision. A new Pratt & Whitney engine, the R-2800-8W, was then available and was installed. This gave a performance similar to standard F4U-1s but the water injection feature of the '8W' provided additional bursts of speed in an emergency. While the F4U-1Cs were under construction additional contracts were let to Vought and Goodyear for the '1D' version. These were in all respects the same as the '1C' model,

F4F-4 Wildcats, the major production version of this type. Over 200 of this model, designated F4F-4B, were delivered to the Royal Navy, the first being delivered to Speke, England in November 1942. A few were also used by No. 590 Squadron R.A.F. at Mombasa. For over a year after Pearl Harbor, the Wildcat had to combat technically-superior Japanese aircraft, but several Navy and Marine Corps pilots ran up high scores during this period.

Like the Wildcat, the Corsair was sub-contracted. While Vought-Sikorsky was the main producer, large contracts were placed with the Goodyear Aircraft Corporation of Akron, Ohio and the Brewster Aeronautical Corporation. Their respective products were distinguished in designation by the suffix letter to the type; Goodyear had the letter 'G' and Brewster, previously assigned 'A' for the Buffalo, retained this letter. By the combined efforts of Vought, Goodyear and Brewster, the Corsair series became the greatest of naval fighters, if the largest quantity produced of a type, and the type longest in production, is to be accepted as the criterion.

As 1943 came, tooling up was completed by both Goodyear and Brewster and production was under way at the former by April. By the end of that month the first Goodyear FG-1 (F4U-1) was delivered. Contracts ranged from 597 FG-1s to 300 of this model for the Royal Navy. The FG-1 was practically the same as the Vought-built F4U-1 and had a similar performance; the basic difference

employing the same R-2800-8W engine and the cockpit modifications, but the armament was six ·50 machine-guns and under-wing fittings for up to two 1,000 lb. bombs or eight 5 inch rockets. The bombs were carried on racks just inboard of the landing gear, the rockets on the outer wing panels. The F4U-1D version was a potent, close support aircraft and its range could be extended by a 160-gallon drop tank fitted under the fuselage.

The F4U-1D version was produced by Goodyear as the FG-1D. Chance Vought built 1,777 of these models of which 140 were given to Britain under Lend-Lease as the F4U-1B. Goodyear built 2,007 for the U.S. Naval Services and 475, re-designated FG-1B, for the Royal Navy. The 'B' models were exclusive to the British, being characterised by wings of shorter span, 16 inches less in fact, and were sometimes called the clipped wing version. This was to facilitate wing folding in the limited stowage below decks aboard British carriers. By February 1944, the first of these versions began arriving with the U.S.M.C. squadrons.

Hellcats with long-range fuel tanks and APS radar in wing radomes operating from an American carrier. Altogether 12,272 Hellcats were produced and large numbers were supplied to the Royal Navy as Hellcat Mk. I (F6F-3) and Hellcat Mk. II (F6F-5).

When the last of the F4U-1 series of Corsairs came to an end, 4,700 had been built by Chance-Vought, which included 1,550 original production models, 605 for the United Kingdom and 370 for New Zealand. Goodyear produced 929 FG-1 and 1Ds for Britain, 60 for New Zealand and 2,370 for the U.S.N. Brewster on the other hand had managerial and production problems and produced only 735 F3A-1 Corsairs of their original contract calling for 1,508 models based on the F4U-1. Their first were delivered in July 1943 and a year later, 366 had been sent to Britain and 369 to the U.S. Naval Services. Unfortunately two of their first production models crashed and then production problems which appeared insuperable arose so that the balance of 773 was cancelled. No further large-scale orders were placed and soon afterwards, Brewster closed down.

Two F4U-1s were sent to the Naval Aircraft Factory for conversion in early 1943. A radome underslung on the starboard wing qualified them for re-designation as XF4U-2s. The gross weight of this model was reduced to 11,445 lb. due mainly to the removal of the two wing guns and their ammunition. The entire project was experimental and was kept top secret for many years. Limited success, led to an additional ten Corsairs being fitted with auto pilots and early A.I. radar. Together with the two experimental models which were all then classed as F4U-2s they were sent to the South Pacific as special service status, night fighters.

A specialist group composed of six F4U-2s commanded by Cdr. W. J. Widhelm, formed at Quonset Point, Rhode Island on April 1st, 1943, as VF(N)-75, a land-based night fighter unit. Sent to New Georgia the following September a successful night interception was made over Munda, the next month, completely by radar guidance. A second special squadron formed as VF(N)-101 with the other six F4U-2s but they were unsuccessful in their attempt to operate as carrier-based night fighters. Both squadrons were short lived but their tour of duty provided data for the subsequent successful F4U-4N and F4U-5N Corsair series.

High altitude performance became more and more important as the war progressed. Combats were taking place above the 30,000 feet mark and on some occasions up to 35,000 feet. As special high altitude prototypes, three F4U-1 Corsairs were converted as XF4U-3s. These had R-3800-16 engines fitted in March 1942, for which the first of the Type 1009A turbo-superchargers had become ready for installation. Technical problems, however, plagued the engine development and it was not until 1946 that the project was completed and an example was fully test flown. By that time the need was over.

Curtiss, the first name in naval aviation, was being pushed into the background as far as fighters were concerned, although they were making an effort to develop a large, powerful, heavily armed, high-altitude fighter for which a contract was awarded on June 30th, 1941, for two, as XF14C-1s. The specification called for a high altitude, high performance single seat, shipboard fighter to be designed around the 2,200 h.p. Lycoming XH-2470-4

An F6F-5 Hellcat (BuAer No. 77593) shown with full operational load on February 20th, 1945. Visible are three of the six 'Holy Moses' 5″ high-velocity air rockets, a 'Tiny Tim' 11·75-inch air rocket, 156 gallon jettisonable fuel tank and one of the two JATO units.

liquid cooled V-12 engine, to be armed with four to six ·50 machine-guns which changed to four (two in each wing) 20 mm. cannon.

A wind tunnel model was tested at the Aerodynamical Laboratory, Navy Yard, Washington, in late October 1942. This test proved somewhat contradictory to the estimates by Curtiss. The brake horsepower was shown as only 1,750 h.p. for take-off and propeller efficiency would be only 86 per cent. effective. A year elapsed before the tests were evaluated and the conclusion was reached that the XF14C-1 would not meet the specification, and that the hypothetical Curtiss estimate (374 m.p.h. at 17,000 feet and climb to 10,000 feet in three and a half minutes with a 30,500 feet service ceiling) was unfounded. The Lycoming engine was looked upon with disfavour, as all in-line engines appeared to be in American naval circles, although in Europe the liquid cooled in-line engines powered the most famous of the fighters, the Supermarine Spitfire and the Messerschmitt Me109. The Lycoming had great potential as a power-unit but it was a case of overcoming the Navy's preference for an air-cooled engine as well as problems in development. The XF14C-1 project was therefore cancelled in December 1943.

during construction, when America was gaining an upper hand with the fighters already in production, the need diminished. Engine troubles including excessive vibration was another factor that hastened the cancellation of this second project. Curtiss proposed a similar large fighter with a pressurised cabin armed with four 20 mm. cannon in the wings as per the XF14C-2. The Navy ordered two, assigned XF14C-3 to the project, but never did follow it up with a development contract and there the matter lapsed, purely as a paper project.

The Navy kept a close eye on development in the U.S.A.A.F. Only one type of theirs offered promise as a naval fighter in the war and that was the famous North American P-51 Mustang. One P-51A specially built for the Navy was received on May 17th, 1943, and test evaluation followed throughout 1944. Trials included carrier landings and take-offs for which the Mustang was fully navalised, including the cockpit arrangement and carrier gear. It was proved that the P-51 could be adapted for carrier use as a naval fighter, but considerable modifications would have to be incorporated which might adversely affect performance. The in-line engine was a factor against the Mustang,

A night fighter version of the Grumman Hellcat, the F6F-5N distinguished by the airborne radar on the starboard wing. Seventy-seven of this model were supplied to the Royal Navy. The example shown is from Naval Air Station, Anacostia.

The basic design of the aircraft was not abandoned. The Navy requested that a second prototype be re-designed for a Wright air-cooled radial engine embodying a turbo-supercharger and the experimental XR-3350-16 radial of 2,300 h.p. for take-off was later fitted. With this the Navy had hopes of achieving an operational altitude of 40,000 feet.

Trials of this model, the XF14C-2, commenced from July 1944, but there was a series of delays. Only the XR-3350 engine could at that time hold any promise of reaching the required performance and this was still at an early stage of development. The estimated performance figures were not met as the maximum speed at sea level was approximately 300 m.p.h., well under the estimated 317–325 m.p.h. The best speed attained was 398 m.p.h. at 32,000 feet when 424 m.p.h. had been expected. A service ceiling of 39,000 feet was relatively good, but not up to expectations. Weights of 10,582 lb. empty, 13,405 lb. gross and 14,582 lb. in overload condition gave rise to some concern about getting this heavy fighter aboard a carrier. During development the tactical need for such a type had increased, but

not merely from the prejudices of the past, but because of the logistic support a completely new type of engine, employing a coolant, would entail for carriers spread out over the vast expanses of the Pacific and Atlantic.

During 1943 a definite pattern was emerging as far as naval fighters were concerned. 1942 had been a year of make-shift as far as operations went, with the Wildcat as the mainstay of the defence. Then in 1943 the Corsair had come on the scene and as it was thought unsuitable for carriers was used mainly by the Marines on land bases while the Hellcat was in large-scale production for Navy squadrons with the fleet. Until they were available in quantity, the escort carriers continued to use Wildcats. These three names, Wildcat, Corsair and Hellcat, stand out as the famous naval fighters of the Second World War.

The Hellcats first went into action less than fourteen months after the first flight of the prototype, which, compared with the development of other fighter types, is most remarkable. The occasion was when Task Force 51 launched nine strike groups from the carriers *Essex*, *Yorktown* (II)

Representative of a large-scale production version of the Corsair is this F4U-4D with wing pylons for carrying bombs. Other versions of this model were the F4U-4C with 20 mm. cannons, F4U-4E with special electronics, F4U-4N night fighter and F4U-4P photo-reconnaissance.

and *Independence* against the Marcus Islands on August 31st, 1943, during the Battle of the Philippine Sea. Their superiority over Japanese fighters was proved without doubt. High individual scores were attained by Navy pilots flying Hellcats, Cdr. David MacCampbell (34), Lt. Cecil E. Harris (24) and Lt. Eugene A. Valencia (23) to mention only a few.

In July 1943, a seaplane version of the Hellcat was tested, similar to the earlier arrangement on the Wildcat. The project evolved from a Bureau of Aeronautics order placed exactly a year after the initial Japanese attack. Specially designed twin Edo floats were fitted which raised the gross weight to 12,617 lb. and so considerably lowered performance. The project was cancelled but the data gained was kept should a need arise for Hellcat floatplanes.

During the occupation of the Gilbert Islands, standard F6F-3 Hellcats from the *Enterprise* were used as night fighters. When they took off on the night of November 24th, 1943, it was the first time night fighters had operated from carriers. No interceptions were made, but two days later, two Hellcats and an Avenger led by Lt. Cdr. ' Butch ' O'Hare, made contact and did battle. Although no aircraft were lost by either side, the enemy attack was routed and dispersed before any damage could be done to Task Force 50. Grumman TBF Avengers accompanied the Hellcats as radar equipped hunters for the F6F-3 killers, thus forming the Navy's first night hunter-killer combination. Following this, special night-fighting versions of the Hellcat were ordered and 125 with a radome fitted on the starboard wing, with an APS-6 radar scanner, were designated F6F-3N. Eighteen to this standard plus additional electronic detection gear became the F6F-3E. These radar installations were very effective and were used extensively on many other single-seat, single-engined naval aircraft.

In 1944 the Hellcat was forming the bulk of U.S. Navy fighters on fleet carriers although several hundred did go to the Marines and proved to be a partner equal to the Corsair. Over 250 F6F-3s went to the Royal Navy. Some records loosely describe them and other such types as being delivered to the United Kingdom, but deliveries to the Royal Navy were at times direct to British carriers under repair in U.S. Navy yards and at other times direct to British naval units in Ceylon, India or Australia.

The XF6F-2 Hellcat project was reconsidered in 1944 when a turbo-supercharged R-2800-21 engine was made available. An F3F-3, BuAer No. 66244, was earmarked for the project, which was dropped again in March 1943, and the airframe concerned was converted back to standard and subsequently delivered as the last F6F-3 production model, BuAer No. 43137, in 1944.

The last F6F-3 delivered to the Navy in March 1944 was followed in production by the F6F-5. The same engine was employed and there were no major changes involved, only minor modifications based on service and combat experience. The windshield was modified and streamlined, the undercarriage fairing was modified, the engine cowling was re-designed to give a slightly higher top speed, and armour was increased to 240 lb. During service life the machine-guns were interchanged with 20 mm. cannon, but most of the F6F-5's retained the six wing ·50 machine-guns.

Production was on a grand scale and nearly 8,000 were built, of which 5,529 appeared as the F6F-5N fitted for night fighting. Contracts were also awarded for an additional 1,677, but these were cancelled on V-J Day. Of those built, 529 were delivered to the Royal Navy.

In July 1944, two F6F-5 Hellcats were fitted with the R-2800-18W engine, and were re-designated XF6F-6. This engine provided 2,100 h.p. at take-off and 1,800 h.p. at 21,900 feet and like the final variant of the Corsair, the big engine required a four-bladed propeller for thrust efficiency. Some internal modifications were made and the result was a lighter fighter with a top speed of 417 m.p.h. at 21,900 feet and a service ceiling of 39,000 feet. By the end of 1944 tests were concluded but it was decided that the gain in top speed, ceiling and climb would not compensate for the necessary alterations which would disrupt production.

Although not approved for the Hellcat, the ' 18W ' engine, together with its thin blade propeller, was adapted to the Corsair. This characterised the F4U-4 which also had a new cockpit canopy. The first of these models arrived with service squadrons in October 1944 but it was not until the following year they became fully combat rated. A top speed of 446 m.p.h. was attained at 26,200 feet and 381 m.p.h. at sea level. The service ceiling was 41,500 feet with an initial climb of 3,870 feet per minute. The Navy was still wary of this large bent-wing fighter with its high landing speed of 90 m.p.h. and gross weight of 12,420 lb.; consequently, main deliveries continued to the Marines. Like the F4U-1 series, the '4' was produced in several variants to suit specific needs. The F4U-4 and F4U-4C models were the first to be delivered. The -4 carried the standard six ·50 machine-guns in the wings, with fittings for two 1,000 lb. bombs and an auxiliary underslung fuel tank (rocket racks to accommodate 5 inch HVAR rockets could be fitted in lieu of bombs); the '-4C' models had four 20 mm. cannon in the wings, replacing the machine-guns.

A contract was drawn up for an improved, more powerful version of the Eastern-built Wildcat in late October 1944. A two-stage, two-speed supercharged engine with water injection was proposed and three prototypes as the XF2M-1 were ordered, and then cancelled as the estimated performance figures showed no improvement over existing fighters.

By December 1944, combat requirements had changed drastically. No longer was America on the defensive. The island hopping tactics were bringing a re-conquest of the Japanese gains and threatening the large islands of Japan itself. The new year of 1945 brought the Philippines, Iwo Jimo and Okinawa in American hands through campaigns that had called for close ground support from the air; assistance that could be afforded best by fast manoeuvrable attack or fighter type aircraft. The actions were fast moving and carried out over vast distances. To get the required aircraft within striking distance meant utilising the carriers as usual, but apart from the Navy's Hellcats, the Marines with their Corsairs were called in. When VMF-124 and VMF-213 reported aboard the U.S.S. *Essex* with their F4U-1 and F4U-4 Corsairs just after Christmas 1945 they dispelled the scepticism concerning the operation of Corsairs aboard ship. As the Pacific Task Forces began their succession of strikes the Marines successfully functioned along with the Navy's Hellcats and thus began the first full combat squadron use of Marine Corps fighters from aircraft carriers.

The F6F-5N Hellcats were extensively used in the Pacific by Navy and Marine units. They proved so effective that on occasions they replaced specially developed night fighters; for example at Leyte an Army P-61 Black Widow night fighter squadron was ordered back to the States and replaced by a Marine F6F-5N squadron. During their tour VMF(N)-541 shot down twenty-two aircraft at night from December 1944 to January 11th, 1945, for which they received the Army's Unit Citation—the only Marine aviation unit to be so honoured in the war. During the 'return to the Philippines' campaign, the Marines alone claimed 816 enemy aircraft shot down using their Hellcats and Corsairs.

The first night carrier air group was formed and commissioned as CVLG(N)-43 at Charlestown, Rhode Island on August 24th, 1944. VF(N)-43, with F6F-5N Hellcats, became the first fully equipped night carrier based VF squadron. Thus the Hellcat was effective both as a day and as a night fighter. The top speed of the F6F-5N was satisfactory at 366 m.p.h. at 23,200 feet, but the service ceiling was approximately 1,400 feet under the 38,100 feet of the F6F-3N. The gross weight went up to 13,190 lb. but this was compensated for by a neutral blower with special ram intakes first applied on F6F-5 BuAer No. 78467 for test. Between March 14th, 1945, and August 2nd, 1945, a series of tests were run on this aircraft to evaluate the use of this system on F6F models. Under full ram conditions 2,233 h.p. was obtained at 1,000 feet over 1,985 h.p. at the same altitude with no ram air intakes. This system was consequently employed on all F6F-5 models since trials proved speed could be increased 5 to 7 m.p.h. at all altitudes. Take-off distances were shorter and climb increased.

Several F6F-5 models were modified 'in the field' for photo-reconnaissance work as F6F-5Ps. They were stripped of armament and had additional fuel capacity. The '5P' designation was not officially registered by the Bureau since the changes occurred later at modification depots. One F6F-5P of VMF(N)-533, fitted for photo-reconnaissance but retaining a single gun in each wing, shot down two enemy aircraft in 1945 during the Okinawa landings.

An interesting side-line on the F6F-5N models came from a secret BuAer instruction of March 3rd, 1945, by which F6F-3N BuAer No. 65950 and F6F-5N BuAer No. 70729 were taken to the Patuxent River testing station to evaluate the tactical use of a searchlight installation. Tests were made until June 11th, 1945. Both aircraft had the APS-6A radome installed on the starboard wing with an L-8 carbon-arc semi-spherical searchlight installed on the port wing bomb racks. The unit, including generator,

Developed from the Corsair by Goodyear as a Kamikaze deterrent using a 3,000 h.p. P. & W. R-4360 engine and a bubble-type canopy. The F2G-1 emerged too late in the war for production. The example shown is the first of the five built from the 418 ordered.

The last Boeing fighter for the U.S. Navy, the XF8B-1 all-purpose fighter-bomber which featured an internal bomb bay. Three experimental models were built of which this is the first shown in late 1944. Of interest is the hangar camouflaged with painted windows to give the appearance of an apartment building.

weighed 340 lb. Flights were made at night and in all types of weather with target aircraft ranging from B-17 and B-24 U.S.A.A.F. bombers to F6F and SBD naval aircraft. The technique was to achieve near contact by radar, and then use the searchlight for a visual attack. Presumably the Bureau reached the conclusion that it was impractical, in the way that the Royal Air Force had some three years earlier, for the project was abandoned after thirty-six hours were flown with only moderate success.

To those who flew the Corsair during the last island hopping phase of the Pacific campaign, they were known as 'Okinawa Sweethearts'; to the Japanese pilots, the big bent-wing 'plane was called, appropriately, 'Whistling Death'. To many ground crews as well as pilots it was the 'U bird'. It was in any case the first fighter to bear the name Corsair, as the name was previously given to Vought observation types. To the American Forces it provided a fighter with multiple capabilities at the time it was most needed. During the course of war, Corsairs shot down an estimated 2,140 enemy aircraft for a loss of 189 in air combat; additionally 349 Corsairs were lost to anti-aircraft fire. They flew 64,051 combat sorties from land bases and carrier decks. Pilots like Col. Gregory 'Pappy' Boyington (28 victories including six when with the A.V.G. in China), Lt. Robert M. Hanson (25), Capt. Kenneth Walsh (21), Capt. Donald Alrich (20) and Lt. Jeremiah J. O'Keefe (7) made the Corsair famous. With them VMF-215 received the first Navy Unit Commendation Award in 1944 for 137 Japanese planes shot down in eighteen weeks, 106 shot down in six weeks and eighty-seven Japs downed in one month, producing ten aces within the squadron. It should here be pointed out that the title ace was awarded to those pilots who destroyed five or more enemy aircraft in air combat and did not, for the U.S. Naval Services, include aircraft destroyed on the ground.

Among the famous units using the fighter were those best known by nick-name—the 'Death Rattlers' who did not lose a single man during their combat tour, the 'Black Sheep' and 'Black Mac's Killers'. The Corsair went on to be the last piston-engined fighter to be produced in the United States, although vast orders were cancelled after V-J Day, ranging from a 2,500 order for Goodyear to produce F4U-4s which was completely annulled, to a cut-back to four on an order for eleven F4U-4P photo-reconnaissance specials.

The Japanese Kamakazi menace was first encountered on Sunday, October 15th, 1944, when dedicated Japanese pilots, in a desperate attempt stave off the encircling American task forces, made suicidal attacks by aiming their aircraft loaded with bombs at Allied ships. That day the U.S.S. *Franklin* (CV-13) was put out of commission and 772 men of her crew were lost. The menace grew and altogether, until the end of the war, 474 hits were made for 2,500 aircraft expended. To meet this threat a fast, low-level fighter was urgently needed. Pratt & Whitney had a 3,000 h.p. engine, their R-4360 Wasp Major, under development for the general pressing need for more power, and this was immediately earmarked to provide a counter to the Kamikaze threat. Goodyear were asked on March 22nd, 1944, to marry a FG-1 Corsair airframe to take the R-4360 engine. They found that this needed a taller fin and rudder to counteract the additional torque of 3,000 h.p. This aircraft, the XF2G-1, was later used extensively by Goodyear for power plant development at the N.A.T.C., Patuxent River which, since August 1943, had been relieving Ancostia of testing work.

Over 4,000 F2G-1s were ordered, but construction was subject to the success of a pre-production four. They were armed with four ·50 machine-guns and had provision for the carriage of two 1,600 lb. bombs on wing racks. To avoid delay, these first models, intended for use by Marines from newly acquired island bases, were not fitted with carrier gear, which came with the second five designated F2G-2. The later version featured split rudders to effect full control under conditions that would appertain if a wave-off was received during carrier approach. However, by the time they were ready, the Japanese surrender had been received and the need had lapsed. The original order had been cut back to just over 400 and then to fifteen, and finally only ten appeared, which were sold as surplus the following year.

Another design around the R-4360 Wasp Major engine came from Boeing, the XF8B-1. This utilised the twenty-eight cylinder, air-cooled R-4360-10 which developed 3,520 h.p. at 2,700 r.p.m. with water injection. It was, for an aircraft engine of limited weight, the most powerful engine in the world and probably still is. It drove, in this case, a six-bladed, contra-rotating co-axial Aeroproducts propeller of $13\frac{1}{2}$ feet diameter. Although classed as a fighter the XF8B-1 was exceedingly large for a single-seat, single-engined aircraft and was built for multi-purpose roles including dive-bombing and torpedo-bombing. For these purposes the fighter—so called—had an unusual feature for its class, an internal bomb bay in addition to external wing fittings, which allowed various combinations of stores to be carried to a maximum bomb load of 5,200 lb. Provision was also made for six ·50 Browning machine-guns or six 20 mm. cannon or a combination of each, but for trials only machine-guns were fitted. Three prototypes were ordered on May 4th, 1943, BuAer Nos. 57984–6, but only two were delivered to the Navy as No. 57985 went to the U.S.A.A.F. for evaluation.

An important factor in the evolution of the XF8B-1 was a standing agreement between the Army and Navy from July 1942. The Navy was in need of long-range land-based bombers, the U.S. Army Air Force too was needing expanded bomber production, particularly of the B-29 Superfortress, which was then in the final tooling stage at Boeing's Renton plant. The Navy agreed to relinquish its cognizance of the Boeing plant to Army orders and it would limit its orders from the Consolidated Aircraft concern to PBY Catalinas and so avoid interference with B-24 Liberator production. In turn, the Army off-set from their production B-24 Liberators, B-25 Mitchells and B-34 Venturas for the Navy's needs as their PB4Y-1/2, PBJ-1 and PV-1 aircraft, respectively. This delayed the Navy's original order for the XF8B-1 so that it was mid-1945 before the first model appeared.

In the background an aircraft project with a pre-war lineage was being pressed forward. The building had commenced in 1940 and by early 1942 the aircraft was ready for flight but first it was wind-tunnel tested at the N.A.C.A. Laboratories at Langley Field to prove the design of this circular, flying-wing aircraft. On November 23rd, 1942, the maiden flight was made from the Stratford Plant of Vought, over five years since Chance Vought had hired C. H. Zimmerman to turn his design into a practical proposition. Two 80 h.p. Continental flat-four air-cooled engines, driving two three-bladed Hamilton standard propellers, powered this aircraft which was built mainly of wood, fabric covered. During the remainder of 1942 and most of 1943, the Vought test pilots, Richard Burroughs and Boone Guyton, flew a total of 131 hours in this Vought V-173.

Since the object was vertical ascent and descent, with hovering capabilities without affecting normal speed, the Navy were vitally concerned. Actual tests of the yellow painted prototype showed a speed range of 30-150 m.p.h.

Many problems were encountered including engine over-heating, pilot visibility and of accommodating the pilot in this radical design. Unlike the first concepts where prone position piloting was envisaged, this single-seater had an upright seat with conventional controls. Because of the high angle of attack in landing and take-off, the nose was extended slightly and a clear plastic window was incorporated to allow the pilot to see forward and downward between his feet.

Following tests the Navy awarded a contract for a twin-engined, single-seat fighter capable of operating from restricted carrier decks as the XF5U-1. Perhaps it was early to class this as a fighter, but at least this designation gave no hint of a radical design, which may have been a security consideration. A Vought-engineered process, Metalite, was used for the surface structure. This was a sandwich composition of light alloy sheets bonded over a thin core of balsa wood. Power was supplied by two 1,200 h.p. Pratt & Whitney R-2000-7 radial air-cooled engines placed within the wing structure, lying flat with bevel gearing and shafting out to the propellers forward of the wing. The propellers were inter-connected, so that in the event of an engine failure, a single engine could drive both propellers to facilitate a safe descent. Originally F4U-4 four-bladed Hamilton standard hydromatic airscrews were used when the XF5U-1 was engine tested under strict secrecy late in 1944, but standard blades put an undue strain on the shafts

Forerunner of the XF5U-1 was the Vought V-173 experimental flying wing proposed by C. H. Zimmerman which first flew on November 23rd, 1942, and was flown regularly for about one year after this. Built of wood and fabric, and powered by two 80 h.p. Continental engines it is at present held in storage by the U.S. Navy.

and set up a disrupted airflow over the wing, which in turn caused serious vibration. Specially built propellers 16 feet in diameter were made of compressed, impregnated wood attached to steel shanks. They were made by the Freedman Burham Engineering Corporation of Cincinnati, Ohio and were mechanically inter-connected so they turned at the same speed. A two-speed gearing system was employed so that the propellers could turn for the greatest efficiency at either low or high speeds, but as before, special clutches permitted either engine to drive both propellers in an emergency. Standard Corsair undercarriage legs were utilised. Armament was not fitted, but proposals were for six ·50 Colt-Browning machine-guns with 400 r.p.g. or, alternatively, four 20 mm. cannon. Provision for the carriage of two 1,000 lb. bombs, underslung externally was planned, which could be replaced with two 125 gallon auxiliary fuel tanks as operationally necessary.

The XF5U-1 was called many names, 'Flying Flapjack', 'Flying Hamburger', 'Flying Carpet', 'Flying Pancake' or 'Flying Crab', but none were official. It never actually flew, and only successfully completed engine trials. Flight tests were finally planned at Edwards Air Force Base, California in 1948, but the jet age had arrived and the concept was outdated in its existing form. A crane swinging a heavy steel ball reduced the fighter to a pulverised mass of metal on March 19th, 1948. It was survived by the original V-173, which is in storage at the U.S. Naval Air Station, Norfolk, for eventual exhibition in the proposed new National Air Museum, near Washington, D.C.

The Tigercats and Bearcats of post-war, evolved during the war years, indeed the Tigercat was conceived before America entered the war and the Bearcat was planned in 1943 as a Hellcat replacement. These Grumman aircraft evolved on divergent lines, the F7F as a heavy multi-purpose fighter, the F8F as a lightweight fast climbing interceptor.

The twin-engined fighter requirement followed on from the XF5F-1 Skyrocket design with a contract on June 30th, 1941, for two XF7F-1 prototypes, BuAer Nos. 03549–50, incorporating the trial recommendations of the Skyrocket. In spite of the many commitments of the Grumman concern, the first of the two aircraft made its initial flight at Bethpage on November 3rd, 1943 to be followed by the second a few days later.

Of all-metal construction, with extremely short span shoulder wings, the XF7F-1 was the first U.S. naval fighter to feature a nosewheel undercarriage. Its long pointed nose was reminiscent of that of the U.S.A.A.C.'s XP-50 version of the Skyrocket. Power was provided by two Wright R-2600-14 engines driving three-bladed Curtiss-Electric propellers with large spinners. The original XF7F-1, No. 03549, crashed on May 1st, 1944, shortly after official trials had got under way. Testing of the second prototype, however, led to a production contract. With a gross weight of 19,500 lb., a maximum speed of 398 m.p.h. at 19,000 feet was recorded. Some difficulty with engine cooling was experienced and subsequent modifications included the removal of the large fairing spinners.

The XF7F-1 proved to be a dependable Grumman design, easy to fly, manoeuvrable and rugged, but its weight and size relegated it to land-based operations. It was subsequently assigned to the Marine Corps, although all the Tigercats were, initially, fully equipped for carrier operations and stressed for deck catapult launchings. During service, however, hooks and catapult points were removed since they were land based. The design and capabilities of the F7F made it a natural choice for night

The first of the Tigercats, the Grumman XF7F-1 BuAer No. 13549. This type in the U.S. Navy re-introduced the twin-engined fighter concept, was the first type with tricycle undercarriage to be accepted for service and was the heaviest armed fighter to date.

fighting, and this concept had been borne in mind when the machine was designed.

Two pre-production models of an order for thirty-four were delivered in April 1944 as F7F-1 Tigercats. Their appearance differed little from the prototypes, but new power plants were installed, 2,100 h.p. Pratt & Whitney R2800-22W engines. On production aircraft, the engine cowlings were further modified, and Hamilton-Standard hydromatic propellers were used. With four ·50 machine-guns on the underside of the nose and four 20 mm. cannon in the wing leading edge, the Tigercat was one of the most heavily armed fighters of its day. Further service evaluation trials were made on the two pre-production models and radar gear was installed in the nose. The third pre-production model, BuAer No. 80261, appeared as the XF7F-2. This became the trials aircraft for the F7F-2 series which was to incorporate a rear seat for a radar operator and be known as the F7F-2N.

Prototype of the Bearcats, the XF8F-1. This fighter was designed as a lightweight intercepter armed with four ·50 machine-guns, but it had provision for the carriage of two 1,000 lb. bombs. The prototypes were distinguishable from production models by the absence of a long fin fairing.

Top speed of the F7F-1 was around 400 m.p.h., with 425 m.p.h. claimed as top speed at 20,000 feet using water injection and the aircraft in clean condition, i.e. without underwing stores. In addition to the standard version, radar-equipped F7F-1Ns were built, their numbers being increased by the modification of many of the earlier aircraft to a similar state. By September 1944 all the F7F-1 Tigercats had been delivered.

In July 1944 No. 80261 modified as a two-seater, but still with Pratt & Whitney R2800-22W engines, was delivered for tests. Production versions of this model were to be powered by R-2800-34W engines and successful trials led to an order for sixty-five, known as F7F-2Ns. Outwardly they resembled the '1' model, but a 156-gallon fuel tank immediately aft of the pilot, had been removed to make way for the radar operator. The entire aircraft was strengthened and performance was improved. To make room for the radar gear, the machine-guns in the nose were deleted. Thirty were delivered in 1944, the balance by March 1945.

The third variant of the Tigercat, the F7F-3, was actually ordered before the '2' model. Of the original order for 1,650 Tigercats to be powered by R-2800-34W engines, only 189 materialised of which 175 were delivered by the end of 1945. The single-seat F7F-3 was similar to the F7F-1 but for its power plant. Further examples were ordered as two-seater night fighters with a nose 18 inches longer as F7F-3Ns.

An F7F-3E variant was envisaged with improved electronic gear, but the end of hostilities led to the cancellation of the tentative order for 150.

Of the Tigercat series, the F7F-3 in its single-seater form had the best performance. Its top speed was 435 m.p.h. at 22,200 feet. Due to its exceptional range of 1,200 miles a few examples of the type had automatic cameras installed in place of the extra fuselage tank and were designated F7F-3P for photo-reconnaissance. These had the nose machine guns, but the wing cannon were removed to reduce weight.

The first Tigercats reached the Marine fighter squadrons in April 1944 and these units were still in training at the end of the war. None were assigned to overseas duties before hostilities ceased. The first '2N' models joined the Marines in November 1944, and in March 1945 the '3' models began to arrive but like the earlier models did not see war service.

The Bearcat was designed both as a Hellcat replacement and with an eye directly on outclassing the premier Jap fighter the 'Zeke'. This meant a rugged fighter of minimum weight to facilitate a good rate of climb.

On November 27th, 1943, a contract was awarded for two XF8F-1 prototypes. The first, BuAer No. 90460, made its first flight on June 25th, 1944, just six months from the date of the contract. The design was typical of Grumman—

First production model of the Tigercat, the F7F-1 single-seat version with nose armament of four ·50 machine-guns. This twin-engined fighter was designed primarily for operation from 45,000 ton carriers of the Midway class although it has operated from the rather smaller Essex class.

short and stubby—in general somewhat similar to the Hellcat, but more compact and having cleaner lines. A bubble canopy was chosen, a feature that soon became a world-wide feature on fighters.

The Bearcat employed the Pratt & Whitney R-2800-22W with a normal rating of 1,600 h.p. at 16,000 feet. Tests were conducted for almost a year, and embraced all likely tactical uses. During initial trials the first XF8F-1 was destroyed in a crash and trials continued with the second. Instability led to several modifications on production aircraft. Vertical tail surfaces were more rounded, and a short dorsal fin was added. The wing span was increased by 4 inches and fuselage length shortened by 5 inches. Due to a large four-bladed propeller (12 feet 4 inches diameter) the Bearcat sat at a rather nose-high attitude on the ground. A top speed of 424 m.p.h. was reached by the XF8F-1 with 393 m.p.h. at sea level. A spectacular feature of its performance, was its initial climb-rate of 4,800 feet per minute.

The first production order for F8F-1 Bearcats was placed on October 6th, 1944. By February 1945 the first of the twenty-three had been delivered and an additional contract for 2,135 placed. A further order was given for 4,000 but as the Bearcats began rolling off the Grumman lines and squadrons were forming up with them, the war ended and all but 765 were cancelled. Due to the large orders separate contracts were signed with General Motors (Eastern) to produce 1,876 as F3M-1s but the war ended before any were built and the contract was terminated. In the final analysis 899 F8F-1s were produced, including three experimental machines.

First deliveries of the F8F-1 series began in February 1945. In the production form standard military items had increased the gross weight to 9,385 lb. and the performance figures fell slightly to a top speed of 421 m.p.h. at 17,300 feet and 382 m.p.h. at sea level. Due to the use of an improved blower on the R-2800-34W engine, the service ceiling rose some 5,000 feet, to 38,700. With two wing drop tanks, the range could be extended to 1,700 miles. The series had four ·50 machine-guns in the wings and could carry bomb loads of up to 2,000 lb. or four 5 inch HVAR rockets.

With the dropping of the atomic bombs on Hiroshima and Nagasaki, the Japanese, who had been fighting a losing battle over the past two years, finally surrendered. During the years since December 1941, U.S. Navy and Marine Corps aviation had grown manifold. Returns for the day before the Pearl Harbor attack, December 6th, show that the Navy had seven large and one small carrier, 5,241 pilots, 5,029 aircraft of all types of which approximately 295 were VF class. By V-J Day statistics for the Navy's air arm show that 13,756 single-engined VF types alone were in service and 184 twin-engined VF types—the F7F-1 Tigercats.

The Marine Corps had expanded similarly. At the end of 1941 they had two Aircraft Groups of thirteen squadrons of which four, VMF-111, VMF-121, VMF-211 and VMF-221 were fighter squadrons. These rose to twenty-nine aircraft groups of 132 squadrons including thirty VMF and four VMF(N) squadrons by August 31st, 1945. During the same period personnel strength rose from 610 officer pilots, twenty-eight student pilots, forty-nine enlisted pilots and sixty-seven enlisted student pilots to 10,005, 104, 44 and 116 respectively.

The Bearcat, planned as the superiority naval fighter of the war, evolved just too late and large-scale production was cut. Those which were built soon went to the Reserve and this Bearcat became a ' stooge '. Painted chrome-yellow over-all it was the ' sitting duck ' target in a ' Blue Angels ' act.

The naval air war had been carried by three main fighters, the Wildcat, Corsair and Hellcat which had also been used by the British Navy. Because of the pre-war administrative policy the Royal Navy had no first-rate naval fighters of its own, except possibly the Seafire which was not a true naval fighter but an adaption of the famous Spitfire to sea duties. The Seafire could fight well and did well against the Japanese Zeros, but as Vice-Admiral Sir Philip Vian, the Flying Officer Commanding the First Aircraft Carrier Squadron of the British Pacific Fleet said, the robustness of Corsairs and Hellcats, their reliability and long endurance, showed up in marked contrast to British types, particularly the Seafires. The British fighters had in fact a higher crash-rate and were hard put to maintain the sortie rate of U.S.N. aircraft. It was a tribute to the U.S. naval aircraft fighter types that the Royal Navy not only used American fighters because their own productive resources were limited, but that they preferred the American fighters.

CHAPTER ELEVEN

Jets—The New Generation

One of the few night fighting F8F-1N Bearcats (BuAer No. 94819) ready for testing with its radar pod slung under the wing rather in the manner of a fuel tank. Bearcats were armed with four 20 mm. cannons with 200 r.p.g., but this particular version had four .50 machine guns.

When the world again came to an uneasy peace in 1945, the Allied powers were confronted with an overabundance of military strength. The war was won, and its warriors were anxious to return to civilian life. Contracts for arms were drastically cut, and demobilisation of the Services was demanded as imperative. The Naval Services were again to experience limited procurement funds and had need to be frugal, at least for the first few years following the war. Their first step was the ' moth-balling ' of useful aircraft which included fighters like the Hellcats and Corsairs, and many carriers were similarly decommissioned and put into a state of preservation. This trend actually began in 1945 before the cessation of hostilities. Old airship hangars were filled, warehouse space was taken up and open fields utilised for storage. By early 1947 some 3,800 aircraft had been put into storage. On V-J Day the Navy had 43,116 aircraft on its charge. Plans called for a reduction to 27,473 by July 1st, 1946 and decommissioning of many formations commenced from October 1st, 1945. Between July 1st, 1946 and June 30th, 1947 the number of aircraft on naval charge fell from 23,156 to 15,371.

Many of the surplus war-weary Hellcats, Corsairs and Wildcats were sold to a limited civilian market. The high operational costs of fighter types precluded extensive use. Some were donated to air-minded towns and cities and a few went into storage ear-marked as museum pieces. A number of Hellcats, sprayed red, were used as radio controlled drones or were used as experimental radio-controlled glide-bombs. However with the rundown of the regular forces came a building of Naval Reserve for which Corsairs and Hellcats among other types were made available.

Of the many contracts axed after the war, those placed with Grumman and Vought were the least affected as these two concerns were producing fighters still in demand to maintain the peacetime front line. Grumman were particularly fortunate with their fighters. Although the last Hellcat, an F6F-5N was delivered in 1945, the F8F Bearcat remained in production and when the amended F8F-1 order was complete, it was followed on the lines by 100 F8F-1Bs. Appearing in May 1946, this type had four 20 mm. cannon in lieu of machine-guns.

Two of the original version, BuAer Nos. 94812 and 94819 were completed as XF8F-1Ns to evaluate the Bearcat as a night fighter. This led to a limited production order for twelve models known as F8F-1Ns which retained the original machine-gun armament.

A post-war day-fighter Corsair, the F4U-5 which as the last production version of the series for the U.S. Navy appeared in 1947. Many, such as the example shown, served with the Navy Reserve.

A post-war night-fighter Corsair, an F4U-5N shown on March 16th, 1951. This version had a re-arranged cockpit and like all the F4U-5 series, with R-2800-32W engines, featured intakes on each side of the cowling.

The F8F-2 variant, besides internal changes and taller vertical tail surfaces, had a revised cowling over its R-2800-34W engine. All of these employed four 20 mm. cannon in the wings, except for sixty F8F-2Ps produced for photographic reconnaissance for which two guns were deleted to lighten the craft, and extend its range and improve its speed. Two experimental XF8F-2s, Nos. 95049–50, preceded a production order for twenty-nine as day fighters and a dozen F8F-2N night fighters. This model showed a slight improvement in the top speed of some 7 m.p.h., but the climb rate was affected by the increased all-up weight.

F8F-1s were initially delivered to VF-19 of the Pacific Air Command on May 21st, 1945, some three months before the end of the war, by which time other Navy squadrons had them ready to operate from carriers. This tubby little fighter formed the backbone of the VF squadrons during the immediate post-war years, giving way only to the jet fighters. Some years later the Marine Corps acquired some Bearcats for training purposes at Quantico.

Some particular Bearcats deserve mention. One F8F-1, BuAer No. 94804, was diverted to the U.S.A.F. for evaluation trials at Wright Field. In a lightened Bearcat Lt. Cdr. M. W. Davenport claimed a national climb record at Cleveland, Ohio on November 20th, 1946. Using an F8F-1 he took off after a 115 foot run, and climbed to over 10,000 feet in 94 seconds. When at the end of 1945 the FH-1 Phantom was under test at Patuxent, it was pitted against an F8F-1. At maximum speed the Phantom left the Bearcat standing, but the jet was still on its way up when the F8F made runs at 10,000 feet. Further attest to the Bearcat's excellence came in 1948, when an F8F-1 was released to the stunt pilot Major Al Williams. His specially modified de-militarised machine, registered NR-3025, had a specially prepared R-2800-34W engine which, with water-alcohol injection, could deliver 2,800 h.p. for short periods. Additionally it was 1,300 lb. lighter than the military version. This ex-Marine flyer who had flown F3Fs christened his acquisition 'Gulfhawk IV', and wrung 500 m.p.h. from it at 19,000 feet. He claimed that he could reach 10,000 feet in 100 seconds in the aircraft from a standing start.

In May 1949 the last of 1,264 Bearcats left the production lines where they were followed by F9F Panthers. The Bearcat's useful life was not by any means over, since many went to reserve flying units. As some of the early versions were retired they were delivered to other air forces. Under the designation F8F-1D, armed with four ·50 machine-guns, they were sold to France and used in Indo-China. By 1954 100 F8F-1Ds had been transferred to the Vietnam Air Force, and twenty-nine F8F-1Bs with 20 mm. cannon were in the hands of the Royal Thai Air Force, which still uses them. Like the Corsair, the Bearcat soldiers on in foreign service whilst the Navy is equipped with aircraft that, when the Bearcat was being designed, seemed a mere pipe-dream.

The F7F Tigercat was produced concurrently with the Bearcat by Grumman, and by early 1946 the most widely used model, the F7F-3, was being flown by the Marine Corps. The last version of the series was the F7F-4N of which only thirteen of the order for fifty-seven were built. No. 80548 served as the pre-production test model, following directly after the last F7F-3. The model '4N' had improved radar, electronic gear and a strengthened structure. Additional radar gear was subsequently added in a retracting chin position.

The Corsair from Vought was also still in production. This was the F4U-5 with an R-2800-34W engine which was considered more suited than the turbo-supercharged engine which had finally got underway just after the war. The Corsair used for tests had a four-bladed propeller with all other features as the F4U-1D version, and was designated XF4U-3. The gross weight was 11,650 lb. and empty, 9,039 lb. At 30,000 feet developing 2,000 h.p. the top speed was 412 m.p.h. with 314 m.p.h. obtained at sea level, while the efficient cruising speed was found to be about 180 m.p.h. The service ceiling was 38,400 feet and the initial climb was 3,000 feet per minute. It was indeed an improvement over the war-time versions but just too long in the development stage. XF4U-3 models were altered, put into service as F4U-5s and the project terminated.

When surplus military aircraft were sold after the war the P-38 Lightnings, P-39 Airacobras, P-40 Hawks and P-51 Mustangs became favourites at the National Air Races resumed at Cleveland in 1946. Navy aircraft were not so popular because it was thought that they could not compete with such fast types as the Mustang. Cook Cleland proved differently in the 1947 and 1949 Thompson

Races, when his clipped wing F2G-1 won—almost uncontested by Mustangs—at average speeds of 396·131 m.p.h. and 397·071 m.p.h., respectively. Cleland had put three F2G-1 Corsairs into the 1947 Race; he flew one himself and the others were piloted by Richard Becker and Tony Janazzo. While Cleland won, Becker came in second in spite of five exhaust stacks blowing causing him to reduce speed. Janazzo crashed in the seventh lap due to carbon monoxide trouble. In the 1949 races all pilots wore oxygen masks to prevent a re-occurrence of the trouble, which had already claimed the lives of two race pilots. That year witnessed the last of the classic Thompson Trophy Races for piston-engined aircraft. The celebrated Bill Odom was killed in his much modified Mustang and this incident led to the decision to make the Thompson Race a speed dash over a measured course for military jet aircraft.

The F2Gs were extremely powerful machines. Service tests showed that they had 3,000 h.p. available for take-off, and that their top speed was 431 m.p.h. at 16,400 feet. Cruising at 190 m.p.h., their range on internal fuel load was 1,100 miles. An attestation to the reliability and stamina of the Pratt & Whitney R-4360-4 engine is indicated by its military rating which allowed it to be run at 3,000 h.p. for as much as three minutes. During the 1947 Race Cleland pushed the Wasp Major to give 4,000 h.p. for 48 minutes of the race, through the use of an alcohol-petroleum mixture, along with hydrogen peroxide in the water injection system.

In several navies the Corsair became a perennial. As the last F4U-4 left the Vought-Sikorsky line at the end of June 1947, a newer version took its place. Certainly the Corsair was a ' die hard ' to the coming of peace and the jet era. Its continued popularity was due to its adaptability for a variety of roles and its suitability for action ashore or afloat. While the F4U-4 is considered a post-war development, the first really post-war variant emerged in 1947 when the production F4U-5 series appeared. In July 1945 two F4U-4s, BuAer Nos. 97296 and 97415, were converted to use the later Pratt & Whitney R-2800-32W engine rated at 2,300 h.p. at 26,000 feet. No. 97296 crashed during flight trials on July 8th, 1946 and was replaced by another F4U-4, No. 97364. These aircraft were re-designated XF4U-5, and their performance showed an all-round improvement, since they had a service ceiling of 44,100 feet and an initial climb rate of 4,250 feet per minute. At gross weight their top speed was 460 m.p.h. at 31,400 feet.

The F4U-5 model was ordered into production on 1947 funds, 233 being built as day fighters. A further order called for seventy-five F4U-5N night fighters carrying a radome on the outer section of the starboard wing. Thirty photo-reconnaissance F4U-5Ps were also constructed. In 1948 an additional 240 F4U-5Ns were ordered and some of these participated later in Korea. These had a true radar scanner, a new cockpit arrangement, slimmer fuselage and a new cowling incorporating double air scoops on the lower sides. Armament was four 20 mm. cannon in the wings in addition to which the type could carry two 1,000 lb. bombs on wing pylons as well as eight 5-inch rockets under the outer wing sections. In lieu of these external stores, two 150 gallon drop tanks and one 150 gallon belly tank could be fitted. In spite of its radome and equipment the night fighter had the best performance and a range of 1,500 miles.

The promising Boeing XF8B-1 fighter project of the war, faded out in peace, but not without due consideration. Trials had shown that the aircraft had great potential and following an Army Air Force request for a test article, BuAer No. 57985 was loaned to them on February 15th, 1946. This was tested first at Wright Field, then sent to Elgin Field, Florida for armament and bombing tests. Meanwhile, the Navy tested the other two at the Patuxent Naval Aircraft Test Center. The two services collaborated on the evaluation.

Assessed in its several roles, varying conclusions were reached. As a long range combat aircraft, it failed, unless bombs were sacrificed for extra fuel tanks, for although the internal capacity was 385 gallons the big Wasp Major had an exceedingly high rate of consumption. For skip and torpedo bombing, it proved controllable, safe and stable, and either internal or external bomb racks could be used; as a dive bomber, it became unstable when diving in excess of 60° and it did not have dive brakes. Fighting was its primary role by designation and in that respect it proved more capable. Low level strafing could be accomplished at 350 m.p.h. with complete stability and excellent visibility was given by the 5° nose-down flying attitude. At altitude, the performance was equal to the best fighters and the service ceiling was 37,500 feet. It reached a speed of 411 m.p.h., and when cruising at 200 m.p.h. the range (with 300 gallons in extra fuel tanks in the bomb bay) could be

Lack of markings on this Bell YP-59A is consistent with the rigid security imposed upon the introduction of jet aircraft. The model shown, BuAer No. 63960 ex-U.S.A.A.F. 42-108778 was used for naval evaluation.

stretched to 3,000 miles—if pilots could stand the excessive fatigue that would be encountered in 15 hours flying.

The XF8B-1 had a basic weight of 13,693 lb., with a maximum weight in any loading combination of 23,000 lb.; trials were conducted at about 18,000 lb. load. The weight was a profound problem at first for carrier operations, but as a land based aircraft it promised to have no equal. Some of the early fears were dispelled when flight trials proved that take-off could be accomplished in approximately 800 to 1,000 feet. Good directional stability was inherent at slow speeds, take-off and landing, and a 'Go around' could be accomplished without difficulty.

By mid-1946 the general concensus of opinion was that an aircraft of this size and weight should have the added safety of two engines. The Air Corps completely rejected the design on account of the high speed performance and manoeuvrability being sacrificed for low speed landing stability and control; to which the Navy pointed out that an ' Air Force Approach ' was not possible in landing aboard a small, bouncing aircraft carrier deck. Finally, due to the protracted trials, the Boeing was outclassed by the Tigercat outmoded by jets, and the project was terminated late in 1947.

Serious study of the possibility of a practical jet engine for aircraft was first made by Frank Whittle in 1933, but not until the war did he receive the necessary backing to bring success to his theories. However, as early as August 29th, 1939, the Germans had flown their Heinkel 178 powered by a crude He S-2 jet engine. In America, an early connection with the jet engine came in 1940 when the Bureau of Aeronautics initiated a contract with Professor H. O. Croft of the University of Iowa, to investigate turbojet propulsion for aircraft. The findings of the professor's team gave recommendations that conflicted somewhat with the Army's investigations. While the idea was considered feasible, it was deemed unsound where carrier based aircraft were concerned. The jet engine was considered to have a slow reaction on take-off, and it was thought that it might prove hazardous during a ' wave-off ' from landing. Since take-off runs were likely to be long, some form of catapult assistance was considered necessary. Fuel presented quite a problem for, although kerosene was cheap, all carriers would need storage tanks for this in addition to petrol for conventional aircraft. The decision was eventually taken, after much deep consideration, to convert jet engines for naval use to run on low value 55 to 60 octane gasoline.

Aircraft performance suffered due to this decision, yet it was proved during trials, that the jet engine could run on gasoline, just as well as kerosene. While the Air Force was forging ahead in its application of jet engines, the Navy was encountering a series of technical problems that did not affect the Air Force with land based operations. Questions were even asked as to whether the Navy really needed jets; indeed, it was even suggested that the Navy's role should rest with patrol and scouting—familiar words, echoing from the past! The old cry of the vulnerability of the carrier was countered by the plea that it was a mobile airfield. The result of all this controversy was that although the Navy initiated jet research at a commendably early date, the actual development lagged much behind the Army Air Force. Considering the facts involved, no valid recriminations can be placed.

Still waging war in two great Oceans, not to mention the Mediterranean, the Navy in 1943 had laid down its first jet fighter specifications. The first step towards actual manufacture of an American jet engine had come on September 22nd of that year when a design study presented to the Westinghouse Electric Corporation was amended to read ' authorisation to construct two Type 19A axial flow turbojet power plants '. By early 1943 the first all American/Navy jet engine had been constructed and on July 5th, 1943 the 19A successfully completed its 100 hour endurance test.

The Navy acquired two YP-59A Airacomets from the Army. Tested during January 1944 at Patuxent, they showed a top speed of around 380 m.p.h. with two

An object of great interest. The first carrier landing by a U.S. jet aircraft, the McDonnell XFD-1 (later XFH-1) piloted by Lt.-Cdr. James Davidson, after landing aboard the U.S.S. Franklin D. Roosevelt (CVA-42) on July 21st, 1946.

The twin-jet Phantom was the first of the U.S. naval jet fighters to reach service status. It is represented here by McDonnell FH-1 BuAer No. 11793. The fairings on the nose cover the barrels of the four ·50 machine-guns.

engines each of 1,620 lb. thrust. Armament tests were conducted with one, and later two, 37 mm. cannon in the nose in addition to three ·50 machine-guns. Trials were in general satisfactory and, with modifications, it was believed the P-59 might be adaptable for shipboard use. The general arrangement of these aircraft was felt to be highly suitable as regards naval needs; particularly was this true of the positioning of the engines close in and amidships. This feature was already in mind when the first specification for an operational carrier based jet fighter had been drafted.

Unofficially the Army fliers called the P-59s ' Swish ', but the Navy gave to them their usual name of Airacomets. In succeeding years, three Airacomets were acquired by the Navy, one YP-59B delivered in 1945 and two in 1946. To them were allocated the BuAer Nos. 64100, 64108 and 64109. Three Lockheed P-80As were also evaluated.

The ex-Army jets were pace-makers for the Navy's own designs which evolved post-war from contracts placed during the war. A letter of intent had been sent to the McDonnell Aircraft Corporation of St. Louis on January 7th, 1943 to a proposal for the design of two experimental fighter aircraft to be jet propelled. The requirements, both technical and structural, were exacting and it was August 30th, 1943 before the design was approved. Two prototypes were then built as the XFD-1 and in October 1944 the first prototype flew at the E. St. Louis plant, where it crashed the following month. The second machine, after initial tests, was turned over to the Navy whose tests at Patuxent were commenced on January 26th, 1945.

The designation XFD-1 must not be confused with that of the old Douglas biplane of 1933. Douglas had since been removed from the list as a supplier of fighters to the Navy, and the letter ' D ' was allotted to McDonnell in 1943. To avoid possible confusion when Douglas became a supplier again, McDonnell was later given the letter ' H ' as its manufacturer's designation. Thus the XFD-1 became, retrospectively, the XFH-1. Its production successor, the FD-1, was similarly re-designated FH-1.

The XFD-1 was powered by two Westinghouse WE19-XB-2B engines each giving 1,365 lb. static thrust. Following successful trials at Patuxent Lt. Cdr. James Davidson accomplished the first tests with a jet aircraft aboard a carrier, the U.S.S. *Franklin D. Roosevelt*, on July 21st, 1946. Davidson made several landings and take-offs with the XFD-1, and it appeared that quite a few of the problems of introducing jet aircraft to carrier operations had been overcome. The XFD-1, with its straight wings and the added safety of two engines, proved that a high speed jet could be built, which could perform at the low speeds necessary for carrier landings. Thus the first Navy jet came into being, McDonnell and Westinghouse having combined to produce the first of a series of successful Navy fighters.

Even before sea trials, an order was placed with McDonnell on March 7th, 1945, for 130 FH-1 fighters. Due to engineering delays, the engines were slow to reach production and it was not until January 1947 that the first FH-1 Phantom was ready for service use. Production models had the improved J30-WE-20 engines delivering 1,600 lb. thrust. These boosted the top speed to 487 m.p.h. at 40,000 feet. The service ceiling was around 43,000 feet.

Shortly after VJ-Day the order for Phantoms had been reduced to sixty and in May 1948 the final example came off the production lines. The first Phantoms were delivered for operations with VF-17A (LANT) whose initial duties were familiarisation with the aircraft, followed by qualification flying for carrier duty. In three days the squadron completed trials uneventfully aboard the U.S.S. *Saipan* off the Atlantic Coast and, once qualified, the squadron became the first regular shipboard jet fighter squadron in the world.

The FH-1 Phantom was the first jet fighter to be used by the Marine Corps, Squadrons VMF-122 and VMF-311 being first to equip when their F4U-4 Corsairs were replaced in 1948. Major Everton and three pilots from VMF-122 formed a precision aerobatic team at Cherry Point; as the ' Marine Phantoms ', they toured the U.S.A. for two years.

Thus the Phantom had become the Navy's first jet fighter to go into service in quantity. Several other designs had been considered including a composite power fighter.

The Ryan Aeronautical Company of San Diego commenced work on an ambitious project in February 1943. A contract was placed for a single-seat fighter, known later as the XFR-1 Fireball, to be powered by a Wright R-1820-56 Cyclone piston engine in the nose, and a General Electric I-16 jet engine in the rear of the fuselage with the efflux in the tail. Slow landings, short take-offs, fast climbs and high speeds at altitudes all seemed to be offered by the design. In the event of a wave-off on a carrier approach, the piston engine would be of great value, while the

The Fireball, of which this is the XFR-1 prototype was the first aircraft with composite power units to reach production. A Fireball is reported to have made the first carrier landing on November 6th, 1945 on the U.S.S. Wake Island off San Diego following an engine failure.

jet engine would boost performance at high altitudes.

Externally, the Fireball appeared to be a conventional low-wing monoplane of clean design. The jet engine air intakes, situated in the wing roots, were barely evident and looked to the uninitiated to be merely cooling air inlets.

The Fireball made its first flight on June 25th, 1944, so that in less than eighteen months it had been designed and flown, which was no small achievement considering its complexity. In retrospect the Fireball was looked upon as a compromise, as if the Navy had tried to get the best of two worlds, but in fact the placing of the vastly differing engines eased the centre of gravity troubles, so that the wings could be moved well aft, which afforded a very good view for the pilot. The wings were fitted with automatic folding devices.

An engineering feature on the FR-1 was a fast-feathering Curtiss Electric propeller system. An automatic propeller control worked in direct conjunction with jet engine thrust output and altered the pitch automatically. Both engines could be operated together and it was found, during trials, that speeds varied little up to 25,000 feet, where the aircraft performed most economically, as they cancelled out each other's operating deficiencies. The high safety margin offered by this combination had great appeal to the Navy. Three XFR-1 prototypes were ordered, all basically similar but with different tail units in order that the design which allowed the best control on approaches might be selected as standard for production. The first, No. 48232, unfortunately crashed soon after its initial flight trials began at San Diego. Shortly after the third prototype had made its first flight it was involved in a mid-air collision with PB4Y-2 No. 59836 over San Diego leaving the second machine to undergo carrier trials which proved successful. A production order placed in December 1943 for 104 was followed later by a second order calling for an additional 600, some of which were to be to a new standard. In the event the latter were never built for in November 1945 the initial production order was reduced and the project cancelled, the sixty-sixth FR-1 being the last.

It was during February 1945 that the first production FR-1s reached the Navy. These were fitted with a 1,350 h.p. Wright R-1820-72W water injection engine which could be boosted to 1,425 h.p. In March VF-66 (Pacific Air Command) became the first, and only, squadron to have a full complement of FR-1s, and immediately went into combat training, which called for new tactics due to the radical nature of the design. Squadron VF-66 was combat ready when the Japanese surrendered.

For development purposes, two Fireballs, BuAer Nos. 39661 and 39665, were taken from the line to become the XF2R-1 and XFR-4 respectively. The former had a different nose power plant, a General Electric XT31-GE-2 turboprop engine. This proved to be the only occasion when the Navy called for a turboprop to be fitted to a fighter aircraft and was the second aircraft to fly in the U.S.A. with a turboprop fitted, the first being the Air Force's XP-81. Ryan were able to claim that their machine was the first to make a cross-country flight of any length on turboprop power in the U.S.A. when it flew from Muroc to San Diego for the Company's trials. As on the FR-1, the air intakes for the jet engine were in the wing roots. To compensate for the larger nose and increased torque the XF2R-1 had an additional large fin and slightly increased fin area. The new engine combination gave the machine a combined horsepower of about 4,800 with which it attained a

The XFR-4 was another version of the Ryan Fireball with an engine change. This model had a Wright R-1820-74W radial engine in front with a Westinghouse J34-WE-22 jet engine behind with side intakes along the fuselage sides.

top speed of 500 m.p.h. at sea level and a climb rate of 10,000 feet in two minutes with both engines running. Its service ceiling of 39,100 feet could be reached in seven minutes.

Production of the XF2R-1 was never anticipated and, as with the XP-81, it was found that the turboprop was not really suitable for fighters. Compared with the XFD-1, the Fireball would clearly pose more maintenance problems and its performance was inferior although rather as an interim type its introduction to service posed fewer problems from the point of view of actual combat. On balance the pure jet was obviously the better choice, hence the cancellation of development and production in 1945.

The other experimental machine, No. 39665, was retained by Ryan for possible reworking as the XFR-4 in which a Wright R-1820-74W and a Westinghouse J-34-WE-22 were to be fitted. Apart from this major change, flush mounted air intakes on the side of the nose ahead of the wings were to be fitted. A great overall improvement in performance followed the engine change, but it was un-

crashed before it reached naval hands. The second two were equipped for carrier trials and had catapult fittings. During naval tests several unpleasant traits were apparent including instability at low landing speeds. Consequently the two aircraft were returned to the manufacturers for modifications, including an entirely new tail assembly. The vertical surfaces were enlarged and the horizontal stabiliser was raised to the top of the fin. Thus came into being one of the first 'T-tail' units. Performance trials showed a top speed of 450 m.p.h. at 25,300 feet using both engines at full power; with only the piston engine, the maximum speed fell to 373 m.p.h. at the same height. Cruising at 163 m.p.h. on full internal fuel load the aircraft had a range of 1,300 miles.

As with the Fireball, the two engines running off the same fuel supply, automatically adjusted themselves as heights and speeds changed. The considerations that led to the rejection of the Fireball as a major service type, also called a halt to the XF15C-1. Unlike the Fireball the Curtiss machine had major faults so on October 10th, 1946 the

A composite jet/piston-engined fighter, the Ryan XF2R-1 was virtually an FR-1 Fireball with engine change and a modified tail. The single example built, BuAer No. 39661, known as the Darkshark, is shown at Edwards Air Force Base.

balanced in power since the jet thrust was too much to permit efficiency when the engines were used in combination.

Space aboard carriers is strictly limited and any equipment that can be reduced in size and weight, yet retain its usefulness and the same rate of serviceability, is to be favoured. There is not room aboard ship to spread aircraft out for overhauls and adjustments. For this reason pure jet aircraft, with their relative simplicity had much to commend them over piston-engined aircraft and certainly more so than a complicated aircraft like the Fireball, with composite power units—thus the Fireball went out.

The final Curtiss naval fighter was also built to evaluate the mixed power plant concept. On April 7th, 1944 three prototypes were ordered and designated XF15C-1. These were larger than the Fireball, the specification having called for a more powerful fighter, for use at high altitudes. The two selected engines were a Pratt & Whitney R-2800-34W of 2,100 h.p. output and an Allis-Chalmers-built de Havilland H-1B turbo-jet giving 2,700 lb. static thrust. Of all-metal construction, the low-wing XF15C-1 had a nosewheel type undercarriage and a bubble type canopy perched high on its fuselage. An unusual feature was the placing of the jet engine under the fuselage with the orifice amidships, making the tail surfaces sit high above the jet orifice.

At the start of 1945 the first example was tested, but

Navy terminated its contract with Curtiss and in so doing it severed its long link with the firm who had supplied A-1 the first U.S. naval aircraft. Ironically, the XF15C-1 was the ninety-ninth Curtiss design, and soon after the failure of this the firm withdrew from airframe manufacture.

Chance Vought entered the jet fighter era with a contract placed by the Navy on December 29th, 1944, for a single-engined day fighter. Three prototypes were, as usual with jet types, ordered. The designation XF6U-1 was given and the first aircraft of the type flew at Muroc on October 2nd, 1946. Designed as a minimum weight fighter, answering needs similar to those which produced the Bearcat, it proved a disappointment to the Navy as well as Vought, who had made special use of their 'Metalite' construction, designed for use in high speed aircraft as it offered strength with a very smooth finish. The relatively small machine was of conventional appearance with straight wings and nosewheel landing gear.

Initial tests at Muroc led to some consternation among company officials. The aircraft was safe but only moderately stable and clearly underpowered. In the hands of a former Navy Commander, Edward Owen, the XF6U-1 Pirate went through its paces. It was then returned to the Vought plant for a number of modifications.

At low speeds, control was extremely sloppy, and all

Curtiss Design No. 99 and the last of the Curtiss fighters for the U.S. Navy the composite powered (D.H. jet engine and P. & W. piston engine) XF15C-1. This type did not advance beyond the prototype stage and the example shown is the second of the three built.

manoeuvres were considered far below the level of attainment expected of a post-war jet fighter. To improve handling a large dorsal fin was added. Tip tanks were of special value on the aircraft since they proved of valuable assistance in stability and only slightly detrimental to speed. The major problem was still the matter of boosting the available power.

In 1947 when the XF6U-1s were undergoing tests, America had available three types of axial flow turbojets. Two were in production, and a third fully tested awaiting introduction. The latter, the Westinghouse 24C, showed great promise, and under responsible engineering seemed likely to prove extremely successful. In this respect the British were ahead in the turbojet field, but they had always made their knowledge available to the United States Government and, of course this mutual collaboration has far-reaching effects today.

In 1945 the Navy had issued to Ryan a contract for further study into the field of re-ignition gases and exhaust at the aft end of the jet engine. The so-called 'afterburner' had been developed, and one of the first uses of it was made in the Vought XF6U-1. This process added as much as 30 per cent. additional power to the thrust of the engine. When the modification was applied to the Pirate it resulted in the aft configuration being made more rounded, leaving a straight contour to the rest of the fuselage. Performance showed an improvement, particularly the rate of climb.

Short bursts of speed resulted when the after-burner was brought into use, but five minutes limit was imposed on its use, since it made a rapid drain on the fuel supply. The Westinghouse 24C engine in the Pirate was now delivering 4,000 lb. thrust. Some instability was still encountered at low speeds and this resulted in two small fins being placed on the tailplane of the thirty production aircraft. These machines each had a gross weight of 9,200 lb. and were powered by the Westinghouse J-34-WE-22, previously known as the 24C.

Manufacturing space was limited at Vought's Stratford plant so all facilities were moved to a new factory at Dallas, Texas. The move came just as Pirate production was about to get under way and in part explains why the contract took three years four months to fulfil. Not until August 1949 was the first F6U-1 delivered; the thirtieth and last was rolled out in February 1950. As early as December 1946 the prototypes had been delivered to VX-3(LANT), a development and experimental squadron. Classified as second line combat types, their use by VX-3 and further service use on an experimental basis resulted in the decision not to introduce them into front line service being taken. Its continued overall poor performance caused the Pirate to be relegated to various research units. With these relegations the Phantom was left as the first successful serviceable jet fighter of the United States Navy.

Vought's first jet fighter venture was the XF6U-1 Pirate of which this is the first of the three prototypes. Four 20 mm. cannon were grouped in the nose beneath the cockpit. Metalite of thin alloy sheets bonded to a balsa wood core, was used in its construction.

CHAPTER TWELVE

Finding a Fighter for the 'Fifties

Scheme followed scheme for an all-round fighter and as in pre-war days, there was more attention paid to experimentation than to production as funds, once again, became limited. A conventional jet fighter was in service early in the post-war years, but even before the war ended radical changes in planform had been envisaged and the evaluation of German research in 1945 had brought startling projects to light. Without the urgency of war, there was time to consider new designs that needed a long programme of development. It was in fact a case of finding a fighter for the 'fifties and with 1950 came another severe test of U.S. naval fighters—the Korean War.

North American proposed, on January 1st, 1945, a fighter based on the same requirements as had led to the Vought Pirate. This single-engined machine was designed to be suitable for carrier or land use and North American had, at this time, considerably more knowledge of jet fighter design than Vought. The Army's XP-86 was underway by May 1945, and much was being learned about the

early September taxying trials began followed by first flight tests at Los Angeles International Airport. On September 10th the prototype was flown by Al Conover to Patuxent for Navy acceptance trials.

The first flight from Patuxent came the day after delivery and trials continued for several weeks. There was trouble when, during final spin tests, the canopy came adrift and Conover's head was jammed against the cockpit side. His strong helmet saved his life and, although knocked unconscious, he regained his senses and partial control at 1,400 feet. Pulling out from $8\frac{1}{2}$ 'G' and climbing again, Conover blacked out and came to after two turns of a slow roll at 16,000 feet. Battered, bleeding and acting on instinct, Conover brought the Fury safely back to Patuxent. From there it was sent to Muroc for further trials. On one of the first dives performed for Navy officials, the recovery stresses wrenched the elevator controls loose, and the aircraft had to be landed with trimming tabs acting as elevators to provide the necessary effectiveness.

Newcomers to the field of naval fighters, North American introduced the highly successful Fury series which paralleled development of the U.S.A.F.'s Sabre. The second of the three XFJ-1 prototypes is shown before it flew on August 1st, 1946.

design and aerodynamics of jet aircraft. After careful study, Ed Scmued, the firm's assistant Chief Engineer who was in charge of the design, came to two conclusions. Firstly, due to Navy requirements a straight wing design was deemed necessary as it afforded several requisites not least of which was stability at the low speeds, necessary for carrier landings. Secondly, to employ the General Electric J-35 engine, with which straight line ducting would be used. The single duct was to extend directly, and uninterrupted from nose to exhaust without a split entrance and expanding duct. This was an arrangement peculiar to this aircraft. The General Electric was not immediately available and it was in fact built by Allison as the J35-A-2.

The basic design proposed was submitted to the Navy in May 1945 and subsequently three such fighters were ordered as XFJ-1s. North American named their machine Fury, which became officially adopted. It was the first machine designed for the Navy by the firm and the progenitor of the long line of Fury fighters. In overall appearance it resembled a bloated version of the Army's XP-86. The first example was completed by February 1946 but it was nearly five months before the engine was available. In

In the XFJ-1 the Navy realised they had a potent fighter, and before full acceptance trials were complete a contract for 100 FJ-1 Furies had been signed. This contract was subsequently reduced to thirty in keeping with immediate post-war needs. Fine though the Fury was, by the time it was available in production form far superior machines were flying and it came to be looked upon as little more than an interim type until North American could improve further on their promising, if not perfect, design.

Final trials for the Fury really started in November 1947 when the first production models were put in service with VF-51(PAC) under Cdr. Evan 'Pete' Aurand. In the hands of VF-51 pilots, the Furies broke three records on February 29th the following year. In speeding along from Seattle to San Diego, Cdr. Aurand flew the 1,025 miles from Seattle to Los Angeles in two hours twelve minutes fifty-four seconds, while Lt. Cdr. Bob Elder averaged 492·6 m.p.h. and covered 690 miles in one hour twenty-four minutes. On March 10th, 1948, Aurand and Elder started carrier suitability trials aboard the U.S.S. *Boxer* off San Diego. Practice landings ashore had given VF-51 pilots some idea of what to expect during sea-going operations,

First production version of the Fury, the FJ-1. Originally one hundred of this version were ordered on May 18th, 1946, but this was later cut to thirty. This type was the first jet fighter to operate in squadron strength from a carrier.

and by coming in at 120 m.p.h., and subtracting wind and carrier speed, a landing was made around 95 m.p.h. This was not considered to be too high, and on his first carrier landing, Aurand hooked the first arrester wire. Elder followed him in, snatching the third wire.

These were not by any means the first jet landings on a carrier. A re-worked and strengthened YP-59A Airacomet had earlier been tried in addition to tests with Fireballs, but these had represented little more than experiments. When VF-51 put their two FJ-1 Furies aboard the carrier and flew them off, it was the first occasion U.S. naval jet fighters from a front line squadron had thus operated. Six weeks of intensive training of pilots and deck crews followed, but these operations concerned only two aircraft, and cannot be considered in the same light as VF-17's experience with the FH-1 Phantoms. VF-17, it will be recalled qualified to become the first fully operational all jet carrier based squadron in May 1948. The knowledge gained with the two Furies had been passed on to the crews of VF-17.

VF-51 only used the FJ-1 for a short time, for both the early Furies and Phantoms had shown the need for larger carriers equipped with improved arrester gear, new style catapults and revised methods of aircraft handling. The FJ-1s were eventually turned over to the Naval Reserve; many served at Oakland, California where they were used as trainers.

On March 2nd, 1945 a contract had been signed with McDonnell for an improved version of the Phantom. It was to be larger and more powerful but retain the same basic configuration as the FH-1 including straight wings and engines buried on the wing roots. Designated XF2D-1 upon inception it was to be known as this until after the first prototypes began flight trials, by which time the revised McDonnell titling caused the new fighter to become the XF2H-1. All subsequent McDonnell navy fighters had the 'H' designation and production examples of the new fighter began life as F2H-1s.

The XF2D-1, named Banshee, made its maiden flight at the Company Plant on January 11th, 1947. Originally it was planned as a day fighter, but before its career was completed it was employed in various roles. The prototypes were powered by Westinghouse J34-WE-22 engines of 3,000 lb. thrust. Initial difficulties encountered concerned the high gross weight of 13,000 lb. which was then considered too high, but later Banshees were to rise another 6,000 lb.! The prototypes reached a top speed of 585 m.p.h. and an absolute ceiling of 50,000 feet, but production aircraft proved to be slightly faster with the same engine type. The climb rate from take-off was 9,000 feet per minute.

A production order for fifty-six F2H-1s was placed in May 1947 following successful trials. March 1949 saw the entry of the Banshee to service, initially with VF-171 (LANT).

On August 9th, 1949 Lt. Jack Fruin of VF-171 flying an F2H-1 in the vicinity of Walterboro, South Carolina, was performing aerobatics at 30,000 feet when the Banshee went out of control. He pulled the ejector seat release which functioned properly, but Fruin was injured when he struck the water. Ejection tests had previously been performed by volunteers over the Pacific, but this was the first emergency use of an ejector seat in America.

Two other types that were to see combat over Korea had their origins in orders placed in 1946. On April 3rd Douglas received a contract to develop a night fighter which became the Skyknight, and Grumman were similarly asked to tender such a design on April 22nd. By the middle of the year, Vought had contracted to build a more radical design, an all-wing fighter which became known as the Cutlass. It was close on four years before any of these designs reached fruition and in the case of the Cutless, nearly nine years. They were ordered at a time when funds for the Navy, and of all America's defence services, were very restricted. Only 926 naval aircraft of all classes were procured in 1947, which was barely enough to keep up with the attrition rate. Some 1,600 were tentatively ordered but lack of funds prevented procurement. Not until the necessity for the Berlin Airlift warmed the Cold War did Congress vote money for firm procurements. Meanwhile, throughout 1947 and 1948,

One of the twenty-eight first production models of the Skyknight, an F3D-1, with its hydraulically operated air-brakes in operation. The crew of two were accommodated side-by-side in a pressurised cockpit with an escape chute.

operations were conducted by less than 3,000 front line Navy aircraft and 900 in the hands of the Marine Corps. Some 6,130 aircraft were on charge altogether, 2,200 being second line types. Additionally in storage there were 3,800 second line machines. By the close of March 1948 funds permitted the operation of three large CVB carriers, eight CV long-hulled carriers, two CVL light carriers and seven CVE escort carriers. A new build-up was underway but this time was the Navy's post-war nadir.

The Douglas contract of 1946 called for two prototype night fighters, XF3D-1s, carrying highly complex radar. Aptly named Skyknight, it was one of the world's first jet night fighters and came to be the first true all-weather carrier based fighter.

Preliminary design had actually started in September 1945 by E. H. Heinemann, before the actual contract was placed. Design requirements to be solved by Douglas were twin-jet power, side-by-side seating for pilot and radar operator, 500 mile combat range, an operating altitude of 40,000 feet, a top speed of over 500 m.p.h.—and the ability to operate from a carrier. Armament was to consist of four 20 mm. cannon.

The Navy had felt that airborne radar interception gear needed to be handled by a second crew member for maximum effectiveness; in the Skyknight this meant the addition of another 200 lb. weight. Side-by-side seating, then considered best in operational conditions, also posed the problem of increased frontal area. By the same token it necessitated a fuselage of exceptionally large diameter which was partially utilised to carry large quantities of fuel.

The first XF3D-1, BuAer No. 121457, flew from El Segundo on March 23rd, 1948 with the Douglas Chief Test Pilot, Russell Thaw, at the controls. Later in the year, in August, the second XF3D-1, No. 121458, flew and company tests continued until October when both aircraft were flown to Edwards Air Force Base for full flight trials before going to Naval Air Test Centre. From the latter they were taken to a carrier, and during mock interception tests they theoretically destroyed jet fighters above 40,000 feet, and were considered to be faster than any enemy bomber of the immediate future.

The prototypes had J34-WE-22 engines and a top speed of 510 m.p.h. They landed at 96 m.p.h., and their combat radius was 600 miles. But tests showed that in the vital fast climb, the Skyknight performed most unsatisfactorily. It began at 1,780 feet per minute and then tailed off to 540 feet per minute at 40,000 feet. Considering its intended role, this was just not good enough although in other respects the Skyknight met the specification.

Since the XF3D-1 had, in most respects, met the minimum requirements set forth by the Navy in 1945, Douglas received a contract for twenty-eight F3D-1s Skyknights in June 1948, purchased with 1949 funds. The first production model went to VC-1, a composite service test squadron at Moffett Field, in December 1950. Major improvements were in the more powerful J34-WE-32 engines of 3,250 lb. static thrust. More electronic gear was installed, raising the

The McDonnell F2H-1 Banshee seen here is one of fifty-six ordered in 1947, and thus was one of the first American jet aircraft built in quantity. First unit equipped with this type was VF-171 in 1949. Although no engine change was involved, production aircraft were faster than prototypes.

gross weight to 27,362 lb. as against 22,000 lb. for the XF3D models. The range was unaltered, but landing speed rose to 107 m.p.h. in spite of the fitting of air brakes on the rear fuselage. The WE-32 engines raised the top speed by 25–30 m.p.h. and improved the climb rate. Extensive tests were conducted by VC-1, who considered increased performance necessary. The majority of the other F3D-1s, twenty in all, were transferred to the Marine Corps, and when an improved version became available they were used as training aircraft.

An interesting feature of the Skyknight was the procedure for abandoning the craft. Originally upward—or even downward—ejector seats were contemplated, but these arrangements proved heavy, bulky and impracticable with the seating arrangement. After careful study, it was decided that the easiest and safest means of escape was to slide, feet first, down a chute so arranged that its bottom door panel acted as a wind screen during exit. By pulling a lever, the two seats swung backwards and outwards to lie flush against the cabin sides. At the same time the forward half of the belly door swung open and the rear half was jettisoned.

The improved and more powerful version, the F3D-2, was initiated for night interception duties with the Marine Corps in November 1950, and the first example was flown

An F9F-2 Panther (BuAer No. 127153) of Marine Squadron VMF-311 utilised as a bomber from K-3 Airstrip in Korea, 1953. Note bombs on wing pylons of this aircraft and another in the background of this photograph. The F9F-2 was the only model of the type to use the J42-P-6 engine.

on February 14th, 1951. The re-design entailed incorporating new electronic and radar equipment, improving air conditioning, providing for air to air rocket weapons, thicker bullet proof windscreen, wing spoilers to better the rate of roll and fitting a General Electric G-3 auto-pilot. The F3D-2 was scheduled to employ the Westinghouse J46 engine of 4,080 thrust, an improved variant of the J34. The engine nacelles tucked under the aircraft were enlarged to accept the increased size of these engines. But as it was, extensive development difficulties beset the J46 engine and production deliveries were delayed and failed to reach the F3Ds in time.

The well-known Grumman Panther was conceived in 1946 as the XF9F-1 night fighter. Design studies were undertaken in 1945 at about the same time as work on the Skyknight began. The Grumman XF9F-1 was, however, to be a far more complicated proposition. Preliminary specifications called for a monoplane night fighter utilising four Westinghouse 24C engines with a proposed output of 2,700 lb. each, which could be boosted to 3,000 lb. Such an engine was far from being a reality at this time, but it became available as the J34 by 1947. Meanwhile project studies were made on the airframe and engine combinations. A Douglas venture involving two of the engines posed many technical problems, and Grumman's problems, with *four*, seemed likely to be insuperable. The XF9F-1 project was therefore considered too ambitious and the original design contract for two prototypes was reconsidered. The Bureau of Aeronautics recommended abandonment of the XF9F-1 in October 1946 in order that Grumman might turn their attention to a single-seat day fighter. The XF9F-1 design was shelved and for the most part still remains classified material.

The decision to produce a day fighter meant a complete change in Grumman's plans. Certainly the XF9F-1 in no way remotely resembled the new fighter which bore a designation indicative of a mere new model—XF9F-2. The decision to change the entire concept also led to a change of engine to the Rolls-Royce Nene, a very powerful engine which the British, rather oddly, did not favour for their own aircraft. It was to prove highly successful in the new Grumman fighter, and the basis would appear to have been used by the Russians for the MiG-15. Initial interest in the engine was made via the Bureau of Aeronautics through the Taylor Turbine Corporation of New York. Two Nenes were sent to America in December 1946 for a standard 150-hour test run. The Taylor Corporation negotiated licence and sales agreements to erect, manufacture, test and maintain the Rolls-Royce engines in America.

Tests run at the Naval Air Material Centre showed a thrust of 5,000 lb. which was much higher than the thrust from available American engines, but provision in the design was made for the time when an American engine could take the place of the British Nene. Two fully tested production Nenes were imported for the prototypes while negotiations were underway for American production. The Taylor Corporation, unable to cope with large-scale building of the engine, arranged with the Pratt & Whitney Division of the United Aircraft to put the engine on to its assembly lines, and this version became the J42 series.

In June 1946 studies of the XF9F-2 design were still informal, but some conclusions had been reached. Formal submission of the design was worked out by August, and two Grumman engineers went to Britain to study the Nene engine. October saw the commencement of wind tunnel tests which continued into 1947. First tests proved the basic design satisfactory and in November 1946 it was decided to proceed with the mock-up which was inspected by the Navy in 1947 and approved. Work commenced on the prototypes early that year. The first two Nenes, flown over in an American Overseas Airlines freighter arrived from London in July, and in November the prototype was ready for tests. It was flown by twenty-seven-year-old Corwin 'Corky' Meyer, chief engineering test pilot for Grumman. Engine tests were started on November 20th, taxying trials next day and three days later the Panther went aloft on her maiden trip from Bethpage. The second prototype took to the air two days later. Such was the initial success of the type that it was ordered in quantity before trials were completed. The P-80 Shooting Stars then in Air Force use required as much as a 4,000 foot run for take-off, the Panther took 800 feet, or 430 in a 30 knot wind—the equivalent of a carrier steaming into the wind. Much of this was due to the incorporation of leading edge wing slats providing the closest approach to a variable camber wing then achieved. Use of these slats coming down from the wing leading edge, allowed the XF9F-2 to land at 85–90 m.p.h. Initial speed runs indicated a top speed of around 500 m.p.h. and a climb rate of about 9,000 feet per minute.

In appearance the F9F Panther was short, stubby and

pot-bellied—a typical Grumman design. It had squared off wing-tips to the straight planform. The egg-shaped nose housed four 20 mm. cannon. A third prototype was ordered in which an Allison J33 engine was easily interchanged with a Nene, and one of the first proven J33-A-8 engines was flown in the third XF9F-2 and was found to have characteristics similar to those of the British engine. Production machines could therefore alternate between two powerful engines, and production bottlenecks which had beset so many post-war types were likely to be avoided.

Pratt & Whitney made ready the Nene for American production in an extremely short time. Indeed they were producing the J42-P-6 as fast as Grumman was building airframes. First production Panthers had the J42-P-6 but this was soon superseded by the 5,750 lb. s.t. J42-P-8. The first contract called for 437 F9F-2s and the type was issued first to VF-51 (PAC) on May 8th, 1949. This was the squadron that had pioneered the use of the jet fighter in naval service with the FJ-1. Most of the F9F-2s went to the Navy squadrons, but the Marine Corps equipped Squadrons VMF-311, VMF-115 and VMF-451 with Panthers.

An interesting side note on the Panther occurred in June 1950, involving some of the first F9F-2s. During large scale exercises held that year on Vieques Islands, the Navy, Marines, Army and Air Force all participated. In the final assessment of the results it was found that VMF-115 had theoretically 'destroyed' ninety F-84 Thunderjet fighter bombers with a loss of only nine Panthers.

When the Bureau invited bids to produce an interceptor fighter in the 600 m.p.h. at 40,000 feet class, Vought, Grumman, Curtiss and Douglas all submitted designs. The offer remained open from April to June 1946 after which the Vought project was selected for development. An order for three experimental aircraft was placed, and not long after, word leaked that this was to be a tail-less aircraft. The original design called for use of two straight flow Westinghouse J34 engines, but tests with the F6U Pirate suggested the value of after-burners. The design staff decided to add two huge Solar after-burners to a XF7U-1 a change that gave an odd appearance to the rear of the aircraft. Engines finally chosen were two J34-WE-32 turbojets of 3,000 lb. dry thrust, 4,200 lb. using the after-burners.

As on the XF5U-1 and F6U-1, 'Metalite' sandwich construction was employed, the skin being of painted aluminium alloy. Elimination of a tailplane and rear fuselage would considerably reduce drag, but longitundinal stability and control was likely to suffer. Extra control surfaces would therefore be needed to be built into the wings, and Vought also fitted twin fins and rudders. Theoretical low drag/maximum lift coefficients were skilfully computed and the use of a 35° wing sweepback added to speed and performance, but also added more airflow problems. The use of full span leading edge slots, however, solved some of the problems of air spill over the high speed wing. Vought also carried the fin leading edges well forward to assist stability. Thus the F7U was not truly an all-wing aircraft; control problems resulted in a compromise rather than a solution.

John K. Northrop, a pioneer of all-wing design, proved that the flying wing could be built without vertical surfaces and still be stable, but Vought and the Navy were not prepared to risk the skidding that might occur at slow speeds with such a design, which could be hazardous for carrier landings. With no horizontal tail surfaces, elevators and ailerons had to function simultaneously, in conjunction with, as well as separate of, each other. Northrop had called these 'elevons', Vought scrambled the words 'aileron' and 'elevator' together and produced 'ailevators'.

The first prototype made its maiden flight at the Vought plant at Dallas, Texas on September 29th, 1948. It was rather a hair-raising event. The swept wing machine, sitting at an angle of 9°, raised its nose on take-off to 15° before it left the ground. This high angle of attack was to provide a good lift for take-off, but the attitude of this 10-ton fighter was frightening to those watching this new take-off routine. Once airborne, the Cutlass was a great performer, but there were many troubles to shed. Three prototypes were complete by the end of the year and full evaluation trials had begun. Testing could not be performed at Dallas since the runways there were too short.

The three XF7U-1 Cutlasses were flown to Patuxent for official trials early in 1949, where, in May, the third aircraft disappeared under somewhat curious circumstances. On a routine test, William H. B. Millar flew the aircraft into a cloud bank at about 7,000 feet and was never seen again. A few days later parts of the aircraft were found floating in Chesapeake Bay, with no indication as to the cause of the accident. During May the other two prototypes made the cross-country flight back for further modifications. These were undertaken at Fort Worth, some twenty miles away from the Vought plant.

For two years, Navy and Company test pilots put the bat-like machines through trials, which culminated on a carrier. In June 1948 an initial procurement order for fourteen F7U-1s was placed. Due to the time consumed in

The Skyknight was the first nightfighter to shoot down a jet at night. This version, the F3D-2, first appeared in early 1951 and the example shown, BuAer No. 125824, is shown serving with VFAW-3 at San Diego on August 29th, 1959.

tooling, it was not until March 1950 that the first delivery was made and then considerable time elapsed before it reached a unit.

A design competition for a delta wing fighter was held in 1947. Douglas, Grumman, and other companies entered and received contracts. After reconsidering the designs and changing recommendations, Douglas received a development order in June 1947 and a contract was signed early in 1948. The Grumman design was for an all-weather interceptor replacement for the Panther. Due to a large number of designs projected by Grumman, the final version was not schemed until the closing months of 1948. Grumman engineers, pointed out that the thick delta aerofoil requested by the Navy was not in keeping with the speed and flight characteristics they demanded. Similar problems faced the Cutlass, and the forthcoming Douglas F4D Skyray. Both companies had to compromise on the pure delta form. Successful as that form has proved since, on land based aircraft like the Delta Dagger and Delta Dart fighters and the Hustler bomber, the true delta has not proved practical for carrier based aircraft.

Confronted with technical problems Grumman proposed a compound swept wing. This, it was felt, would decrease the weight, yet retain the delta configuration. Engineers felt that they could work out the aerodynamics of such a radical design, and that the mechanical problems would be mastered in a matter of time. The proposal was accepted by the Navy and a contract was let for two experimental aircraft, designated XF10F-1 and known as the Jaguar. After the initial announcement of the contract being placed, little was heard of the project partly for security reasons, and partly because the design stage was not running at all smoothly and it was after the Korean War before the Jaguar was seen in the open.

Doctor Alexander Lippisch, well known for his all-wing and delta designs in Germany, had proven theories, by models and war-time research, that led to the Messerschmitt Me163 rocket fighter. The Navy displayed interest in his designs and consultation with Doctor Lippisch prompted the Navy to delve deeper into the delta design. The study led to the contracts which Grumman and Douglas received, although as recounted, the XF10F-1 was re-designed so that the original delta planform was dropped in favour of the swept wing. Douglas, at first, accepted the challenge, but finally produced a modified delta design, albeit a most successful one.

It was June 1947 when Douglas received the contract for the preliminary study of a delta winged interceptor fighter. The entire project was handed to the Douglas Chief Designer, Mr. Edward H. Heineman, and their Project Engineer, Charles S. Kennedy. By December of the following year the basic engineering concept was laid out to enable inspection and a contract was awarded on December 16th, 1948 for two prototypes to be known as XF4D-1s. Douglas was now faced with the problems associated with the delta wing that had beset Grumman. One of the requirements of the Navy specification was the ability of the fighter to 'intercept enemy bombers within five minutes of an alert'.

Douglas had chosen the prefix 'Sky' for the names of their military aircraft, and the XF4D, on account of its 'sting-ray-like' wing shape, became the Skyray. The design was under development for two years, which was not unduly long for such a challenging project. Douglas make it quite clear that, after thorough evaluation, the Skyray was not a true delta design. In all respects it was a swept wing aircraft of low aspect ratio. Its wing of triangular planform had its corners rounded. Although at first glance unusual, the Skyray was relatively conventional, with a thin high speed swept wing and a well faired fuselage.

By January of 1951 the first prototype XF4D-1 had been loaded aboard a trailer and transported to Edwards A.F.B., Muroc, California. The small compact fighter, its wings folded, fitted with ease on to a special lorry for its journey by road from El Segundo. Test pilot Robert O. Rahn took the machine on its maiden flight on the morning of January 23rd, 1951. A few days later the second prototype arrived and a series of tests followed. During these trials the characteristics peculiar to the delta form were closely investigated and the Skyray, with further development, promised to be in a class of its own as an interceptor.

The delta form or swept back wings were favoured in the 'fifties.

The XF7U-1 Cutlass over the airfield of the Chance-Vought Division at Dallas. This revolutionary aircraft was the first Navy jet fighter to be designed from the drawing board to employ an afterburner; on earlier aircraft this had been introduced as an afterthought.

CHAPTER THIRTEEN

Korean Kinetics

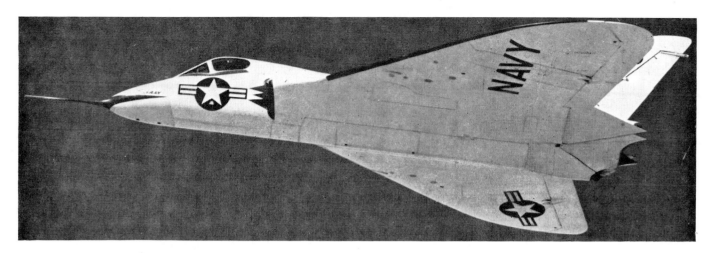

The prototype of the only naval aircraft to have the distinction of participating in the Continental Air Defense System of the U.S.A., the XF4D-1 Skyray which first flew early in 1951. This aircraft is aptly named from its ' sting-ray' appearance and its sting is four 20 mm. cannon, rockets and Sidewinder missiles.

The Korean War stirred fighter development in two ways, firstly by bringing into prominence operational requirements and secondly by securing the funds so necessary for development. With the need and the funds, forces were soon set in motion.

Early on the morning of June 25th, 1950, North Korean Forces invaded the Republic of South Korea. The United States Government who until recently had been caretakers in South Korea asked the United National Security Council to consider measures to stop this invasion. The Council adopted a resolution calling for the cessation of hostilities and for North Korean Forces to withdraw to their positions north of the 38th Parallel, a request they ignored as, backed by Russian material, they advanced well into South Korea.

President Truman, on June 27th announced that he had ordered sea and air forces in the Far East to give support to the Republic of South Korea. An order was sent to the Seventh Fleet off Formosa to prevent a possible invasion of that Island. Later that night the United Nations adopted a resolution to assist the Republic in repelling the armed attack upon its territory. Without a formal declaration of war by either side, and without a full Senate consenting vote, the President, in keeping with the U.N. resolution, ordered the U.S.A.F. into battle and the U.S. Army to engage in ground warfare, while the Navy was deployed to blockade the entire Korean Coast.

On July 3rd, the carriers U.S.S. *Valley Forge* and H.M.S. *Triumph* went into action under the U.N. flag. Air Group 5 aboard the *Valley Forge* launched strikes against supply and transport targets and airfields around Pyongyang. VF-51 flying F9F-2 Panthers delivered the first strikes and before the day had passed the U.S.N. had made its first kills of the war. This was of course the first real test of the Panthers, and Lt. (j.g.) L. H. Plog and Ens. E. W. Brown each shot down a Yak-9 over Pyongyang.

The British carriers with their piston-engined Fireflies and Sea Furies may well have felt out of it until American carriers arrived with a veteran aircraft—Corsairs, which the Fleet Air Arm had not used since 1946.

VMF-214 flying from the escort carrier U.S.S. *Sicily* started combat operations in close support role on August 3rd. The U.S.S. *Badoeng Straight*, with VMF-323 aboard, joined the action three days later thereby bringing their F4U-5 Corsairs into action. The first Marine aircraft to enter combat had done so on July 4th when two F4U-5Ps of VMJ-1 made a reconnaissance sortie from the *Valley Forge* which took them over the Ryangyang area. One was destroyed by fire from anti-aircraft guns.

Both the Navy and Marine Corps air elements used the F4U-5 and 5N Corsairs in quantity in Korea, but in the face of the Russian built MiG-15 they had to withdraw. However, on low level attacks the Corsairs performed extremely well. Over 80 per cent. of the support missions flown by the Navy and Marine Corps were made by Corsairs. On one such mission five MiGs jumped two Corsairs, which fought back and destroyed one of these jets. Top cover and fighter patrols were flown over the Corsairs by Grumman F9F-2 Panthers.

The Panther was one of the Navy's mainstays during the war and was still being developed. The Allison J33-A-8 was delivering 4,600 lb. thrust by 1950, and with water injection this could be raised to 5,000 lb.; one F9F-2 was taken from the assembly line to become the XF9F-3 with this installation. While overall performance was little improved, a considerable order for F9F-3s thus powered was placed, because of the easy interchangeability with the J42 engine. After a while it was found that they were rather inferior to the earlier aircraft, so the engines were removed and they reverted to being F9F-2s.

First of the major changes in the type came with the F9F-4, of which there was no prototype, since it was introduced on production. A 2-foot extension was built on to the Panther's fuselage to accommodate the J33-A-16A engine guaranteed to give 5,500 lb. thrust with the aid of an after-

In a jet age, conventional aircraft played an important role in Korea and fighters famous in the 1939-1945 war, such as the P-51 Mustang and F4U Corsair, were used operationally in strike roles. Here, flying along the Korean coastline are F4U-5 Corsairs, accompanying Douglas AD Skyraiders.

burner. To improve slow handling qualities, a taller, more pointed, tail was fitted. One hundred and nine ordered were eventually phased into F9F-5 form which had the Pratt and Whitney J48 engine. Of the F9F-5, 655 were built, and a total of 1,388 Panthers had then left the assembly lines when the final F9F-5 was delivered in December 1952.

It was the F9F-2, 4 and 5 models that saw use in Korea. Among squadron users of the F9F-4 were VMF-334, VMF-122 and VMA-334, while the F9F-5 was operated by VMF-223 (VMA-223) and VMJ-3 (VMA-323). The F9F-5 variant had by far the best performance, on a J48-P-6 of 6,240 lb. thrust. Maximum speed at 40,000 feet was 575 m.p.h. with a service ceiling of over 45,000 feet. Cruising range with wing tip tanks fitted, was 1,200 miles. Several F9F-2s were modified for photographic reconnaissance and were designated F9F-2Ps.

To evaluate control during carrier landings, Grumman and the Navy withdrew one of the F9F-4 models and fitted wings for boundary layer research, and late production F9F-5s were subsequently modified as a result of findings. Boundary layer research had been under study some time, various contracts for study being awarded to American Universities. Many of their experiments had been carried out on light private aircraft, and in this connection, work by Mr. John Attinello of the BuAer Engineering Department, led to later developments.

With the coming of the first winter it was found that Corsairs needed special modifications to combat the cold. This led to the F4U-5NL version for after-dark missions in cold weather. Rubber boot de-icers were fitted to the leading edges of all wing and tail surfaces. Windscreen shields had defrosters added, and propllers sported de-icers; extra oxygen supplies, dual radio equipment and a superior and more efficient heating system was also installed.

As the war in Korea progressed the close support role became increasingly important and brought into existence the final Corsair variant for the Navy. A heavy offensive armament was prerequisite; no longer was a good high altitude a special requirement. Maximum all-up weight rose to 19,398 lb., and additional armour and fuel were carried. External offensive load could be 4,000 lb. bomb load or ten rockets. This new Corsair began life as the XF4U-6, test bed for which was a converted F4U-5, but in production it was known as the AU-1.

Ordered jointly with the F9F-2 Panther, which was powered with a P. & W. J42-P-6 engine was the F9F-3 Panther to be powered with an Allison J33-A-8 engine and an early example of the latter, BuAer No. 122564, is shown aboard the U.S.S. Franklin D. Roosevelt (CV-42). This photograph was taken before the modification of this carrier which was completed in 1956 at a cost of $48 million, some 53 per cent of the original cost.

The first of 111 ordered was delivered in January 1952, and the type entered service with the Marine Corps in Korea being the first aircraft to join the Navy under the new 'Attack' designation.

A see-saw battle raged in Korea and turned into a type of 'brush war'. The carriers U.S.S. *Badoeng Strait*, U.S.S. *Sicily* and U.S.S. *Philippine Sea* were despatched for duty. The U.S.S. *Boxer* arrived with a load of P-51 Mustangs for the U.S.A.F. The U.S.S. *Princeton* and the war veteran U.S.S. *Essex* also joined the fray. But normally not more than four carriers were involved at any one time.

The first major air strikes delivered by the U.S.N. were made on November 9th, 1950 when bombing runs were directed against the bridges crossing the Yalu River. The first noticeable opposition to the U.N. Forces was also encountered here. Russian built MiG-15s roamed the area and on that day the Navy first battled with them. In the ensuing fight, Lt. Cdr. W. T. Amen, Commander of VF-111, flying an F9F-2, scored one victory and thus became the first Navy pilot to shoot down a jet while flying a jet.

nose. Provision was made for two 500 lb. underwing bombs to be carried. The FH-1 and FJ-1 were, incidentally, the last Navy fighters to have ·50 calibre guns as main armament, subsequent to which the rapid fire 20 mm. cannon came into general service. The fourteen F2H-2N night fighters also had the 20 mm. cannon with limited radar fitted in the nose.

When the need for a superior all-weather fighter version arose, the F2H-3 came into being. Classified as a long range all-weather fighter the '3' Banshee was further elongated, this time to accommodate improved electronic gear in the fuselage and a bulbous fibre-glass nose accommodated a radar scanner. The armament was re-positioned further aft, along the fuselage sides. First of the 250 F2H-3s ordered were introduced to squadron service in April 1952. Although increased thrust with J34-WE-36 engines giving 3,600 lb., were fitted, the top speed remained around 590 m.p.h. at 10,000 feet; possibly this was because the gross weight was up to 21,300 lb. The Royal Canadian Navy purchased thirty-nine from the U.S.N. in 1955 as all-purpose

The Douglas F3D-2 Skyknight was a refined version of the original F3D-1. Refinements included improved air conditioning, new type auto-pilot, thicker bullet-proof windscreens and wing spoilers for improving lateral control. It was the standard nightfighter in the naval services for several years.

The Banshee was combat ready when the campaign opened and later it was brought into the War. Altogether Banshees served with twenty-nine squadrons of the Navy and Marine Corps in the U.S.A., the Mediterranean, Atlantic and Pacific, and it is still used by reserve units.

The production lines at McDonnell were never idle, since as the last F2H-1 rolled out, so production of the F2H-2 came underway. Ordered in June 1948, this Banshee was basically similar to the first, and came to be the primary Navy and Marine fighter for all-purposes, including night combat duties, throughout the mid-'fifties. It had improved Westinghouse J34-WE-34 engines delivering 3,250 lb. static thrust. First flown in August 1949 it incorporated such improvements as a lengthened fuselage accommodating an additional 147 gallons of fuel, and 200-gallon fixed tip tanks which increased the range from 1,500 to 1,800 miles with little change to the overall performance. The initial order called for 174 F2H-2 and fourteen F2H-2N night fighters. A further 160 F2H-2s were ordered in April 1954, since the type was so easy to adapt to varying roles. With an elongated nose housing six cameras, eighty-nine Banshees were built as the F2H-2P reconnaissance version.

Armament of the Banshee was improved over that of the Phantom, since four 20 mm. cannon were fitted in the

fighters. They operated from H.M.C.S. *Bonaventure*.

The last Banshee variant was the F2H-4, of which 150 were procured. Again, improved engines, Westinghouse J34-WE-38s were fitted and although power output remained unchanged these turbojets used less fuel, and were more reliable. Yet again the fuselage was lengthened, this time by five feet over the F2H-1 and an entirely new tail surface was incorporated in a generally cleaned-up design. At last the Banshee topped 600 m.p.h. and the combat range was extended to 200 miles. In-flight refuelling was provided, and again the radar was improved. Further development of the type was however deemed inadvisable and in August 1953 the final Banshee left the production line, the 894th of the type. Now it is referred to as the safe and reliable 'Old Banjo'.

In the first months of the Korean campaign bad weather, shortage of forward airfields, insufficient numbers of adaptable aircraft and political restrictions all hampered engagements. By February 8th, 1951 the Marine fighter squadrons had returned to support operations around Pusan airfield. F9F-2 Panthers fitted with bomb or rocket racks were then available; these were known as F9F-2Bs. Strikes by two F9F-2Bs of VF-191 were carried out on April 2nd, each aircraft being loaded with four 250 lb. and two 100 lb. G.P. bombs. The machines had been catapulted

Banshees of the first limited production version, the F2H-1, at Glenview, Illinois. Note that white paint has given a false impression of tail shape against the sky background.

from the *Princeton* for an attack on railroad bridges near Songjin and on this occasion the U.S.N. was using jets as fighter-bombers for the first time. The process became routine, and in part led to the development of aircraft specially devised for this role. Squadron VMF-311 equipped entirely for it in 1952. The arrival of the U.S.S. *Essex* loaded with fresh supplies, equipment and men in August 1951 also brought the first F2H-2 Banshees to the theatre, where they were flown by VF-172 as part of Task Force 77. On August 25th alongside F9F-5 Panthers, the Banshees provided escort for B-29 Superfortresses on a high altitude bombing mission on the extreme north-east of Korea. By December, the southward advance of the Communist Forces forced the evacuation of airfields, and the Marine Corps retired to Japan to re-group. A push into the sea seemed imminent, but the Pusan perimeter held.

The F7F-3N Tigercat too, was also early on the Korean scene. VMF(N)-542 was called to duty in Korea and in October 1950 made its first combat sorties, operating Tigercats by day and night. In May 1951, flying Tigercat No. 80575, M./Sgt. Barney Olsen and T./Sgt. Frederick together destroyed a complete convoy; for which Olsen won the D.F.C. Later the Tigercat squadron returned to El Toro in California for transitional training to the F3D-2 Skyknight with which they later returned to Korea.

By 1951, the conflict brought a need for a night fighter of the F3D Skyknight type, so production was ordered in spite of the delivery failure of the J46 engine. Some improvements were made to the existing J34, boosting the output to 3,600 lb. Seventy Skyknights were ordered from 1950 Fiscal Year Funds, and fitted with J34-WE-36 engines.

Although the larger J46 engines were not produced in sufficient quantity for installation in the type, the F3D-2 with the revised J34 showed an improvement in performance by the top speed rising by about 25 m.p.h. and climb rates increased to around 2,000 feet per minute at sea level. What was more important, the F3D was in the hands of the Marine Corps for use in the Korean War, albeit the closing months. To train in two-man radar equipped night-fighters, the U.S.M.C. procured twelve ex-U.S.A.A.F. P-61 Black Widows as F2T-1s.

While the war was on attempts were being pressed for a superiority fighter by swept wings. Douglas engineers realising that the Skyknight had reached ultimate development in its present form turned to the possibility of a swept-wing version. The Navy's first study of the swept-wing aircraft was as early as 1943, but actual flight evaluations did not get under way until 1946 when two ex-A.A.F. P-39Q Airacobras were transferred to the Navy as F2L-1K target drones. Because of their configuration and aerodynamic design with the engine located amidship, they were suited for swept wing experiments. At least one of the Airacobras was subsequently fitted with varying forms of swept back wings, with performance and stability being carefully analysed. Test reports quote this aircraft as the 'L-39 swept-wing Cobra'. A later purchase, a P-63 Kingcobra labled L-39-2, provided additional performance data on the swept-wing design. High speed research found the wing inefficient at low speeds, and led to use of leading edge slats, vertical fins and greater wing areas to meet naval requirements in this field.

A swept-wing version of the Skyknight was projected as the XF3D-3, utilising J46-WE-3 engines which could guarantee a combined thrust of 8,160 lb. The overall

Apart from an engine change, the F2H-3 Banshee differed greatly from the early model shown above, with re-designed nose and tail, longer fuselage and, visible on the right, the probe for in-flight refuelling.

dimensions would be somewhat less than earlier versions, and estimated figures promised a substantial improvement in performance. Combat range increased to 860 miles, the initial climb to 2,640 feet per minute and the maximum speed to 515 m.p.h. at 40,000 feet. Gross weight would rise to the all-time naval fighter record of 34,000 lb. The 1951 procurement programme provided for two experimental models of the aircraft and 100 production machines. Extensive development work went forward on this sweptwing design, but the likelihood of delays during development and production, caused a complete cancellation in February 1952. Changing policies and tactics at the end of the Korean War placed more emphasis on aircraft suitable for several roles. As it was, the conventional F3D-2 delivered on March 23rd, 1952 was the last of the Skyknights built.

While the Skyknight was 'out' the Cutlass was definitely 'in'. The first F7U-1 was assigned to the Advanced Training Command at Naval Air Station Corpus Christi for evaluation and further service use in January 1952. The remaining F7U-1 Cutlasses were relegated to training duties, to N.A.T.C. for trials, and to various squadrons to

Red Forces began to fall back in the face of services whose training was vastly superior and equipment more than their equal. The air campaign developed into a rigorous routine upon which the Commander of Task Force 77 commented, ' a day-to-day routine where stamina replaces glamour, with persistence against perseverance '. The Korean war was a small war, for at no time were more than four carriers in action simultaneously. Fighter groups went into action were relieved and returned to battle. On August 28th, 1952 news of the first use of explosive laden drones was announced, the first of six attacks being launched from U.S.S. *Boxer* by an F6F-5K drone, under the control of two AD Skyraiders of Guided Missile Unit 90. This unit also directed successfully drone Hellcats on to a railroad bridge at Hungnam.

With the recapture of Kimpo airfield, the main landing field in South Korea in mid-1952, further value was placed on fighter aircraft for close support and for fighter-bomber missions. Late in 1952 the first F3D-2 Skyknights arrived in the hands of VMF(N)-513 and flew from K-2 airstrip near Taegu. By the end of the war they had logged 12,669 combat missions. A second unit, VMF(N)-542, commenced oper-

The F2H-2 model of the Banshee, which introduced integral wingtip tanks, was produced in several versions, two for bombing, eighty-nine for photo-reconnaissance and fourteen for night-fighting designated F2H-2N of which an example is shown—BuAer No. 12300 bearing the name of the pilot, Lt. A. Newman, by the cockpit.

report on its potential and adaptability. The last F7U-1 was modified to become the only XF7U-2, although it is believed that the designation on the actual aircraft was never changed. It was to be modified to the extent of having taller vertical tail surfaces, revised intakes, and two Allison J35-A-21s of approximately 4,000 lb. thrust. The basic configuration was unchanged and indications were that the engineers were on the right track to improve the Cutlass.

As trials of the first F7U-1s were being conducted, an extensively re-designed and more powerful version was in the mock-up stage. J46-WE-9A engines were to be used, and the new larger version was approved without a prototype order, since the Cutlass had proved itself after years of flight trials. By December 20th, 1951, as the last of the F7U-1s were coming down the assembly lines, the new F7U-3 made its first flight. The J46 engines were not installed in this or the first few production models until a later date.

During 1952 the tide of war slowly changed and the

ations with F3D-2s in 1953 and set up a similarly impressive record. On November 2nd, 1952 a lone F3D-2 of VMF(N)-513 destroyed a Yak-15 over ' Mig-alley ', and this was the first time in history that an American jet had, at night, shot down an enemy jet.

Four F4U-5N Corsairs were deployed to Kimpo to operate under the U.S. Fifth Air Force on June 25th, 1953. Night harassing attacks made by the enemy were known as ' Bed check Charlies '. The Corsairs soon put a stop to their tactics. It was with VC-3 a composite fighter group using Corsairs that the Navy had their first and only ace in the Korean War, Lt. Guy P. Bordelon, Jr., of Ruston, Louisiana. Flying F4U-5N ' Yokosuka Queen ' No. 12442, Unit No. 24, Bordelon made five kills—mostly Yak-15s, during the course of fifteen missions.

Final variant of the Corsair was the F4U-7, a development of the AU-1 for the French Navy. Although carrying Navy serial numbers, they were built specifically for the

French and were delivered direct. When the order for ninety-three was completed on Christmas Eve, 1952, the last piston-engined fighter had been produced in the United States of America.

Throughout 1951 and 1952 while the war in Korea went on the two XF4D-1 Skyrays were under trial and appeared to be meeting the specification called for in the original contract. In conjunction with the Navy, Douglas officials agreed to work up the design making detail modifications and changes in the power plant. The first XF4D-1 had the proven Allison J35-A-17 turbojet of 5,000 lb. s.t. installed, the second used the new Westinghouse XJ40-WE-8 of 6,500 lb. s.t. which offered 11,600 lb. thrust when utilising the Solar after-burner. After initial flights, the Allison engine in the first Skyray was replaced by an XJ40 engine, and during final evaluation both prototypes were flying with Westinghouse engines. An initial order had been placed, but delay was incurred with the termination of the Westinghouse J40 engine development programme. In its place came the Pratt & Whitney J57-P-2/8 of 9,700 lb. s.t. (13,000 lb. with after-burner), which entailed extensive re-design before installation. Emphasis was also placed on improving the radar and armament production. Meanwhile during November 1953 carrier trials were successfully craft using J40-WE-8 engines, of 11,000 lb. static thrust.

While the Skyray went on to the 'sixties, another project under test during the Korean war period came to nought. This was Grumman's XF10F-1 Jaguar which took five years to progress from design to first flight. A variable sweep wing was not entirely new, for the U.S.A.F. had incorporated this feature on the Bell X-5 of 1951, which was in turn largely based on the Messerschmitt P1101-V1. So, whilst the feasibility had been proven in theory, Grumman engineers had to adapt it to meet naval requirements.

A power control system was employed in the machines, and through mechanical linkage, the pilot's control column operated a small horizontal vane on a streamlined boom which protruded ahead of the all-flying tail. This small surface would, in turn, actuate the tail surfaces proper, providing longitudinal control at high speeds without excessive stick forces, and reducing the weight of hydraulic systems for these controls. Wing spoilers were used to improve lateral control. The main undercarriage folded into the lower sides of the fuselage, and was of similar geometry to earlier Grumman designs such as the F3F and F4F series. Complex and unusual controls were naturally involved on this aircraft to govern the forward and aft movement of the type of wings employed.

The Jaguar was the only variable sweep wing naval fighter. Thirty-two were ordered, two only were built and only the one shown flew. The swept wing was adjustable in the range between 20° and 40°.

completed aboard the U.S.S. *Coral Sea* and the type was accepted for service.

The outstanding performance of the prototype XF4Ds led to an attempt on the World's official speed record, termed as 'part of the flight evaluation test programme'. Using the second Skyray No. 124587, Lt. Cdr. James B. Verdin captured the International Three Kilometre Course record, after making four passes at an average speed of 753·4 m.p.h. This record, on October 3rd, 1953, at Salton Sea, California, was an amazing achievement for a carrier based aircraft in combat state. At this period the speed record was passing to and fro between the U.S.A.F. and Britain, with the Navy out of the running until the XF4D-1 made its mark.

With this record in hand, the Douglas test pilot, Bob Rahn, decided to try for the 100 kilometre closed circuit record currently held by Britain. On October 16th, 1953 the XF4D-1 averaged 728·11 m.p.h. at Muroc and the record passed to America. Both records had been set up by air-

By the first of May 1953 the first prototype was ready for testing. It was carefully rolled out via the back-door of the assembly shed between two buildings, hidden from prying eyes, at the Bethpage plant. After careful inspection it was crated and despatched to Muroc for flight trials, where it first flew on May 19th. From the start, the intricate control systems proved a source of trouble.

The second prototype, which was also sent to Muroc was externally similar to the first, but incorporated some exterior, and many interior, technical modifications. The delta tail surface was changed on both aircraft at Muroc in order to incorporate a more conventional swept design with rounded tips. This improved flight characteristics, but it was evident that more problems were arising than could be readily coped with.

Loaded, the XF10F Jaguar weighed 33,000 lb. Power was supplied by one Westinghouse J40-WE-8 engine of 7,200 lb. static thrust and 11,600 lb. with after-burner. The top speed estimated, since the aircraft were never flown

In the same way that F9F-2 and F9F-3 Panthers were subsequently merged to a common F9F-2 series, so the F9F-4 and F9F-5 Panthers were merged into the later series. This F9F-5 of Minneapolis Navy Reserve started life as a F9F-4.

really fast, was 720 m.p.h. with wings in the fully swept position. The leading edge of the wing root actually slid into the fuselage and moved 40 to 42 inches as the angle of sweep increased from 10 to 50°.

Flight trials, which concerned the first prototype only, indicated that the design had possibilities, and could perhaps be adapted for carrier use. The Navy placed an order for thirty pre-production models but as tests continued and problems increased the whole project was terminated.

The first Jaguar was deliberately destroyed testing a new type of crash barrier at Johnsville Naval Air Station. Pilotless, the machine was sling-shot into the barrier to prove the capabilities of the device to withstand the force of a jet crashing into it on a ship. The barrier withstood the shock—but not the Jaguar. The second Jaguar, which was never flown but used for static tests and mechanical evaluation, was stripped of its engine and items of value then sent to the Aberdeen Proving Grounds, there to serve as a target for tank and artillery gunfire practice.

A project that developed during the war that not only became reality but saw some war service before the end was the Cougar. Grumman and North American were in the forefront of swept wing design, so when the first contracts for fighters for the Navy incorporating this feature were issued, they naturally went to these two concerns. To Grumman on March 2nd, 1951 came a call for a design study and one experimental model to be known as the XF9F-6. Six days later North American received a contract for a navalised version of the highly successful F-86 Sabre and this new aircraft was to be known as the XFJ-2.

Grumman utilising the earlier F9F Panther as a basis for their design, proceeded to apply the new aerodynamic features while retaining what was suitable from the old design to facilitate rapid production. Taking an F9F-5 Panther fuselage, a 35° wing sweep was tried with little trouble. The engine remained a single J48-P-6A as used in the Panther. Grumman considered the change in planform warranted another name and this version of the Panther became the Cougar. After six months of design, research and construction, the prototype made its first flight in September 1951 from Bethpage. Showing an overall improvement in performance and stability over the Panther, the type went into production as the F9F-6, and the first example left the line five months after the completion of prototype acceptance trials.

In the F9F-6, the problem of the high landing speed was solved by increasing the chord of the leading edge slats, as well as increasing that of the trailing edge flaps. These refinements were also incorporated in the Panther. When the F9F-6s went into service with VF-32 (LANT) in November 1952, they became the first swept-wing fighters in service aboard any carrier. These Cougars and subsequent models were also the first military aircraft to be equipped with tubeless undercarriage tyres.

Production of the F9F-6 was limited, but it was available as planned and in use towards the end of the Korean war. The majority of the early Cougars were turned over to the Marine Corps, and besides serving as fighters, a number were modified into F9F-6P reconnaissance aircraft. By 1958 remaining examples of the F9F-6 had reached the hands of reserve flying units.

The F9F-7 model followed the F9F-6. This new type

While new types and new models of basic designs made their appearance, older types were converted to other roles during the Korean emergency. Here an F9F-2B Panther of VMF-115 takes off in Korea with a full bomb load.

had an Allison J33-A-16A turbojet of 6,350 lb. thrust (7,000 lb. with water injection). Performance differed little, the top speed being about 650 m.p.h. Unlike the ' 6 ' this second Cougar model was produced only as a day fighter. Most of the ' 7 ' models went to Navy squadrons, and many are now used by reserve units. It was during the front line service days of the F9F-6 and 7 that the old midnight blue overall colour scheme for U.S. Navy aircraft was changed to the present seagull grey and white.

The Cougars were not the fastest fighters of their day by any means, but they were rugged, reliable and manœuvrable. Some regarded the swept wing aircraft as unstable fire-power platforms, but the generous wing area and the method by which Grumman handled the design, resulted in the Cougar being a reasonably stable aircraft at all speeds and altitudes.

The North American FJ-2 Fury may be considered as the counterpart of the Grumman Cougar. The contract date for the prototype XFJ-2, was March 8th, 1951. Somewhat similar in appearance to the FJ-1 Fury, this new design had swept main and tail plane surfaces and was virtually a navalised version of the F-86 Sabre used so successfully by the U.S.A.F. So, like the Cougar, the FJ-2 was built around an existing design. While the U.S.A.F. version had six ·50 machine-guns the Navy version had four 20 mm. cannon. Its engine was a General Electric J47-GE-27 offering 6,000 lb. thrust. Additionally the Navy variant had a completely redesigned cockpit interior, retractable barrier guard, barrier pick-up, ' Y ' shaped arrester hook and catapult points. An improved Navy gunsight was fitted and wing-folding was arranged for carrier stowage. Although the two prototypes were listed as XFJ-2s the actual aircraft, Nos. 133754 and 133755, were marked ' FJ-2 ' only. The first flight came immediately after the 1951 Christmas holiday at the Los Angeles plant. Its performance was not quite the equal of the land based counterpart, but then it had much additional gear, without a more powerful engine to compensate. Two hundred FJ-2s were ordered together with design studies on a more powerful version. With this order in hand North American negotiated the purchase of the old Curtiss-Wright plant at Columbus where the Fury —and also the F-86H for the Air Force—was produced. In later years North American have used the plant for production and research work.

In mid-1952 the FJ-2 qualified for carrier service, but it did not prove to be particularly well suited for the role, which influenced deliveries to the Marine Corps who used them as land based aircraft. The same held true for the next series of Fury, the FJ-3, which differed in that it had the more powerful Wright J65-W-2 engine based upon the British Sapphire and also had an ' all flying ' power boosted tail control system. From both the FJ-2 and FJ-3 the deck gear and associated equipment was sometimes removed. Although developed quickly the Furies were rather too late to have any impact on the Korean war like their Sabre counterparts of the U.S.A.F.

As the Korean war dragged to an end, the U.S. Marine Corps was highly elated when, on July 11th, 1953, Lt.-Col. John F. Bolt became the first jet ace in the Corps history. Col. Bolt downed his fifth and sixth MiG-15s while leading a four-plane flight east of Sinuiju. On temporary duty with the U.S.A.F.'s 51st F.I.W., he was flying an F-86 Sabre. This brought Bolt's score to twelve enemy aircraft since he had downed six confirmed Jap aircraft during the war.

If the contributions of Naval and Marine aviation forces in Korea are not fully appreciated then the following facts, quoted from *Naval Aviation News* of January 1961, are worthy of the reader's attention. ' In comparison with WWII, the total Naval Aviation force employed in Korea was small, but its achievements in some respects was higher, the combat employment of carriers was on a more continuous basis and the ordnance expenditure not only exceeded the per flight delivery but force expenditure was also higher. Navy and Marine aircraft flew more than one-third of all the combat sorties flown by the United States air forces in Korea '.

In spite of the Korean War and its diversions, the Navy's development programme was not interrupted, indeed it was given impetus by the funds that became available, while the Pacific and Atlantic Fleets and the Carrier Task Force in the Mediterranean went about their lawful tasks uninterrupted by operations in Korea.

An FJ-2 Fury being arrested aboard the U.S.S. Hancock (CVA-19) off San Diego during initial trials of the steam catapult. This Fury, BuAer No. 131979 flown by Maj. F. C. Thomas, U.S.M.C., bears the squadron markings of VMF-235 which are of white stars on a red background stripe.

CHAPTER FOURTEEN

The Radicals

The first of the pre-production F7U-1 Cutlasses on test. This machine first flew on March 1st, 1950 and all of its batch of fourteen pre-production models were extensively tested and served only in composite units. Four 20 mm. cannon were fitted and Mighty Mouse rockets could be carried.

Following the Korean War came a series of radical designs, some that had been conceived years before, others during the war. Much research on high speed aircraft had been conducted and new ideas were mooted.

One really radical fighter was already in service and new versions were appearing. This was the Cutlass. The F7U-3 followed the F7U-1 down the production lines and by December 1954 the first were being assigned to Squadrons VF-81 and VF-83 of the Atlantic Fleet and VF-122 and VF-124 of the Pacific Fleet. Several were assigned to newly-formed attack squadrons or composite squadrons such as VC-3. The Marine Corps then acquired their only Cutlass, BuAer No. 128466, for tests of its high speed minelaying capability. On March 12th, 1956 VA-83, fully equipped with F7U-3M Cutlasses carrying Sparrow I missiles, departed from Norfolk aboard U.S.S. *Intrepid* for duty in the Mediterranean. Thus, VA-83 and the Cutlass became the first squadron and first naval aircraft type to be armed with missiles. The Cutlass was an extremely potent fighter and more suited for an intercepting role, but served mainly as an attack aircraft, although the appropriate designation was not applied. Most of the F7U-3s had been relegated from front line status by 1957 and were then considered as second line combat ready attack aircraft. A few went into storage, others to reserve units. On March 2nd, 1959 the last F7U-3 in active service, used by the Naval Parachute Unit at El Centro, California, was retired.

The 290 F7U-3s were produced in three versions; the F7U-3 fighter-interceptor, F7U-3M missile carrier, and F7U-3P reconnaissance variant with a nose lengthened by 25 inches to incorporate camera installations. The series were larger than the original Cutlasses and offered the pilot improved visibility and, unlike earlier models with cannon fitted just above the air intakes, two on each side of the fuselage; the F7U-1s had the cannon in the nose. Except on photographic versions, this compartment was reserved for the radar gear.

F7U-3s were capable of carrying 5,500 lb. of bombs or external fuel tanks, missiles and rockets. With the two J46-WE-8A engines of 4,600 lb. thrust (6,000 lb. using the after-burner) a top speed of 700 m.p.h. at 40,000 feet and 610 m.p.h. at sea level was recorded. In clean configuration top speeds of 650 m.p.h. were reached low down. Its tremendous initial rate of climb of 13,000 feet per minute was awe-inspiring to witness. The Cutlass was capable of flying on one engine, and even doing so on 'wave-offs' in which case the after-burner had to be 'cut in' on the operating engine. Additional armament was placed on the late model Cutlass in the form of a bellypack housing. The Cutlass weighed 18,000 lb. empty and with normal combat gear, 27,350 lb.

In 1953 the Cutlass qualified for carrier operations aboard the U.S.S. *Coral Sea*, a record of seventy-two flights being made in twenty-nine days. As it became operational the Cutlass received two distinctions. It was the only tailless turbojet fighter of non-delta configuration to achieve service status, and the U.S. Navy's first fighter designed from the outset to employ after-burners. It was unfortunate that the F7Us were so long in the development stage, and the design was too involved for successful continued use as a service aircraft. Although very stable at high speeds the Cutlass had frightening spin characteristics and was difficult to control. Although stressed for 12 'G' manoeuvres careful handling was a necessity. Throughout its service use continual tests were made to correct vicious characteristics. Drag shutes were employed, additional stabilising and correctional surfaces were experimentally tried. All proved in vain and the Cutlass had to be given only to well trained units with experienced pilots.

As the last Cutlass rolled off the line at the Dallas plant, in December 1955, a great change took place in company policy, planning and management. Before the next and last Vought fighter series came into being, the Chance Vought Division of United Aircraft broke away from the parent company to form a separate organisation, the Chance Vought Corporation. It was evident that the long haul of bringing the Cutlass into production had put a great deal of strain on the plant's facilities, not to mention the drain on time and money. On the drawing boards were several designs, one in particular that seemed as far advanced as

117

Called 'Pogo', the Convair XFY-1 vertical take-off fighter made many flights from Brown Naval Auxiliary Air Station near San Diego, in the hands of J. F. 'Skeets' Coleman who is Convair's engineering test pilot. A gimbal-mounted seat rotates outward 45° for vertical flight and locks into the conventional position for horizontal flight as demonstrated here. To exact full value from the tests a stenographer took down every word Coleman spoke into his radio microphone during tests.

any Navy fighter could be, by slide-rule computations giving figures that would revolutionise the naval fighter concept. When the Navy expressed interest in the design, Vought broke with the United Aircraft Corporation and pinned their hopes on the new aircraft. How successful this venture was, is evident by the present-day Crusader.

The most unusual Navy fighter designs ever to fly were the vertical take-off Lockheed XFV-1 and Convair XFY-1, orginated in 1951. In March of that year a true vertical take-off design was proposed to industry. Such a fighter was envisaged, from the Navy's point of view, as being able to take-off and land on the stern deck of a destroyer or even operate from a freighter. In such a case each ship in a convoy would be able to carry its own fighter escort on a lengthy voyage. Such a craft would also be able to operate from restricted land areas. The use of a pure jet power was considered, but after some study was dropped in order to ease the problems confronting the designers, and also because suitable engines did not then exist. The programme was initiated on March 31st with a development award to Convair for a propeller-driven vertical take-off (VTO) fighter. Design and mechanical problems would clearly be great, and additional lengthy discussion arose from a delta wing design suggested by the Company. Lockheed's design proposals were somewhat more conservative, but of equal promise, and so three weeks later a contract was signed with that firm for their design.

These craft were not helicopters, convertiplanes or STOL (Short Take-off and Landing) aircraft, they were pure vertical take-off aircraft. A most powerful engine was needed and the Navy approached the Allison Division of General Motors for a turboprop engine of more than 5,000 s.h.p. Both machines were planned to use the same engine.

The technical problems involved were considerable, and it was 1954 before flight trials were ready to begin. Allison also had difficulties with the power plant which was finally solved by harnessing two T38 engines together and devising 'twin power section controls'. This allowed either engine to be used separately or for the two to work together. This package engine was known as the Allison T40-A, and had a guaranteed minimum output of 5,800 s.h.p. It became available for use late in 1953. Both the airframe manufacturers and Allison knew that large contra-rotating propellers would be necessary to offset the powerful torque set up by the large engine and special ones were built.

Lockheed's XFV-1 had short straight wings with normal control surfaces combined with an 'X' shaped tail unit which could be moved to effect directional changes. Convair settled on a delta wing configuration for the XFY-1. Unlike the XFV-1 it was intended to take-off in the

Taxying on the water in San Diego Bay is the first of the Convair XF2Y-1 Sea Darts. There are two significant dates with this aircraft, initial launching, which was December 16th, 1952, and first flight, which was April 9th, 1953.

vertical attitude and four small castoring wheels were provided one each at the extremities of the tail surfaces. After vertical take-off the XFY-1 was flown into horizontal flight. In landing the aircraft was lowered vertically and backwards to the ground, hanging, as it were, by 16 feet diameter, six-bladed propellers. Although both aircraft used the T40 engine it was realised that it could be used for only limited periods in any one position. The engine was housed in a bulbous fuselage with the cockpit placed high above to ensure good visibility. To ease the strain on the pilot his seat was so fitted that it could tilt in order that he should not sit in too uncomfortable a position for take-off. Along with this the controls had to retain their position relative to the pilot.

Emergency abandonment of the aircraft needed special attention. For level flight the usual type of ejector seat sufficed, but to provide a good chance of safely leaving in any emergency a special system of sideways ejection was developed. Tests showed that ejections could be made from as low an altitude as 25 feet, with a good chance of success, even if the approach speed was as high as 150 m.p.h. Normally approaches were to be made at around 45 m.p.h.

The 'roll-out' of this aircraft, popularly called 'Pogo' took place in March 1954. Pre-flight tests were made at Brown Field, near San Diego, before the XFY-1 was moved to Moffett Naval Air Station in July 1954. At Moffett the machine was tethered to a rig in a hangar where, years before, airships had been housed. On August 1st it proved its ability to rise vertically and settled safely back. This was the first VTO flight. J. F. Coleman, Convair's chief of flight testing, raised the aircraft first to 20 feet and then up to 60 feet on later flights, which was as high as the roof would safely allow. After many such tests—getting on to 300 in fact—Convair decided it was time for 'Pogo' to be tested free from such restrictions.

The XFY-1 was returned to Lindberg Field, San Diego, where instruments were recalibrated in both vertical and horizontal positions, and a special rate of descent indicator was installed. Further trials were flown on November 2nd, 1954, when 'Skeets' Coleman took it up on a twenty-minute flight in the course of which the first transition to level flight was made. The following day a public demonstration of its capabilities was made including low level passes and steep banking. The transition meant that the XFY-1 was to pass into history as the world's first VTO fighter. In the following weeks more flights were made.

Lockheed's approach to flight testing differed from Convair's. A special ground handling rig was constructed to contain the aircraft and turn it from the vertical to horizontal position. Additionally auxiliary wheels, much the same as conventional landing gear, were fitted for taxying tests in the horizontal posture. Using this landing gear rig, the XFV-1 made its first flights at Edwards Air Force Base late in 1955. Herman R. Salman made the flights, but the XFV-1 never accomplished a true vertical take-off and landing. All flights were conventional, although Salman hovered nearly motionless and held vertical positions for some seconds. By the end of the first series of trials the Navy and Lockheed agreed further tests should be postponed until a suitable engine was developed for the project. Tests had normally been flown at around 10,000 feet and the XFV-1 exceeded 500 m.p.h. easily. A second example was partially completed and was used for static tests and to check out pilots. Recently it was with the Hiller helicopter company who are studying VTO aircraft at Palo Alto in California. The original XFV-1, No. 138657, is in storage currently but will be exhibited at the new National Air Museum in Washington, D.C.

Sea Dart, first fighter to feature a retractable hydro-ski undercarriage, which was developed jointly by Convair, the U.S. Navy, the N.A.C.A., the Edo Corporation and engineers of All-American Airways. The YF2Y-1 is shown with the single-ski version under test.

An announcement was made in January 1956 that the development programme on both VTO fighters had been terminated, and that findings with these two aircraft would be filed for future use. Their evaluation almost certainly encouraged the U.S.A.F. to award a contract to Ryan to develop the X-13 Vertijet and aroused lasting interest in VTO aircraft. The Lockheed XFV-1 was the company's first and only fighter for the Navy. Apart from the XFY-1, Convair's only Navy fighter was the XF2Y-1 ordered June 17th, 1951 and its derivative.

By 1950 research on high speed aircraft had progressed to the point where 'designs of the future' had a chance of materialising. The idea of a seaplane fighter was certainly not new, nor was the jet fighter seaplane, for the British Saro SR/A1 fighter flying boat had shown a remarkable performance some years previously. Convair's machine was a rather different proposition for it married a flying boat hull with a supersonic fighter design—and delta winged, at that. Long take-offs, now essential for most fighters are unacceptable for carrier operations, but a water based fighter would have ready made take-off and landing facilities.

When first the Navy showed interest in a water based fighter designed to expand perimeter defence at sea and on land, several companies submitted proposals. Most planned around a heavy hull with tail surfaces placed high to keep them clear of the spray. In general these resembled scaled down, rather conventional, flying boats. Convair produced several designs, the most promising of which somewhat resembled the F-102 Delta Dagger being built for the U.S.A.F. The project was finalised and on June 17th, 1951 the contract to develop the fighter-seaplane went to Convair. A host of wind tunnel and hydro model tests were run before the final configuration was accepted. Several radio controlled models were constructed and hundreds of flights made by

Built to the same requirement as the Convair XFY-1 was the Lockheed XFV-1 shown here in its test rig. Both had a common factor in the Allison T40-A engine, of 5,800 horse-power.

them. Unorthodox as the project was, the XF2Y-1 was felt to be practicable, in spite of the engineering difficulties which would obviously arise in the construction of such a fighter.

Two prototypes were to be built of the XF2Y-1, which was to become better known as the Sea Dart. Throughout 1952 the prototype was being carefully built. In planform its wings were of delta shape, and its long nose protruded well forward of the wing. Fin and rudder formed another triangular shape, and the two Westinghouse J34 engines, each to deliver 3,000 lb. thrust, were placed side-by-side on top of the fuselage with their intakes aft of the wing leading edge, and so placed as to reduce spray intake to a minimum. The exhaust orifice was on either side of the rudder above the trailing edge of the wing, and just above the water line. The bullet-like waterproof fuselage had a 'V'-shaped planing bottom for its aft section.

Perhaps the most unusual feature was the use of hydroskis, this being the first use of these on any combat aircraft. The prototype utilised two which protruded from the lower side of the hull, and were actuated hydraulically. Two such skis it was felt provided good stability on take-off and landing. During flight they were retracted into the lower hull sides thereby conforming to the fine contours of the aircraft. In addition to easing stability and spray problems on the water, the skis also allowed for a faster take-off and a shorter run.

On the morning of December 17th, 1952 the first XF2Y-1 was successfully launched in San Diego Bay and with her engines idling she wallowed around, as a spectator described it, like a killer whale on the prowl. On the days following many trials were undertaken, including taxying directly from the water on to the beach, and vice versa. On April 9th, 1953 the XF2Y-1 Sea Dart made its first flight, of ten minutes duration, over San Diego Bay.

Sea Dart progress was rapid, and two additional prototypes were also ordered. On the second machine the double skis were discarded in favour, experimentally, of one large ski placed centrally on the forward underside of the hull, in the belief that this would reduce vibration and improve stability. Trials with this aircraft also proved successful and the Navy decided to put the aircraft into the service trials test category, the designation thereby being changed to YF2Y-1 for this aircraft. Flight trials of the single-ski gear showed that less spray was thrown up than with the dual gear, but that stability was decreased. Albeit, in skilful hands the performance of the two types proved to be about equal. Jet orifices were also redesigned and placed slightly further back on the hull. Extensive tests continued at San Diego in the course of which the Navy ordered three more Sea Darts as YF2Y-1s Nos. 135763–135765.

Ill fate was to plague the Sea Dart for during a public demonstration flight on November 4th, 1954 the second machine, No. 135762 which was by then designated YF2Y-1, blew up in the course of making a high speed fly past at about 50 feet. In a blinding flash it disintegrated in front of the horrified crowd. The accident was traced to a fuel leak.

Sea Dart performance data is still restricted, but it is known that the YF2Y-1 before its crash exceeded Mach 1 on August 3rd, 1954 in the hands of C. E. Richbourg during a shallow dive at 34,000 feet. The original XF2Y-1s powered by two J34 engines later had these replaced by J46-WE-16s of 4,000 lb. thrust (6,500 lb. with after-burners). J46s were fitted into the three YF2Y-1s from the start. Intensive trials with 135765 proved the twin ski pattern to ultimately show superiority, and these replaced its single ski. Convair had under study a redesigned version powered by a single Pratt & Whitney J57 or Wright J67 in the 10,000–12,000 lb. thrust class, but the problems of maintenance and service operations coupled with changing requirements brought about the cancellation of the project on November 4th, 1954. Of the four remaining Sea Darts, Nos. 135763 and 135765 are in storage at the Miramar Naval Air Station in excellent condition while No. 135764, also converted with twin skis, is currently in storage at Norfolk Virginia, for the National Air Museum.

CHAPTER FIFTEEN

Fighters in Fine Fettle

The fighters of today did not materialise overnight or even in a year or two. Their design was conceived from five to ten years ago and in this final chapter the famous modern front-line naval fighters in service are traced from their beginnings to service.

Production of the well-known Skyray commenced early in 1954; the first with R. O. Rahn at the controls, made its maiden flight from El Segundo on June 5th, 1954. No troubles were experienced and supersonic speeds in level flight were achieved over the Pacific shortly after take-off. The initial order was soon completed by which time further Skyrays were ordered, the last of the 420 production aircraft being built during December 1958.

The first Skyrays went to VC-3 of Pacific Air Command at Moffett Field in April 1956. Several were handed over to the Marine Corps in 1958 and the first feel of the Skyray prompted Maj. E. N. Faivre, a World War II and Korean

Production of the F4D-1 Skyray was completed early in 1959, but it still is an important operational fighter to-day. This F4D-1 of VFAW-3 shown at San Diego has the standard two 300-gallon drop tanks which can be carried in addition to more potent stores.

War veteran, to attempt the Official World Climb Record. In May he actually broke five records—four formerly held by France—at the Naval Air Missle Test Centre, Point Mugu, California.

These records, led to the Skyrays being assigned to VFAW-3 (All-weather Fighter Squadron Three) for use at the San Diego base of the North American Air Defense Command. This was an honour for both Douglas and the Navy as well as VFAW-3 since the Command normally comprised U.S.A.F. formations. The Skyray is still the only Naval type to receive such distinction.

Since the F4D-1 remains in widespread front line service with both the Navy and Marine Corps much detail about the machine remains classified. It attains more than Mach 1 above 10,000 feet and 750 m.p.h. at sea level, it can carry an assortment of armament which usually includes four 20 mm. cannon and a weapon load of up to 7,000 lb. Additionally six underwing fibreglass pods can be carried each containing seven 2·75 inch unguided folded fin air rockets, or four larger pods each carrying nineteen 2·75 inch unguided missiles. These are aimed by directing the aircraft itself towards the target. Sidewinder infra-red missiles may also be carried. VFAW-3 have forsaken

cannon and machine-guns for the rockets, their load usually comprising two Sidewinders and one pod of nineteen unguided missiles and two pods each containing seven rockets. Two extra-long range tanks are frequently carried.

The tactical versatility of the Skyray led to a pre-production contract for a more powerful version having superior all-weather interception capabilities. Refinements were to be increased armament and improved electronic auto-pilot and revised control and weapon firing systems. F4D-2 was the allocated designation, but to incorporate the modifications required such extensive redesign that the contract was changed to call for four new models as XF5D-1s. They received the name Skylancer. The new machine was larger and sleeker than the F4D, for to accommodate a Pratt & Whitney J57 engine the fuselage had to be lengthened and to improve aerodynamic form, the nose was lengthened. The wing thickness/chord ratio was reduced, and a taller fin and rudder of increased area fitted. A substantial all-round improvement in performance was forcast, due to the various improvements which included an increased fuel capacity.

Rahn, the firm's test pilot, took the first Skylancer on its maiden trip from the El Segundo plant on April 21st, 1956 during which it exceeded Mach 1. On June 30th the second Skylancer became airborne, and at this time the Navy placed an order for nineteen Skylancers to replace the F4D Skyrays of VFAW-3.

During trials a top speed of 1,098 m.p.h. at 10,000 feet and 990 m.p.h. at 44,000 feet was recorded. The ceiling was 55,000 feet. Climb rates surpassed the Skyray figures, and it seemed that this new form would replace its predecessor. However, other new fighters were showing similar performance figures—in particular the McDonnell Demon which had special capabilities as an all-weather fighter. The Skylancer designed basically for the day interceptor role had only limited all-weather capability, and due to this the order was cut to eleven production machines. Later instructions were received to terminate the Skylancer programme with the completion of the four prototypes.

The second two machines, Nos. 142349 and 142350,

A stalwart naval fighter is the F3H Demon which is in fact the standard carrier based all-weather fighter in spite of its inauspicious start. F3H-2N BuAer No. 133550 is shown with the standard white tail surfaces and ailerons.

were turned over to the Ames Laboratory of the NACA/NASA for experimental use and are still flying, together with the first two machines 139208 and 139209. All four therefore are usefully employed. The ' X ' prefix has been dropped and they are now known as F5D-1 Skylancers. The first two machines have also been used extensively at Edwards A.F.B. as trial installations aircraft for radar, instruments and armament.

The Skylancer was to have been equipped, with the Sidewinder missile. This weapon, employed by several air forces, stemmed from the requirements of the Navy, a fact not widely appreciated. An amusing story, often repeated, tells of how a technician at the Naval Test Station at Inyokern, California, was annoyed in his sleep by persistent mosquitoes whose all-weather capability was remarkable. Pondering how the mosquito was able to find its target so readily, he decided to devote study to the insect. It was discovered that it located its target by homing on to the heat emitted, irrespective of whether or not the surrounding atmosphere was warm or cold. The Sidewinder missile functions along similar lines, having infra-red homing and as such is the only Navy missile of its type.

On September 11th, 1953 the first successful interception using a Sidewinder took place at the Naval Ordnance Test Station at Inyokern, as a result of which a radio-controlled F6F-5 Hellcat crashed in flames. Sidewinder has been fired successfully from the fastest jets and was used in the 1959 engagements by Sabres against Chinese Communist jets off Formosa with effective results.

A return to the Navy fighter field was made by McDonnell in 1949 with an entirely new design. Conceived as an all-weather fighter, it was to bring much heart-searching and lead to serious bickering before it became the mighty weapon that it is today. As with so many current service aircraft much of interest about it remains clothed in secrecy. The XF3H-1 Demon encountered serious teething troubles which nearly led to its abandonment as successor to the Banshee. The engine manufacturers too, received a share of criticism levelled at all concerned with the project.

On September 30th, 1949 a contract for four XF3H-1 Demon prototypes was placed and the first machine flew just under two years later, on August 7th, from the St. Louis base. It was powered by an experimental XJ40-WE-6 engine, of 7,200 lb. estimated thrust, which did not come up to expectations. The other three Demons were flown, but none measured up to the requirements of the Navy, for the aircraft was, without doubt, seriously underpowered.

The design was extremely clean and appealing to the eye. Under pressure of the Korean War 146 F3H-1Ns were ordered on August 29th, 1952; the Demon was then seen as an answer to the MiG-15 should the war in Korea drag on. Shortly after the first of the production machines flew,

The Skylancer, as an improved all-weather fighter version of the F4D-1 Skyray, was initially recorded as the F4D-2N. It had the characteristic planform of the Skyray, but the fuselage was larger and more slender. An increased fuel capacity gave it an even better performance than its predecessor. Designated the F5D-1, it was planned to put the type into production but this was cancelled through lack of funds.

it was clear that the modifications already incorporated as a result of prototype flying, still left much to be desired. Banner headlines in the St. Louis papers declared 'Planes won't fly—Blame Democrats' 'Demon failures—Navy's Jet Blunders', and 'Demon explodes over Washington, Missouri'. Although delivery of the first Demons was made in 1953, only eleven had flown by 1955 of which six had crashed taking the lives of two test pilots. All Demons were grounded after the second fatality. By this time the Navy had cut its order to sixty F3H-1Ns of which delivery was to be continued in spite of the crashes, all sixty being complete by June 1955. Apart from eleven production models tested and the prototypes, the Demons awaited their fate at St. Louis, dependant on the outcome of investigations into the whole project at Patuxent. There, it was finally decided that the J40 engine would not be effective for use in the Demon.

Westinghouse was not the only engine builder to encounter such failure, the Douglas A3D and F4D and Grumman XF10F had been linked with similar engine failures. As a result of the state of the Demon, Westinghouse in 1953 were asked to halt further development of the J40, and McDonnell received instructions to redesign the F3H to accommodate the Allison J71 in a new variant to be followed the F3H-1 Demons without a production break. In the event 460 F3H-2s materialised under three model suffix designations. The final machine was delivered on November 17th, 1959, at the close of eight years of production. Included in the production total were the twenty-nine converted from the ill-fated model and when production ended on November 17th, 1959 the last Demon built was an F3H-2M converted from the last F3H-1N grounded years before—and this aircraft was significantly named 'Under-the-Skin'.

Powered by an Allison J71-A-2 engine developing 9,700 lb. thrust and 14,250 employing an after-burner, the F3H-2 had a wing of new design, that proved very successful. Many are still in service with Navy VF and VA squadrons. The first six F3H-2Ns reached VF-14 at Naval Air Station Cecil Field, Florida on March 7th, 1956. Intermingled on the production lines were F3H-1s for conversion, the F3H-2P reconnaissance aircraft and the F3H-2M missile-carrying version which has provision for four beam riding Sparrow III missiles on underwing pylons and exists only as a day fighter. The majority of the Demons built were F3H-2N all-weather fighters armed with four Sidewinder missiles.

The first of the fifty-six initial production Demons of which there were ugly rumours and unfavourable reports. They were delivered as F3H-1Ns as shown, but twenty-nine were converted to the main production standard, F3H-2 with the Allison J71-A-2 engine, which completely vindicated the initial design. Proven in service, it operates in two versions, F3H-2N with Sparrow I missiles and F3H-2M with Sparrow III missiles.

known as the F3H-2N. Of the fifty-four remaining F3H-1Ns, twenty-nine were returned to McDonnell to be refitted with J71 engines and twenty-one, which it was considered uneconomical to convert, were relegated to non-flying roles. The remaining four were converted into prototype models for the F3H-2.

Shortcomings exhibited by the Demon drew a Congressional Investigation at the end of which Defense Secretary, Charles E. Wilson, neatly summed up the whole affair by saying 'None of the parties could really be blamed. The F3H was a crash programme, and whenever several new models are rushed into production you occasionally get a sour one'.

Large orders were placed for the F3H-2 series which

Normal standard armament on production F3H-2 Demons was four 20 mm. cannon. Performance of the different versions varied with their roles, but the top speed was around 700–730 m.p.h. at sea level, initial climb about 12,000 feet per minute and service ceiling approximately 48,000 feet. The F3H was extremely advanced in design, and on reaching the squadrons the re-modelled Demons could readily out-perform many land based fighter types. Unlike previous McDonnell fighters, the Demon used a single engine fed from two air intakes. Under the buddy system of flight refuelling the tanker version of the Demon utilises a 22-foot long external tank fuel containing 300 gallons, from which the second fighter fulfils its need. This same tank can also be used as an external long range tank. All Demons have provision for in-flight refuelling.

An F3H-3 version was proposed by McDonnell using the more powerful General Electric J79 turbojet, but this was dropped in favour of the new design on the drawing boards then known as the AH-1, and now as the F4H-1.

By 1961 Demons were in service aboard seven carriers. They had already gained a number of worthy 'firsts' since it was a Demon that flew from the U.S.A. to England as the first Navy jet to make such a flight. F3H-2s were used to test-fire the first AAM-6-N Sparrow III missiles, and VF-64 aboard the U.S.S. *Midway* was on December 8th, 1958, the first unit on a carrier to fire such weapons. This was the first occasion, too, when a squadron outside continental limits of the U.S.A. had fired these missiles. VF-193 aboard the U.S.S. *Bon Homme Richard* repeated the exercise eleven days later in the Western Pacific where it was serving with the Seventh Fleet.

The first major change in Cougar design and second major change in the F9F series, came with the F9F-8. This version had a fuselage extended by 17 inches to accommodate additional fuel and the larger Pratt & Whitney J48-P-8A engine of 7,250 lb. thrust—8,800 lb. with water injection. The wings retained the 35° sweep back, but many new features were incorporated including a small trimmer at the port wing tip. Wings featured fixed cambered leading

Mach 1·05. Trials with the latest version showed a top speed of 714 m.p.h. at sea level and 556 m.p.h. at 36,000 feet. The climb rate was 9,000 feet per minute initially, an altitude of 40,000 feet being reached in seven minutes. Clean, and at cruising speed, a range of 1,000 miles was possible with the F9F-8, a performance which prompted the development of the F9F-8P reconnaissance variant.

The F9F-8 was immediately put into production in both forms. Normal armament was 4 × 20 mm. cannon in the nose, except for the F9F-8P which had cameras in an extended nose. Six HVAR rockets or four Sidewinders could be carried under the wings on pylons, and although the latter meant the aircraft was missile armed, no 'M' suffix was ever applied to its designation. A 2,000 lb. bomb load could be carried in lieu of rockets, and carrying these external stores a maximum speed of around 705 m.p.h. at sea level was attained.

During 1954 and 1955 the F9F-8s were delivered in quantity to the Navy and Marine Corps, in whose hands they set up a number of records. Three Cougars spanned the U.S. in record time on April 1st, 1954, from San Diego to Floyd Bennett, N.Y. Lt. F. X. Brady made the 2,438 mile journey in three hours forty-five and a half minutes, followed by Lt. (j.g.) J. C. Barrow in three hours forty-seven minutes. Lt. W. Rich, third, had an elapsed time of three hours

Furies were the naval fighters of the 'fifties. The FJ-2 shown paralleled the North American F-86F of the U.S.A.F. except for folding wings, carrier gear and 20 mm. cannon armament. It was followed by an improved version, the FJ-3. Shown, is a standard FJ-2 BuAer No. 132017.

edges, and unlike many aircraft of the type, this Grumman fighter had these built-in and fixed in the extended position, which in turn reduced drag. By this token it provided a low speed wing which improved stability and control at both high and low altitudes, a highly satisfactory combination.

By the elimination of the hydraulic system necessary to operate the movable slats the wing construction was simplified and the space made available provided for thirty additional gallons of fuel in each wing. Lengthening the centre section of the fuselage by 8 inches made room for 80 more gallons, increasing the overall fuel load by 140 gallons and substantially improving the range. All of the Cougar's fuel is normally carried internally, but external tanks can be fitted as necessary.

The F9F-8 followed on the production lines after the F9F-7, and made its first flight on December 18th, 1953. All Cougars were transonic, they were never intended to be more. The F9F-8 series could, however, exceed Mach 1 safely in a dive, but instructions were given not to exceed

forty-eight minutes. En route they made one in-flight refuelling over Hutchinson, Kansas thereby revealing that the later versions of the Cougar had provision for this useful feature. Late model Cougars had the refuelling probe located in the extreme nose, and many earlier machines also were fitted for probe and drogue refuelling. No official recognition was afforded the three Cougars of VF-144 which made the trans-continental, round trip record flight on October 5th, 1956 from Miramar to Long Island, New York and back with refuelling stops both ways at New York and Olathe, Kansas, taking ten hours forty-nine minutes and eleven seconds for the trip. The previous record stood at eleven hours eighteen minutes twenty-seven seconds. An F9F-8P made a non-stop photographic flight across the U.S. in less than four hours while taking a continuous strip of photographs of the terrain.

During 1957 and 1958 the U.S.M.C. perfected the technique of 'loft bombing' with nuclear weapons. These tests were carried out with F9F-8 Cougars for which several

An outstanding adaptable fighter of which nearly 2,000 were produced is the Grumman F9F Cougar, a series which covered the F9F-6 to F9F-8 model range. It was the first swept wing fighter in U.S.N. carrier service. This F9F-8, in orange and black finish, is testing Sidewinders for the N.O.T.S., China Lake.

had the suffix letter ' B ', denoting the change to a bombing role. The Cougar was a most versatile type and served as fighter, bomber, missile-carrying aircraft and attack fighter; some 1,988 F9F-6, 7, 8 models were built before the final variant appeared.

Because of its good overall performance, stability and ruggedness, the F8F-8 series was chosen as the basis for a jet trainer, much needed at the time by the Navy and Marine Corps. A two-seat Cougar was projected and Grumman produced the F8F-8T which, but for the elongated nose and long streamlined canopy, much resembled the F9F-8. The cockpits were duplicated and were to the same standard as those of the fighter. Two Martin-Baker ejector seats were employed. Apart from the fuselage being lengthened by 23 inches other main dimensions remained unchanged. The same Pratt & Whitney J48 power plant was used, and performance suffered little in spite of an increase in all-up weight. The service ceiling of 50,000 feet attained by the fighter, fell to 40,000 for the trainer and combat radius was cut to 280 miles, but the top speed and range were practically unchanged. Armament was reduced to two 20 mm. cannon, but external stores of up to 2,000 lb. could still be carried. On April 4th, 1956, delivery was made of the first of the 400 ordered and the last F9F-8T was delivered on February 2nd, 1960. The total number of Cougars built was recorded as 1,985 from XF9F-6 to the final trainer. Only wartime Corsair, Hellcat and Wildcat fighters surpassed this impressive peace-time production total. The last front-line squadron to use the Cougar was VMCJ-3, still thus equipped in 1959. Although the Cougar series ended with the F9F-8T the F9F designation was used again for the next and last fighter type built for the Navy by Grumman, the F9F-9 Tiger of an entirely revised shape and concept.

As the FJ-2 Furies came from the Columbus works of North American the FJ-3 was, in early 1952, taking shape. The majority of the FJ-2s went to the Marine Corps, VMF-122 at Cherry Point being first to receive them in 1954; a few days later VMF-312 was equipped, at Santa Ana, California. Five Marine Corps fighter units flew FJ-2s: VMF-122, VMF-312, VMF-325, VMF-232 and VMF-334. FJ-2s also served in the ' B ' Bombing category as the FJ-2B, and were usually attached to attack squadrons for this duty. It was during the early service of the FJ-2 that the midnight blue colouring on Navy fighters was discarded in favour of the present grey/white finish; the order was not mandatory for Marine Corps aircraft and for a while many of their Furies had the old colouring. When the order came for the Marine Corps to remove the blue paint, their aircraft were then left in their natural finish.

The first FJ-3s were turned over to VC-3 at Moffett Field for service evaluation, while the first front line squadron, VF-173 at Jacksonville, Florida, received early production models for qualification trials aboard the U.S.S. *Bennington*. Each pilot in VF-173 had to make eight arrested landings, and at least two touch and go landings. Using three Furies, the entire Squadron successfully qualified

The FJ-4 Fury was a complete re-design with a new fuselage, larger wing, taller tail fin and a new undercarriage. It made its debut in May 1955. This Marine Corps Fury is of Squadron VMF-323.

Marine attack squadrons were equipped with FJ-4B Furies. These were equipped to carry Bullpup air-to-surface missiles or bombs, or a 700 gallon fuel tank in lieu of bombs. This version was first introduced in 1956. The example shown is BuAer No. 143497.

with 180 landings within four days. The FJ-3 had met naval requirements and went into full scale production. Most of them went to the Marines, although VF-211 used them aboard carriers as late as 1956. The FJ-3 Furies were particularly tricky to handle on approach to a carrier, and the Navy found they exhibited considerable yaw on glide approach and showed inadequate stall warning on normal landings. Experimental wing fences were built on one machine, but they did little to improve matters. The FJ-3 like its forerunner, was then, mainly land based and went chiefly to units with FJ-2s. By 1959 they, in turn, were being replaced by later types and when this occurred the Furies were switched to the reserve forces.

The Fury 2 and 3 had a normal armament of four MK-12 20 mm. cannon in the nose. Auxiliary underwing fuel tanks each carrying 200 gallons could be dropped in flight. In the fuselage and wings 545·8 gallons of fuel could be carried, refuelling being carried out from a single point. All FJ-2s and the first 343 FJ-3 Furies had movable leading edge slats, but from aircraft No. 36118 onwards, they had a fixed cambered ' wet ' leading edge installation similar to that on the late Cougars. Maximum altitude reached was 50,000 feet with combat ceiling of 45,000 feet. Range at best altitude and engine rating averaged 815 miles. The top speed was around 650 m.p.h. Neither of these versions of the Fury was considered a Mach 1 type, top speed at best altitude being 0·95, and, only in 60° to 70° dives could the sound barrier be broken.

The FJ-3 in final configuration could be used for varied roles. As the FJ-3M, it could carry two Sidewinder missiles on special Aero 15A under-wing launcher pylons. In addition up to 2,000 lb. of bombs or other weapons could be carried. In-flight refuelling probes were installed on late models, being placed about 3 feet from the fuselage on the port wing leading edge. A re-designed rudder was also applied to correct ' rudder buzz '. ' G '-overshoot was experienced with the Furies, for which reason the red line speed limit was fixed at Mach ·95. Once past ·96, this condition disappeared but would be encountered again when speed fell, especially after a dive. Intentional spins were prohibited in the Fury, and also in the Cougar. Once in a spin they were usually normal, topside or inverted, but two types of spin could be encountered—steep slow rotation and rapid oscillating. Two thousand to 2,700 feet could be lost in each turn. A minimum of 7,000 feet was required for complete recovery. Orders were to eject should a spin occur below this height, since the recovery margin was too small. These rather unpleasant traits give some indication of the problems being encountered at this period with high speed aircraft, and applied not only to the Cougar and Fury, the two types in particular with which the Navy and Marine Corps were getting their first real experience of high speed jet aircraft.

In August 1956 VF-211, completely equipped with FJ-3s, departed from the West Coast aboard the U.S.S. *Bon Homme Richard* armed with Sidewinders. This was the first time the Sidewinder had been deployed with the Pacific Fleet. The initial overseas deployment of Sidewinder was in July 1956 when VA-46 using F9F-8 Cougars aboard the U.S.S. *Randolph* sailed into the Mediterranean with the ships of the Sixth Fleet.

On January 27th, 1955 Lt. Cdr. W. J. Manby of VF-33 set up an unofficial record when he reached 10,000 feet in an FJ-3, in 73·2 seconds from a standing start at Naval Air Station, Oceana. This record was beaten by an F3H-2N on February 13th, 1955, 10,000 feet being reached in seventy-one seconds by McDonnell's test pilot, C. V. Brown. The latter record was beaten ten days later when R. O. Rahn reached 10,000 feet in fifty-six seconds in an F4D-1 Skyray.

Applying their findings with the FJ-2 and FJ-3, North American proceeded with the FJ-4, their final fighter for the Navy. It was a completely new design, although it superficially resembled the earlier Fury. The FJ-4 had a taller more pointed fin from which a dorsal spine led to the cockpit canopy. The fuselage was shortened very slightly, and the wing span was increased, but overall height remained the same. Wing area was increased by about 50 square feet, this being brought about by an increase of wing root chord. A thinner wing employed built-in cambered leading edges, barrier stops and airflow fences. A new rudder design, proven on late FJ-3s, was also a feature of the FJ-4.

The engine fitted was a Wright J65-W-16A turbojet delivering 7,700 lb. thrust. Some of the earlier FJ-4B fighter bomber models had a J65-W-4B engine, but had a 16A engine type exhaust cone and thermocouple leads, which gave the engines the same power as the later type of turbojet. The tailplane was of the " all-flying " type. Carrier gear initially installed was similar to that used on the FJ-3, likewise the four-cannon armament and external loads. A new control system was utilised as well as new emergency operation systems and electronic fire control provisions. A re-designed undercarriage gear was fitted, the main gear being based more on " knee action " shock absorption principles, which improved the operation.

Target towing equipment could be fitted and from FJ-4B No. 139303 onwards provision was made for in-flight refuelling for which the probe was fitted 5 feet out from the port wing root. The FJ-4 was designed as a day fighter interceptor and was not suitable for all-weather fighting. An attack variant, the FJ-4B retained the VF designation, as it was merely an adaptation of the standard type. It could be fitted for the buddy tanker role.

On October 28th, 1954 the first of two prototypes, Nos. 139279-139280, was flown and it was in May 1958 that the last FJ-4B came off the line at Columbus. In all 147 FJ-4s and 217 FJ-4Bs were built in addition to the prototypes mentioned above.

One of the first uses of the retractable air-driven hydraulic emergency power control pump was on the late models, and it was subsequently fitted to all earlier models. This unit, located immediately aft of the nose intake on the starboard side of the fuselage, could be extended into the airstream manually to afford power for the control operations in the event of power failure or damage to the hydraulic systems. The turbine, incorporating a speed governor and automatic feathering blades, drove a pump which, in turn, pressurised the No. 2 flight control system which gave approximately the same pressure as that obtained from the engine driven pump.

Normal loaded weight of the FJ-4 was 19,300 lb., the maximum permitted weight being 26,000 lb. Maximum speed fell just short of Mach 1·0 in level flight. A height of 40,000 feet could be reached in nine and a half minutes; it had a service ceiling of 47,000 feet and ultimate ceiling 50,000 feet. Normal range in full combat state was 400-800 miles dependant upon the load. Like the preceding Furies, most of the FJ-4s went to Marine Corps units, VMF-334 and VMF-451 using them as day fighters and VMA-223 for attack and dive bombing, to name but a few.

Two factors played an important part in governing the use of the Fury. One reason for its limited carrier use was that the nose intake shape made ditching procedures hazardous. Pilots were warned that the aircraft would sink within a few moments of ditching. After hitting water in a near level attitude, it would dive violently soon after impact, scooping great amounts of water directly into the intake sufficient to fill the fuselage in a few seconds. The second factor concerned its high landing speed, which placed considerable limits upon its landing weight. Speeds of 125 knots had to be maintained in normal landings in spite of the large slotted flaps and inboard aileron action.

Many FJ-4s went into Navy VA squadrons for attack and light bomber duty. Ten such squadrons used the FJ-4B model at the time when the VF and VA units were being intermixed to use Cougars, Furies, Skyrays and Demons. Aircraft which were multi-purpose would then be operated in whichever role was needed. In the hands of VX-4, a special development squadron, an FJ-4B was, on March 19th, 1958, used to fire the potent ASM-N-7 Bullpup Air-to-Surface missile. This was the missile's first trial, and it was no mean feat for the Fury, since the weapon itself weighed 540 lb., and six were carried.

Squadron VA-212, deployed for overseas duty aboard the U.S.S. *Lexington*, sailed from Alameda to join the Seventh Fleet in the Pacific on March 25th, 1959 as the first fully-equipped FJ-4B Bullpup squadron. During " Operation Cannonball " on October 8th, 1958 the U.S.M.C. Attack Squadrons VMA-212 and VMA-214 landed their FJ-4s at Atsugi, Japan, after a trans-Pacific flight from the Marine Corps Air Station at Kaneoke, with stops at Midway and Guam. The Marines flew their FJ-4Bs in two sections each of twelve aircraft and refuelled from U.S.A.F. Boeing KB-50 tankers near Wake Island and Navy AJ Savages in the Iwo Jima area.

The last Grumman fighter design made its debut in 1954. A new very fast agile fighter was requested by the Navy and from the designs submitted for review the Grumman Type 98 was selected and on April 27th, 1953 six prototypes were ordered. Although this was an entirely new aeroplane, the designation carried on from that of the Cougar, and the F9F-9 prototypes, Nos. 138604-138606, were accepted under that designation, but subsequently this was altered to F11F-1 which gave a more correct imputation—that this was a new type, not merely a new model of an earlier type.

Basically the Tiger was a lightweight air superiority fighter for the day interceptor role. Teething troubles seriously beset the Tiger partly because of the need to satisfy the two main requirements—maximum performance and simplification of design to allow for the fighter to effectively perform its primary mission.

In September 1952 the National Advisory Committee for Aeronautics announced to the American aircraft industry the results of its three years' of study into the area rule principle as it related to aircraft design. Richard T. Whitcomb of the N.A.C.A. is credited with its formulation. On test, it appeared to reduce drag by as much as 25 per cent. Grumman eagerly applied the area rule theory to their Tiger

By installing a new cockpit forward and removing two of the four 20 mm. cannons and accessories, 400 Cougars were produced as F9F-8T trainers.

design. Basically area rule comprised smoothly contouring the fuselage to be concave at the wing attachment zone. The area rule calls for the total cross section area of the fuselage and aerofoil at this point to equal that of a normal well-streamlined fuselage only. Thus the amount of " pinching in " of the fuselage at this point is determined by the wing chord at this point.

Application of the area-rule principle at the start of the design presented the Tiger with a feature which favoured it for becoming a supersonic aircraft. The Tiger was in fact the first aircraft to utilise the rule and to be flown incorporating it. To keep the weight of the wings low they were machined from single sheets of aluminium and a manual folding system was decided upon. The folding area was at the wing tips, and very small. Full span flaps were fitted, with spoilers to work in conjunction with exceptionally small ailerons near the wing tips. Leading edge slats were used to offer better control during landing and take-off. The exceptionally thin wings had marked sweep, and an all-flying tail was chosen. The fin was designed to carry part of the fuel load.

On July 30th, 1954, the first Tiger was flown by Corwin Meyers. No " X " prefix was applied to the designation since the prototypes were really pre-production aircraft.

825 to 950 m.p.h. depending upon altitude, and an initial climb rate of 18,000 feet per minute. With this performance the Navy had its first " faster than sound in level flight " fighter. An initial order was placed for thirty-nine F11F-1 Tigers and from these came the group flown by the famous " Blue Angels " aerobatic team initiated in 1946, and which had subsequently used Hellcats, Bearcats, Panthers, Cougars and now Tigers.

From the first, Tigers had refuelling probes in the nose and also the straight wing leading edge. Several of the early machines were turned over to VX-3 for trials and evaluation in 1957, after which there were trials of suggested modifications. No. 138622 was fitted with a Sidewinder beneath the outer section of each wing and No. 138628 carried special Douglas-built Aero 1-C 150 gallon drop tanks. Normally the armament was four cannon, two on each side of the fuselage, faired into the intake openings. Modifications were made on the intakes and these, among others, were incorporated in a new version which retained the F11F-1 designation, but formed the substance of the second production batch to appear.

In 1957 the suffix letter " F " was added to the model designation which has only been applied to one other fighter, the FJ-4 Fury. It indicates the installation of a

There were three names associated with the Grumman F9F series: F9F-2 to F9F-5 Panther, F9F-6 to F9F-8 Cougar and finally the F9F-9 Tiger. However, as the first aircraft to incorporate the area rule principle, which in effect gives the fuselage an indentation like the popular ' coke bottle ', this was re-designated the F11F-1. This example, the first of the prototypes, first flew on July 30th, 1954.

Early test flights showed serious control and stability problems at both ends of the speed range which were only in part finally corrected. A leading edge fillet was advocated as a feature to improve stability, and this brought some slight improvement in this shortcoming.

Power for the Tiger came from a Wright J65-W-4 (W-7 in the first pre-production aircraft) of 7,800 lb. thrust—11,200 with after-burner on. Originally the Tiger's maximum gross weight was 13,850 lb. and combat weight 17,000 lb. Since the earlier Cougars had weighed around 20,000 lb. in combat state, it is evident that the reduction in structural weight had much to do with its superior performance. During initial armament tests, a Grumman test pilot, Tom Attridge, " shot himself down " by running into one of the 20 mm. cannon shells he had just fired! This odd misfortune took place on September 21st, 1956 off Long Island.

First speed trials credited the Tiger with a top speed of

special power plant. For the ' 1F ' Tiger it denoted the General Electric J79-GE-3 of 12,000 lb. thrust (15,000 lb. with after-burner) fitted to compensate for the weight which would rise as it was fitted out for various roles, which entailed re-positioning of the retractable refuelling probe on the starboard side of the fuselage, a longer nose to house search radar—equipment which was never actually fitted, leading edge fillets to the wings and provisions for a variety of under-wing stores.

When the F11F-1F was completing tests at Edwards Air Force Base it was decided to attempt some record flights. First the speed record was raised unofficially to 1,220 m.p.h., and within three days, on April 16th, 1958, the World's Altitude Record was won when Lt. Cdr. G. C. Watkins reached 72,000 feet. After a day's servicing, the same aircraft raised this record height to 76,939 feet, again with Watkins at the controls. Under normal combat

Models of the F11F-1 Tiger with General Electric J-79 engines were designated the F11F-1F. This Tiger so powered, BuAer No. 141738, has its refuelling probe extended and carries auxiliary fuel tanks and Sidewinder missiles.

conditions the F11F-1F had an absolute ceiling of around 50,000 feet and a service ceiling of 40,000 feet. The new version easily exceeded Mach 1 in level flight.

The F11F-1 went into quantity production, the last one being built in December 1958. When it was delivered the following month it represented the last of that long line of Grumman fighters for the Navy, although not, of course, the last Grumman aeroplane to reach naval hands. Whether or not another Grumman navy fighter will emerge remains to be seen, mention has been made of the XF12F-1 and three prototypes are known to have been ordered. Under Mutual Defence Aid contracts, Grumman offered the so-called " Super Tiger " to Switzerland and Japan in 1958. Eventually the Swiss chose the British Hawker Hunter fighter as being more suited to their needs, but Japan decided to produce the Tiger and negotiations are still pending.

Most of the Tigers in service are in the attack class, although from time to time these units are in part redeployed as fighter groups depending upon operational needs. None of the F11F series reached the Marine Corps, although the Corps did assist in the prototype testing and development work. VA-156 (PAC) was, on March 8th, 1957, the first unit to receive the Tiger and since that time Reserve and Advanced Training Commands have also taken deliveries, for nowadays these second line units are equipped with first line aircraft. On July 30th, 1959 ATU-203 became the first reserve formation to receive Sidewinders for its Tigers.

One of the reasons why the Tiger flew with the VF units for such a short time was that it fulfilled the VA role adequately and was somewhat coerced by the arrival of the powerful Chance Vought Crusader. The Tiger is sometimes referred to as the Navy's counterpart of the F-104 Starfighter of the U.S.A.F. In a sense this is true, since both were conceived around the same time and for a similar need. Both have, too, found employment in roles other than those originally intended, but it must be conceded that the F-104 has found more favour than the Tiger.

When Vought broke their ties with the United Aircraft Corporation and presented the first designs of the Crusader, it was clear that here was a revolutionary aircraft. Vought staked their whole existence on this huge angular fighter—and the Navy backed them. Neither can nurse any regret. The Crusader almost literally flew off the drawing boards, for it was but nineteen months from first prototype to first production delivery.

On June 29th, 1953 two months after the Tiger contract had been placed, Vought received an assignment to build two prototypes of their XF8U-1, which, in competition with eight other firms' designs, had won the competition for a new supersonic day fighter requirement tabled in September 1952. Chance Vought was, incidentally, a division of United Aircraft in 1953, but became independent soon after the design was approved.

February 1955 found the first prototype ready for flight. It was accordingly despatched to Edwards Air Force Base for the first flight which came on March 25th. John Konrad climbed into the cockpit and this hefty fighter, powered by a J57-P-4, took off after a very short run,

This F11F-1 Tiger, BuAer No. 138617, the eleventh production model is being catapulted from the angled deck of a carrier during service with a composite squadron VX-3. The Tiger's very thin wing and slatted edges are shown to advantage in this view.

climbed rapidly, and exceeded Mach 1 before returning for a ' gentle ', carrier style, landing.

Many innovations were applied to the Crusader to ensure that it met the needs of the specification. Its shoulder wings with 42½° sweptback had a sharp saw-tooth leading edge. An extremely large fin led into a dorsal spine. The tailplane, all flying, joined the titanium rear fuselage below the thrust line. The small cockpit canopy was carefully designed to increase pilot's vision and reduce drag. Unusual was the employment of dihedral on the tailplane and an air scoop forward of the cockpit just below the small nose radar-scanner. The landing gear folded into the fuselage. An armament of four 20 mm. cannon was called for in the specification and was fitted on the XF8U-1's and on production aircraft, although later examples have provision for Sidewinders in addition.

The most unusual feature of all was the hydraulically controlled two-position incidence wing. Built as one unit, this pivoted at the rear spar junction with the fuselage to allow it to tilt upwards, thereby acting as an enormous landing flap; this controlled lift allows the Crusader to land at about 115 m.p.h. Not only has it proved to be a feasible idea, it has given hardly any trouble in service. In addition to the changing incidence wing, large trailing edge flaps are fitted and all leading edges can be drooped for additional high-lift at slow speeds.

Where possible, titanium was used in construction of the Crusader; the rear fuselage around the after-burner was constructed of this, likewise a large part of the central structure. To save weight, a new 30 lb. simplified ejector seat was devised. The J57 engine which powers the Crusader is a compound twin-spool turbojet of which the P-4 variant, fitted to the prototype, delivered 10,900 lb. thrust dry and 13,200 with the after-burner on, but production Crusaders use the more powerful P-12 of about 11,200 lb. dry thrust—13,500 lb. with the after-burner firing.

Following successful trials at Edwards Air Force Base the XF8U-1 went for carrier trials on the 1,036 foot long deck of the U.S.S. *Forrestal* shortly after her commissioning at Norfolk, Virginia, on October 1st, 1955 as the first of six ships of her class. The Crusader landed aboard and also showed its amazing facility by landing on the 876 foot deck of the U.S.S. *Bon Homme Richard*.

Quantity production was ordered in December 1955 and in a short time contracts calling for some 592 aircraft had been placed. The first was delivered in the middle of 1956 and on August 21st of that year, Cdr. R. W. Windsor captured the Thompson Trophy by setting a national speed record of 1,015·428 m.p.h. over a 15 kilometre course at the Naval Ordnance Test Station, China Lake, California, thereby snatching the record from an Air Force F-100C Super Sabre which had won it earlier at 822·135 m.p.h. The record breaker was a standard production Crusader with full naval equipment and well loaded. A further indication of the Crusader's capability was given on June 6th, 1957 when two F8U-1s took-off from the *Bon Homme Richard* off the Californian coast and landed aboard the *Saratoga* off the coast of Florida some three hours twenty-eight minutes later.

' Operation Bullet ' on July 16th, 1957 won the Distinguished Flying Cross for Maj. J. H. Glenn, Jr., U.S.M.C.,

Prototype of the first carrier-borne 1,000 m.p.h. fighter, the Chance-Vought Crusader which, as the XF8U-1, exceeded Mach 1 on its first flight on March 25th, 1955; later that year it was ordered into quantity production. An outstanding feature is its two-position variable incidence wing.

now known to the whole world as Col. John Glenn, America's first globe-circling astronaut. At 09.04 hours Glenn took one of the first F8U-1P photo reconnaissance versions of the Crusader off the runway at Los Alamitos, California, and 203 minutes later arrived at Floyd Bennett Field, New York. This journey was made at an average speed of 723·517 m.p.h., and it was the first upper atmosphere—35,000 feet high—supersonic flight from the West to East coast of the U.S.A. Three in-flight refuellings brought Col. Glenn down to 25,000 feet and to 350 m.p.h. to take on fuel from slow propeller-driven tankers. In addition to the speed dash, a photographic record of the flight was made by cameras clicking at set periods providing a continuous strip of photos from end to end of the flight.

In 1957 the ' Collier Trophy ', one of America's highest tributes was awarded to Vought and the Navy, for 1957's most significant achievement in aviation—the 1,000 m.p.h. fighter. On March 25th, 1957 the first F8U-1s went into squadron service with VF-32 (LANT) and as production built up many Navy and Marine squadrons possessed this amazing fighter. By 1961 it was standard first line equip-

The first production version of the Crusader in service, the F8U-1 with Marine Squadron VMF-312. This model had been produced in '1', '1E', '1P' and '1T' versions.

ment in over twenty-eight naval VF squadrons. Production of the ' 1 ' ceased in 1961 when the ' 2 ' took its place.

To the normal armament of four cannon were added two Sidewinders placed on special racks one on either side of the fuselage just below the rear of the cockpit. A retractable missile tray immediately aft of the nosewheel recess, can carry up to thirty-two 2·75 inch Mighty Mouse rockets. Hydraulic power for the operation of much of the mechanical equipment is provided by a ram-air turbine located within the fuselage with intakes for it situated below the leading edge of the starboard wing. The early production models did not have in-flight refuelling, but on September 12th, 1955 the Navy made this obligatory on all fighters. All Crusaders were subsequently so fitted or had in-flight refuelling gear retrospectively applied on the port side of the fuselage immediately aft of the cockpit.

The F8U-1 became, and still forms, the backbone of most VF and VMF squadrons. Perhaps the most remarkable Crusader flight was in December 1960 when a Navy pilot inadvertently took off from the Naval Air Facility at Capodichino, near Naples in Italy, with the wings in the folded position. Although the handbook sardonically states ' It is customary to lower wing tips before flight ' the pilot had apparently overlooked the sarcasm of these words. He did not discover his predicament until he tried to level off at 5,000 feet. His first indication of abnormality was the excessive amount of forward stick needed to control his machine. Noting his upturned wings, he cooly decided to try out the flight characteristics of the Crusader in this unfortunate state while he jettisoned fuel. The F8U remained aloft for twenty-four minutes before making a rather ' fast ', but smooth landing. The speed throughout the flight had remained around 200 m.p.h.—including during the landing! Odd though it seems, the pilot reported that the Crusader was quite easy to handle during the flight. The Crusader has, incidentally, the greatest wing area of any fighter in front line service in the American Forces today.

Several other similar folded-wing flights have taken place in U.S. naval history, but have usually resulted in stalling and crashing. One such flight was made by a Curtiss SB2C Helldiver during the war. It staggered into the air, flew a few seconds, stalled and the pilot successfully landed.

The F8U-1 Crusader series was made in four variants; F8U-1 day fighter, F8U-1E, with special electronics, the F8U-1P photo reconnaissance version and the F8U-1T trainer. The second model of the series, sometimes called the Crusader II, came into existence in 1958. Utilising a standard F8U-1, various modifications were made including the installation of a J57-P-16 turbojet rated at 13,000 lb. static trust and 17,500 lb. with after-burner. Higher speeds and altitudes were reached, but it was deemed necessary to add two long narrow fins under the rear fuselage for added stability in the thin upper atmosphere. Success of these modifications led to a limited production order for the F8U-2 Crusader, and on August 20th, 1958 the first one was flown. A top speed of 1,155 m.p.h. was attained at 36,000 feet representing a Mach 1·75 figure, and an initial climb rate of 16,500 feet was recorded. Overall dimensions were as for the first models, and externally the only noticeable feature was the addition of under-fuselage stabilisers.

Internally many new features were incorporated such as improved fire control systems, revised auto-pilot and cockpit arrangement. Delivery of the F8U-2 to Navy and Marine units was under way by the end of 1958. First to receive the F8U-2 was VF-84 at Naval Air Station Oceana to where the initial delivery was made on April 4th, 1959. After working up, the Crusaders joined the Atlantic Fleet.

Even more spectacular was the F8U-2N with a 20,000 lb. thrust (with after-burner) J57-P-20. An increase of top speed to Mach 1·95 was attained, and many new features

The F8U-1P Crusader, BuAer No. 144608, that made a record trans-continental flight under ' Operation Bullet ', with Maj. J. H. Glenn, U.S.M.C. at the controls. This pilot, is now the famous astronaut Col. John Glenn.

The Crusader II can be distinguished from the earlier I by two stabilising fins under the rear fuselage and airscoops for the afterburner. This II is BuAer No. 145581 of VF-91 aboard the U.S.S. Ranger.

introduced into this night fighter version mainly connected with the larger and more powerful nose radar. The automatic-pilot was further modified to reduce pilot fatigue and allow him to concentrate more on radar interception. The new auto-pilot is able to maintain any given altitude, hold a heading, select a new course and orbit over the target. Fuel capacity was increased by deletion of the nose rocket tray, in place of which two additional Sidewinders can be carried on the sides of the fuselage. The four cannon armament was retained but additionally four of the improved Sidewinder 1C missiles are carried on ' Y ' shaped fuselage pylons.

The F8U-2N first flew at Hensley Field on February 16th, 1960. The loaded weight had risen to 29,000 lb. on this version which has an initial climb rate of around 22,000 feet per minute and a service ceiling of 55,000 feet. First deliveries of a production F8U-2N were made in June 1960 and versions are still in production and are scheduled, at the time of writing, well into 1963. Latest version of this model is the F8U-2NE completed and first flown in July 1961. An improved and enlarged radar nose cone features a small infra-red scanner in a special housing just above the nose. This equipment is used to further improve the guidance of the four Sidewinder missiles. First to use the F8U-2N was VF-111 equipped in May 1961.

Several of the F8U-1s were modified with extra and improved radar gear into F8U-1Es, a variant which cannot be identified externally. None of the Crusaders has, incidentally, any attachment points for under wing stores.

One of the more spectacular feats accomplished by the Crusaders occurred on January 11th, 1962 when a complete Marine Corps squadron flew their F8U-1s non-stop from El Toro, California, to Atsugi in Japan. This was the first overseas deployment of a squadron across the Pacific and was made possible by the aid of in-flight refuelling.

A design contest for an attack aircraft of supersonic speed was initiated by the Navy in September 1953, and a development contract was awarded to McDonnell. The aircraft was to be twin-engined and carrier based, possessing greatly advanced radar and armament control systems. After design preliminaries had been studied McDonnell received a ' letter of intent ' for two of these machines, as AH-1s, on October 18th, 1954. Further changes to apply advanced air-to-air missiles, interceptor capabilities and fighter qualities, led to the Navy changing the concept and designation to F4H-1, a missile fighter, on May 26th, 1955.

By July the necessary changes had been made and the specification finalised. The following month twenty-three pre-production machines were ordered as F4H-1s to serve as prototypes and trials aircraft.

This F4H-1 Phantom II is a highly sophisticated and complex fighting machine, the actual construction of which was started in August 1956. On May 28th, 1958 the prototype made its first flight. Yet already it had a strong competitor in a new Crusader development. Experience by the Navy and Vought with the F8U prompted a redesign of the Crusader in competition with the F4H. Vought presented the improved and more powerful fighter as an operational all-weather Mach 2·0 aircraft and called it the Crusader III. The Navy showed great interest and ordered eighteen pre-production examples in December 1957.

Design of the F8U-3 was extremely rapid, for the first one flew on June 2nd, 1958. Although it resembled the earlier Crusaders it was really an entirely different aircraft. The fuselage was lengthened, the nose was made more pointed and canted slightly upwards. A new Ferri-type forward swept chin air intake was used, within the lip of which was a movable wedge to vary the shock wave angle to suit the engine, which was a J57-P-4 giving 17,500 lb. static thrust and 23,500 using the after-burner. Additionally provision was made for fitting a 6,000 lb. Rocketdyne rocket motor in the rear of the fuselage. This was to be used to increase initial interceptor climb, provide faster acceleration from cruise to combat speed and assist manoeuvres at 90,000 feet—the Crusader III's operating altitude.

As a test vehicle for the auxiliary motor two Fury FJ-4's were modified early in 1958 to become FJ-4F's, the first Navy aircraft to have the ' F ' suffix denoting a special engine installation which was a Rocketdyne AR-1 engine installed over the conventional jet engine tail pipe. The rocket was designed to use standard JP-4 fuel and hydrogen peroxide mixture. The device was used to test the feasibility of the rocket engine for additional short bursts of speed and boosted initial climb.

The wings of the Crusader III retained the same two-position incidence feature and were of the same planform, but had increased span. As with the 1 and 2, the wings folded upwards at the half span mark, but featured blown flaps. The fin was reduced in size but to aid stability at altitude two large retractable fins were added to the lower rear part of the fuselage. For landing and take-off these assumed a horizontal position. All cannon and small armament was forsaken in favour of three Sparrow III AAM-N-6 homing missiles, two being fitted to the fuselage sides and the other beneath the fuselage. An offset nose-wheel allowed the missiles to be fired irrespective of nose wheel position. Of great interest was the push-button system of flying possible with the F8U-3. Vought claimed fully automated all-weather capabilities with this version.

Between July and September 1958 the Crusader III was in direct competition with the F4H-1. Over Edwards Air Force Base and Patuxent River each vied for the Navy's production contract as the future all-weather missile-equipped fighter. The F8U-3 was lighter and a single-seater whereas the more powerful, heavier McDonnell had a two-man crew. During trials the Crusader reached Mach 2·3 (1,520 m.p.h.) and computations rated its endurance at cruising as three hours without recourse to in-flight re-fuelling. A combat altitude of 90,000 feet was claimed with an operating radius of 1,000 miles. But more was involved in this competition than speed, climb rate and the suitability for carrier operations.

All in all, the F4H held the lead during the trials. It could accommodate more internal radar and electronic gear, more external stores and in any case its overall performance was superior to that of the Crusader III. The weight of the two-man crew seemed to have no adverse effect on performance and gave the F4H a decided operational advantage. By November 1958 the Crusader III had accepted defeat. Four had been built, Nos. 146340, 146341, 147085 and 147086, the last of which was never flown. The Navy cancelled the remainder of its order. The third example was the only one to be completely fitted with electronic and armament systems with nose radome fitted, in lieu of a metal cone and flight yaw test boom, although in other respects it was not modified to the latest standards as were the first two. None of them had the rocket motor originally planned and flight tested in the Fury.

The final type of VF aircraft on the Navy list, the McDonnell F4H-1 is now coming into full squadron use. From the final specification acceptance in July 1955, McDonnell have known it as the Phantom II, and the first of two pre-production examples ordered, first flew at Lambert St. Louis on May 27th, 1958. Prior to this, in August 1956, twenty-three other pre-production machines had been ordered. Evaluation tests followed at Muroc and Patuxent as the other pre-production F4H-1s left the lines, few modifications were called for during development.

Initial tests proved the Phantom to be, without question, a Mach 2 fighter and after fleet indoctrination trials aboard the 78,700 ton U.S.S. *Independence*, it was obvious the Navy had a most potent weapon. Successful flights were made soon after from the 41,900-ton U.S.S. *Intrepid*, the U.S.N.'s smallest carrier, after which there could be no doubt that here, was the world's finest all-round fighter/attack naval aircraft. The only thing that holds back the Phantom is the heat barrier, for although much of the Phantom is constructed of titanium alloy over stainless steel honeycomb structure, aluminium still forms the major material used throughout construction.

Somewhat ungainly in appearance the Phantom II will be employed in the dual role of fighter and attack aircraft for use in any weather, by day or night. The extremely thin wings have a modified form of boundary layer control. The saw-tooth wing leading edge is swept back 45°, and the wings fold upwards at the saw-tooth point for carrier stowage. Full span blown leading edge flaps can be deflected to 60°, allowing an angle of attack of 10° during landings, and reducing speed to 145 m.p.h. Polyhedral is employed on the wing the inner portion having 3° dihedral, the outer panels 12°. Inboard wing spoilers are used in conjunction with inboard ailerons.

In the fuselage the crew sit in individual cockpits over which are placed individual cockpit covers and both have British Martin-Baker ground level ejection seats. Provision is made for a duplicate set of controls in the rear cockpit, aft of which the fuselage is crammed tightly with fuel tanks, six in all. Additional fuel is carried in the inboard wing sections integral tanks giving a total internal fuel load of approximately 1,750 gallons. A ventral tank may also be added. The all-flying tailplane is somewhat unorthodox with a 23° anhedral, in order to clear wing tip vortices, thereby improving control.

Two General Electric engines are buried in the wing roots. On the original eighteen F4H-1 Phantom IIs these were J79-GE-2A engines rated at 10,350 lb. static thrust and 16,150 lb. with after-burner on. Current production models have J79-GE-8 turbojets of 10,450 lb. thrust, 17,000 lb. with the after-burner on. The service type will have available over 34,000 lb. thrust to call upon with the water and

A version of the Crusader II with improved radar and a Pratt & Whitney J57-P-20 engine in place of the earlier J57-P-16 is the F8U-2N, shown here with its missile complement of four Sidewinders. This type's more recent variant, the F8U-2NE with improved electronics is now in production.

alcohol compressor injection system. Following successful trials, shape of the air intakes have been modified, production Phantom IIs having variable geometry inlets, and one fixed and one movable entrance baffle with variable throat areas. It seems quite possible that in time the Phantom will emerge with an even more powerful engine type such as the Pratt & Whitney J93.

An all-missile armament has been decreed for the Phantom II which usually is armed with four Sparrow III radar homing AAM-N-6s which have a speed of Mach 3·0 and a range of five and a half miles. They are fitted to the underside of the fuselage and can be replaced by 1C Sidewinders. A combination of missile types may be carried, as well as four additional similar missiles fitted to special 'T'-shaped under wing pylons. A normal mixed load is comprised of four Sparrows on the fuselage and four wing mounted Sidewinders. When the all-Sparrow load is used, the wing pylons accommodate only two missiles since they each weigh nearly twice as much as the Sidewinder.

A wide range of external stores may be accommodated such as additional fuel tanks, one 600 gallon tank under the fuselage and a pair of 360 gallon wing tanks are possible, and they raise the fuel capacity to 3,070 gallons. The attack version can carry twenty-two 500 lb. bombs in clusters on five different racks, as well as four missiles.

The U.S.A.F. have gazed enviously at the Phantom and have finally decided to procure it as the F-110, for this machine is eminently suited to replace the F-101 Voodoo, which incidentaly equips the 81st Wing in Britain. The F4H-1 tips the scales fully-loaded at about 45,000 lb., though this is limited to 40,000 lb. for carrier operations. In its fighter form it has a top speed of Mach 2·4 (1,584 m.p.h. at sea level).

By October 1960 pilot training on the F4H-1 had commenced and McDonnell had received further orders. The F4H-1 introduced a modern feature of many jet aircraft to the Navy pilots, the braking chute to slow the aircraft upon landing. It can also serve as an anti-spin chute, and is carried as standard equipment. It opens out, at the pilot's discretion, from just above the arrester hook in the tail cone and can be used at speeds of below 150 m.p.h. for a rapid reduction of speed during landing.

During the evaluation phase the Phantom broke many records, and these give some idea of its spectacular performance. On December 6th, 1959 during 'Project Top Flight' Cdr. L. E. Flint set a world altitude record when he reached 98,560 feet over Edwards Air Force Base, California. Although officially agreed, the record was not registered with the F.A.I. The Navy believes that the true value of a combat aircraft lies in its ability to manœuvre at high speeds and illustrated this on September 25th when Cdr. John F. Davis averaged 1,390·21 m.p.h. on a 100 kilometre closed circuit course completed in 2 minutes 40·9 seconds. This flight was made over a route less than twenty miles in diameter. This particular record was F.A.I. claimed by a Russian 'T-405' at 1,298·7 m.p.h., but the Phantom had exceeded it by nearly 100 m.p.h.

The twelfth production F4H-1 went to the Marine Corps and in the hands of Lt.-Col. T. H. Miller set a new world speed record for the 500 kilometre course. A blistering 1,216·78 m.p.h. speed was achieved on September 5th, 1960 beating the 1959 record set by an RF-101C Voodoo of 816·3 m.p.h. For his climb and the fifteen minute twenty

Famed fighters in naval service today aboard the U.S.S. Independence in early 1960. In the foreground is an F8U-1E Crusader with a pylon fitted for the carriage of Sidewinder missiles. In the background, on test by personnel of the Naval Air Test Center, is an early Phantom II.

second run over Edwards A.F.B., Miller used 23,000 lb. of fuel. Armament was deleted saving 1,200 lb. of load and three external fuel tanks were carried and were jettisoned before the high speed flight took place. This record was approved by the F.A.I.

On August 28th, 1961 a Phantom of Squadron VF-101 undertook 'Operation Sageburner', an attack on the three kilometre speed record. Lt. Hunt Hardisty and Radar Interception Officer Earl H. DeEsch in rapid runs averaged 902·769 m.p.h. hurtling through the course in seven and a half seconds and breaking a record of eight years standing. The speed run was made four times over the course, never above 500 feet, even during turns, at Holloman Air Force Base, Alamagordo, New Mexico. It was considered hazardous and very exacting since altitude during the straight runs had to be held below 325 feet.

A team from VF-74 was chosen to undertake 'Project LANA' in 1961. This was a subtle name, for 'L' is the Roman numeral for fifty, and ANA represented the initials of Anniversary of Naval Aviation—from Chapter 1 it may be remembered that the Navy received its first aircraft

The large-scale production and lack of modifications points to the F4H-1 Phantom II as a winner from the outset. Lt. John S. Brickner is here making the 1,000th landing recorded by the U.S.S. Enterprise.

in 1911, hence the fiftieth anniversary project in 1961. Although trained on Phantoms, the Squadron had not at the time received them, so borrowed two from Squadron VF-101 and trained for a trans-continental dash. Five Phantoms participated in the special run, comprising three VF teams, and completed the flight in record time. Taking off from Ontario Airport, California, with Lt. Richard Gordon at the controls and Lt. (j.g.) B. R. Young as radar officer, one Phantom flew the 2,449½ mile West–East course to the New York Naval Air Station in 2 hours 47 minutes 17·75 seconds at an average speed of 869·739 m.p.h. beating the U.S.A.F.'s F-101 Voodoo flight of 1957 which had taken 3 hours 7 minutes and 43 seconds. The winning Phantom was missile loaded and carried a 600-gallon belly tank. Taken into account were three in-flight refuelling link-ups en route. For this achievement the Navy team was awarded the coveted Bendix Trophy.

Those who had flown the Phantom realised it could exceed the 1,525·26 m.p.h. at which the U.S.A.F.'s F-106A Delta Dart held the speed record, but it was not until November 22nd, 1961 that an official run to prove this was undertaken at Edwards Air Force Base. Then the F.A.I. approved a record speed of 1606·324 m.p.h. for Lt.-Col. Robinson of the Marine Corps, flying BuAer No. 142260 on the 15/25 kilometre World Speed Record course, and breaking the U.S.A.F. record.

Carrier evaluation of the Phantom was conducted with the sixth from the production lines, in mid-February 1960, aboard the U.S.S. *Independence*, CVA-62. Lt. Cdr. Paul Spencer was the pilot and Cdr. Larry Flint his 'number two'. The first flight was of fifteen minutes duration and trials followed by day and night. The F4H left the carrier fully qualified and by May 1961 VX-5 was using one of the Phantoms for conventional and nuclear weapons delivery trials at the N.O.T.S. China Lake, California. The first F4H-1 Phantoms for front line service were assigned to VF-121 at Miramar Naval Air Station in February 1961, and they were used to train pilots and radar officers in the intricate and sophisticated complexities of the type. By July 1961 five had been transferred to VF-101 at Oceana, Virginia and both bases and squadrons continued to offer training courses on the F4H. All their aircraft were from the initial production batch and were powered by J79-GE-2 engines. On some of the aircraft, the radar operator's cockpit was modified to include a second set of flight controls in this position.

So advanced is the Phantom II that it has set a need for a special training course. Upon its completion, the pilot receives membership of the 'Phantom Phraternity'. Charles Baumeister, commander of VF-101, became 'Phemagler No. 1', and William A. Shryock was 'No. 2' for 'phlying the phenomenal, most pherocious of all phighters—making him phit for Phantom Phraternity membership'.

Production for the F4H-1 was just getting under way by the end of 1961 and in September of that year McDonnell had received a 180·1 million dollar contract for additional Phantoms making a total of $548,996,623 earmarked for this aircraft. In early 1962 the first fully equipped specially trained squadrons joined the fleet.

The first indication of a complete missile weapon system

In competition with The Phantom II was the Crusader III, an enlarged model with push-button controls, all-weather radar, provision for a rocket auxiliary engine and armed with Sparrow III missiles. Ventral fins were lowered for high speed flight.

was made known on December 5th, 1958 when the Navy announced a development contract had been awarded for the 'Eagle' missile. To the Bendix Corporation went an order for this air-to-air guided missile system designed specifically for fleet defence and long range interception missions. Bendix were to develop, produce and instal the missile, and Douglas, who received a contract on July 21st, 1961, were to design the aircraft to launch the weapon. Under the design contract a tentative order was placed, for two machines to be known as F6D-1 Missileers.

The F6D-1 was to be a monoplane employing two Pratt & Whitney TF30-T-2 turbofans each of 8,250 lb. thrust, and to be carrier or land based and of subsonic speed only. The XAAM-N-8 Eagle was also under consideration by the U.S.A.F. in which converted Boeing KC-135A tankers could carry up to twenty-four Eagles and act as airborne defence command centres. But on April 25th, 1961 the entire project was terminated, presumably because of technical troubles with the missile itself, and of changing requirements.

Before temporarily shelving the Missileer, Douglas had gone as far as pre-production drawings, wind tunnel models and scale model mock-ups. Details of the aircraft remain classified, but it is known that it was planned that it should carry eight missiles on wing pylons.

In November 1961 Admiral James S. Russell, Vice Chief of Naval Operations stated that 'the pilot is not obsolete, the missile has not replaced him and the space age has certainly not made him obsolete. Man is the maker, maintainer and operator of machines of war—the best, most versatile and by far the most reliable combination of sensing, deciding, acting devices we can hope to see for a long time'. This is true, but in the not too distant future he will become only the deciding factor.

The VF category in the Naval Services is slowly winding down and being integrated into the attack group, which is trained to be able to deliver a rapid attack which can be repeated again from an entirely different delivery point. The small A4D Skyhawk for example can carry a nuclear weapon with the destructive power equal to all the bombs dropped by the Fortresses of the Eighth Air Force over Europe during the whole of the War; this colossal load being released from an aeroplane considerably smaller than most present day fighters.

Keeping pace with the incredible advancement in naval aircraft have been the equally noteworthy advances in the design of aircraft carriers.

There is an excellent co-operation with N.A.T.O. Allies and very close association with the Royal Navy. Aircraft of the Fleet Air Arm have operated from American carriers and vice versa. Today over 70 per cent of the U.S. Navy's aircraft are equipped with the British made and designed Martin-Baker ejector seat. Many high speed machines have the new ground level escape model first demonstrated to the Navy by Flt. Lt. Sydney Hughes of the Royal Air Force at Patuxent on August 28th, 1957 when he ejected from an F9F-8T.

The twentieth century will be remembered as the backcloth against which was enacted the story of powered flight. The technical and operational progress revealed by a survey of successive generations of fighter aircraft can only engender amazement at their advanced state and speed of development over a period of less than sixty years. In this relatively brief span of time military aeronautics has evolved from a vague idea into the precision of a science.

To test the rocket motor for installation in the Crusader III, two FJ-4 Furies were specially modified and re-designated FJ-4F. One of the two, BuAer No. 139284 is shown with a Rocketdyne AR-1 engine installed and test equipment fitted.

Progress is never easily or cheaply achieved, and this is particularly the case with naval aviation, in which the resolving of purely aeronautical problems has to be simultaneously integrated with the specialised requirements of maritime operations.

The potential of the aircraft carrier, which has, indeed, become the capital ship of the world's navies is a major factor in military affairs, and the fighter aircraft which fly from it now represent the current peak in manned weapons capable of operation far beyond the range of land based types.

Some will argue that the U.S. Navy has come to be *the* greatest deterrent to war. Certainly the Navy can exert its influence over the Oceans of the World and the air above it, on behalf of the Free World. Below surface lurk the Polaris submarines, a mobile unseen force sustaining the deterrent. From the information in this closing chapter of this book, it is evident that the naval fighter of today, with its attack capabilities, is in itself a deterrent force quite apart from its function of fleet defence.

APPENDIX ONE

This very rare and untouched photograph shows a U.S. Navy Sopwith Camel with large white serial number on fuselage, repeated in 3 inch letters on the white stripe of the rudder.

There have been four phases in the colouring and markings of U.S. Naval aircraft; from the dull, uncertain tones of the First World War period through the highly spectacular colours of the inter-war years, followed by the blues of the Second World War until the light grey and white of today. Each change resulted from official directives, but such instructions were not necessarily mandatory; administrative orders may well be interpreted by executive officers in different ways and the availability of the correct paints is a primary consideration in the application of markings. What was decreed can, in general, be regarded as the standard for markings, but there were often anomalies in what was actually applied. In addition, non-standard colours and markings were permissible for such events as air races, air shows, fleet concentrations and speed tests. The United States Marine Corps, with its individual units under separate command, often did not conform to general Navy standards and tended at times, to flaunt their individuality which could perhaps be excused as *esprit de corps*. However, we are concerned here with the general, not the particular, in naval aircraft markings.

The original fighters of the U.S. Naval Services were painted a nondescript grey. The HA Dunkirk Fighters for example, had dark grey fuselages with wings and horizontal tail surfaces clear-doped; while the Curtiss 18-Ts were of a lighter shade of grey with clear-doped wings; tail surfaces were plywood-covered, doped light grey. Unlike land-based aeroplanes, naval aircraft were constantly subject to the elements and careful fresh-water washing and the repeated waxing of surfaces, was essential to maintenance.

At first, the only marking displayed was the individual Bureau number which, logically, started at A-1 with the Navy's first machine. The numbers then progressed apart from a few early exceptions, numerically as each successive one was procured. There was no specific location for these numbers and up to 1916 they were conspicuously applied, generally on the tail surfaces. When America entered the war in 1917, these numbers were applied in large figures, white or black in contrast to their background and positioned centrally on either side of the fuselage. They were usually repeated on the fin or rudder and when rudder striping became mandatory in 1917, as related later, the figures were

This rare flying shot of a U.S. Navy Hanriot HD-1 shows new paint scheme used on this type. Fuselage was grey and wings and tail were painted silver. As on the Camel above, the serial number is in large white numerals on the fuselage side.

First of the First! The new marking system adopted in 1924 clearly identified each aircraft. This Curtiss TS-1 is from the First Fighter Squadron (VF-1) and is the first aircraft of the unit. Note the red pennant on this, the Squadron Commander's aircraft.

reduced to a 3-inch size to permit presentation within the middle band of white, either vertically or horizontally. The large fuselage numbers were standard until 1920 and then optional until December 1924, when a directive was issued for their removal, leaving the rudder as the only position for marking the Bureau number. This location was changed in 1925 from the white stripe of the rudder to the fin, where it remained until 1941; meanwhile, in 1930, the prefix 'A' to the number was dropped. The presentation of 3-inch figures remained until America's entry into the Second World War in 1941, when it was reduced to a 1-inch figure size. Post-war, it reverted to the 3-inch figure presentation and today the size is 8-inch on the vertical tail surfaces and 3-inch on the rear fuselage.

The second official naval aircraft marking and first official National Insignia, was ordered into use on May 19th, 1917. A general order specified that a red disc be placed within a white, five-pointed star on a blue circular field, of a diameter equal to the distance between the leading edge of the aileron and the leading edge of the wing, to a maximum of 5 feet. This insignia was placed at four positions on the wings, inboard by a distance equal to the chord of the wing, with the star point at the leading edge, aligned in the direction of flight. At the same time the order stipulated that the rudder would be striped in three equal vertical bands of red, white and blue, with blue leading. Under this directive the Navy's first fighter, the HA 'Dunkirk Fighter', emerged sporting the rudder stripes, but due to its brief life and experimental status, no markings were applied to the wings.

On February 8th, 1918, the roundel form of insignia replaced the National Star Insignia and was adopted to assist identification of American Expeditionary Force aircraft in the European Theatre by bringing them into line with Allied markings. The roundel was formed by concentric circles of red, blue and a central white disc. Under the same instruction the order of the rudder striping was reversed, red then becoming the leading stripe. This directive affected the Curtiss HAs and 18-Ts and later, the last two HAs sported the large white Bureau numbers painted on the fuselage sides, these numbers also being placed within the white rudder stripe. The 18-Ts however, went without fuselage markings and subsequently underwent many changes of colour scheme during their racing career. The roundel markings were used for some eighteen months, until August 19th, 1919, when the Secretary of the Navy ordered a reversion to the star marking, and rudder stripes with blue to the rudder post. One year's grace was allowed before this order became fully effective and the early type insignia on the Curtiss Fighters was not replaced until a

On Marine Corps aircraft the mission letter was encircled as on this Boeing FB-1. The aircraft was the fifth, of the Sixth Marine Fighting Squadron (VF-6M). Note the unusual tail stripes.

A McDonnell F2H-4 from Cecil Field, Florida, displays large Branch and Unit markings. In addition, individual markings re-appeared at this time and the Boar's Head of the famous "Red Rippers" is evident here.

under surface. New aircraft delivered after July 1st, 1955 were completed to this scheme and it was directed that all naval aircraft would conform by July 1st, 1957. The matt grey on the upper surfaces proved to be the most inconspicuous colour at altitude and the glossy white under surface, provides a reflective finish from the light and heat of nuclear devices. This scheme had the added advantage of assisting flight technicians in spotting oil leaks. There is an interesting parallel here with aircraft of the Royal Navy.

From 1958, aircraft assigned to Reserve Squadrons carried dual branch of service markings 'NAVY' 'MARINES' painted over the orange fuselage bands.

The paints of today are much advanced over those used on early aircraft, they are acrylic-based, adding very little to the weight of the machine, and withstand wear and 'weather' to a remarkable degree. Many of the decorations of pre-war have re-appeared on naval aircraft and individual squadron markings are once again a colourful and attractive feature on them.

The official standard markings in use today, are the National Star Insignia, positioned each side of the fuselage and on the upper left and lower right wing surfaces; a letter code each side of the vertical tail above the last digits of the Bureau number; the full aircraft Bureau number and type identification with large 'NAVY' or 'MARINES' and squadron designation, all appear on the rear fuselage and large individual squadron numbers are positioned each side of the nose and on wing surfaces opposite to the National Insignia. Red equilateral triangles below the cockpit indicate aircraft equipped with ejector-seats, all jet intake areas are marked with a red pennant-shaped symbol and red 'rescue' arrows are placed about the cockpit area for fast cut away of structure to free the pilot in an emergency.

Squadrons are beginning to apply insignia again, the pre-war 'High Hat' has returned with VF-14, 'Red Rippers' VF-11, 'Felix' now 'Tomcatters' with VF-13, 'Striking Eagle' VF-51 and war-time 'Grim Reapers' VF-101, 'Jolly Roger' VF-84 and 'Iron Angels' VF-141.

The 6-inch white-shaded Efficiency 'E' has been revived and may be seen applied to naval fighters of units which have won distinction for gunnery at annual weapon 'shoots', bombing, maintenance, etc. Small suffix letters such as B or M situated below or following the 'E' denoted the particular category in which the efficiency award had been granted.

To sum up this account of Naval Aircraft markings, it may be noted that the paints and colour schemes applied to aircraft have always been functional in that they protect, camouflage and identify, in addition to being used as an aid to safety precautions and maintenance. From the early days, paint and dope were applied as preservatives; through the 'twenties and 'thirties markings were used additionally for identification and decoration; during the Second World War and through the middle 'fifties, the dark blue paints represented the best camouflage at the combat altitudes flown, and today, the ligher shades of colour appear as the most suited for flying in this nuclear age.

Introduced in 1955, the current overall scheme employs a seagull grey scheme with glossy white undersurfaces. Letter codes, Bureau numbers and branch markings are all conspicuously displayed, as are various warning signs, access points and equipment locations.

APPENDIX TWO

Aircraft Carriers of the U.S. Navy

With arrester hook down, this F3H Demon approaches the angled deck of the U.S.S. Bon Homme Richard *(CVA-31) on whose flight deck can be seen Skywarriors in front of the island and Demons beyond.*

That America was late in introducing the carrier cannot be denied, but neither can the fact that today the United States have the largest carrier force in the world.

In spite of a die-hard attitude of some officers in the formative years of American naval aviation, Admiral Sims as early as 1925, when the U.S. Navy had only one carrier in service, foresaw that the first carrier was the capital ship of the future—and so it proved. However, not until 1934 was the first carrier commissioned that had been built from the keel up for that purpose.

Reference letters, indicative of their roles, were allotted to carriers in accordance with U.S. Navy custom for ships, airships and aeroplanes, and carriers were numbered from No. 1 with CV as prefix letters. Up to December 1941 when America entered the war, a total of eight carriers had been commissioned CV-1 to 8.

It was war that brought the carrier into prominence. The U.S. Naval Services went to war in the Pacific in 1941 with three carriers, and in late 1942 they were down to one serviceable—the newly commissioned U.S.S. *Enterprise*. Two years later 100 flat-tops of various types were in service with 5,000 combat aircraft aboard, and apart from those in U.S. Service, others had been supplied to Britain. The *Archer* ex-*Mormacland* was in November 1941 the first of thirty-eight escort carriers transferred to the Royal Navy.

During the war carrier based aircraft of the U.S. Navy sank eight times the warship tonnage of land based aircraft of all U.S. Services, and sank twice as much merchant shipping as land based aircraft; only the submarines sank more over-all tonnage.

Post-war carriers have been used in various ways. The U.S.S. *Midway* CVB-41, in March 1946 went into the Arctic Circle in 'Operation Frostbite' taking F8F Bearcats and FR-1 Fireballs to test, *inter alia*, fighters under extreme cold weather conditions. In a very different way, four months later, the U.S.S. *Saratoga* was sunk in shallow water during the Bikini Atoll atomic bomb tests during which the U.S.S. *Independence* was damaged and was sunk by gunfire— both were too radio-active for salvage. A new U.S.S. *Saratoga* CVA-60 is now in service as it is U.S. Navy policy to re-use famous names, as indeed is customary in other navies, but in the U.S. Navy the reference number, when quoted, avoids any confusion in historical references. American carriers are listed by class in the following pages.

On the carriers, techniques for landing and take-off have been continuously improved. In November 1948, the Mark 7 high energy absorption arresting gear project was initiated allowing aircraft of up to 50,000 lb. at speeds of 105 knots to land safely—and some post-war fighters reached a loaded weight of 35,000 lb.

Later, to replace the former crash barrier of steel cable nets, the 'Davis Barrier' designed by S. V. Davis of the Naval Aircraft Factory, throws up steel cables to engage in the under-carriage while the wings engage in a nylon net.

A number of British techniques have been adopted perhaps the most useful being the angled flight deck, a feature of all modern carriers. An 8° cant allows aircraft to land into the wind and provides room for others to be parked aboard or even to be catapulted as others land on. Trials with the angle deck began on January 12th, 1953 on the converted U.S.S. *Antietam*.

Another important invention for which the British were responsible was the steam catapult; previously catapults were hydraulically operated. Invented by C. C. Mitchell of Brown Brothers & Co. Ltd. of Edinburgh, the steam catapult uses the principle of the slotted cylinder with neither rams nor purchase cables; instead, the large piston is driven by high pressure steam from the ship's boilers, and attached to the aircraft through special pick-up points. Its introduction at sea in the U.S.N. was on the U.S.S. *Hancock* CV-19, first trials being made by Cdr. H. J Jackson using a Grumman S2F-1 on June 1st, 1954.

The mirror landing aid, now standard was first installed on the U.S.S. *Bennington*. F9F-8 Cougars led by Lt. Cdr. H. C. MacKnight were, on August 24th, 1955, the first naval aircraft to land at night with the aid of this British system.

On September 24th, 1960 CVA(N) the U.S.S. *Enterprise* was launched and on December 20th, 1961 she joined the Navy as the world's first nuclear powered carrier—not experimental, but part of the operational group that forms the Atlantic Fleet, carrying Crusaders and Phantom IIs.

Top, an F2H-2P Banshee securely caught by an arrester wire, after landing on H.M.S. Ark Royal *during cross-operations with U.S.S.* Saratoga. *Centre, during the same exercises, this F4D Skyray also landed on the British carrier, and is seen about to be launched from the* Ark Royal's *steam catapult. Note the 'guard' helicopter. Bottom, a fine action shot of an FJ-2 Fury being catapulted from the deck of the U.S.S.* Coral Sea.

FIGHTING FLAT-TOPS

Left column, top to bottom: The first U.S.N. carriers CV-1 to 6—the Langley, Lexington, Saratoga *in 1944 with Hellcats, Avengers and Dauntlesses,* Ranger *on trials 1934,* Yorktown *at Midway where it was sunk and* Enterprise *in 1954. Right column, top to bottom: The* Hornet *CV-8, two views of* Essex *first of its class CV-9 to 21, the* Yorktown *CV-10 of the Essex Class — shown modernised, the* Independence *CV-22 representative of her class covering CV-22 to 30 and the* Leyte *CV-32 of the Repeat Essex Class CV-31 to 40.*

FIGHTING FLAT-TOPS

Left column, top to bottom: The Antietam *with F3D and F9F aircraft on deck, the three carriers of the Midway Class CV-41 to 43* Midway, Franklin D. Roosevelt *and* Coral Sea *respectively;* Wright *CV-49, which together with the* Saipan *CV-48 are the only two of their class, and the new* Saratoga *CV-60. Right column, top to bottom:* Ranger *CV-61 with FJ-4s, F8U-1s and F4D-1s,* Independence *CV-62,* Kittyhawk *CV-63,* Constellation *CV-64 and two views of the world's largest ship, the nuclear powered* Enterprise *CVA(N)-65.*

FIGHTING FLAT-TOPS

Postwar the angled deck and complicated radar array has altered the appearance of carriers. Top: Essex Class Fleet carriers modernised with angled decks, Essex CVS-9 showing its hurricane bow (left) and Intrepid *CVA-11 (right). Middle: Repeat Essex Class changes represented by Kearsarge CVS-33 showing radar array and armament (left) and the angled deck of Shangri-La CVA-38 (right). Bottom: 60,000 ton carriers of the Forrestal Class; the Forrestal CVA-59 (left) and the Independence CVA-62 (right) with Skyraiders, Skyhawks, Demons, Crusaders and Skywarriors aboard.*

FIGHTING FLAT-TOPS

Carriers of today. Top: The old and new in service carriers; Enterprise CVA(N)-65 the first nuclear-powered carrier (left) and Intrepid *CVA-11 shown during service with the Sixth Fleet in the Mediterranean. Middle:* Ranger *CVA-61 in 1959 (left) and* Forrestal *CVA-59 in 1956 (right). Bottom: Refuelling at sea (left) is the* Saratoga *CVA-60 with the destroyer* Miller *DD-535 and cruiser* Boston *CAG-1 and the* Kitty Hawk *CVA-63 (right) during atomic washdown drill at sea, July 1961.*

CURTISS HA

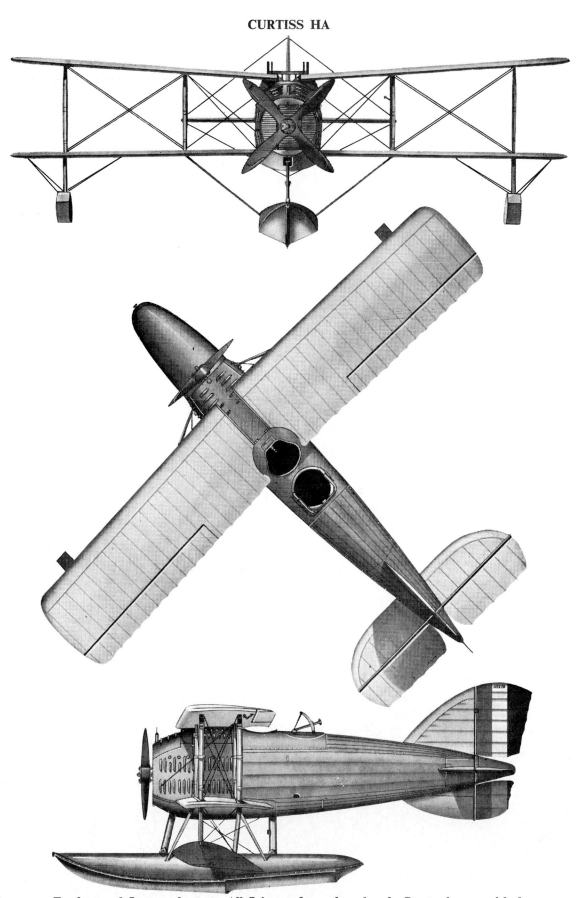

Fuselage and floats early grey. All flying surfaces clear doped. Struts clear varnished.

CURTISS 18-T

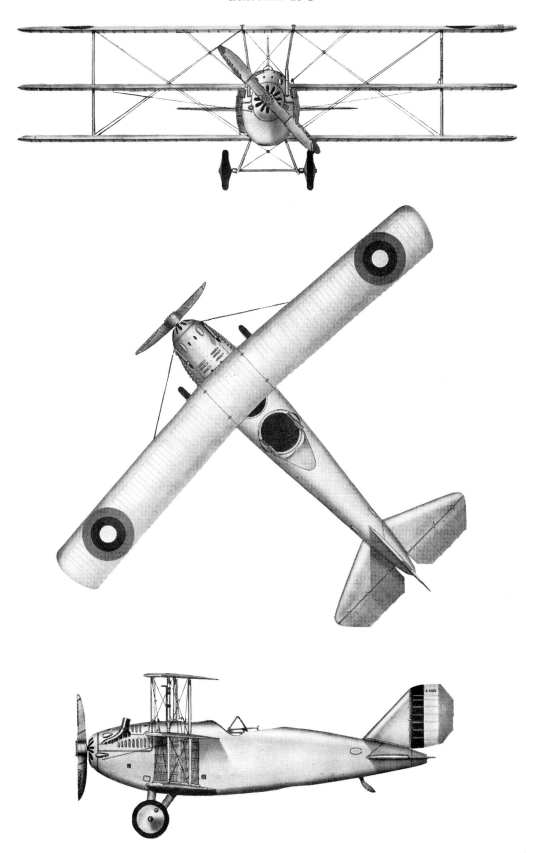

Fuselage and tail unit early grey. Wings clear-doped. All struts clear-varnished spruce.

HANRIOT HD-1

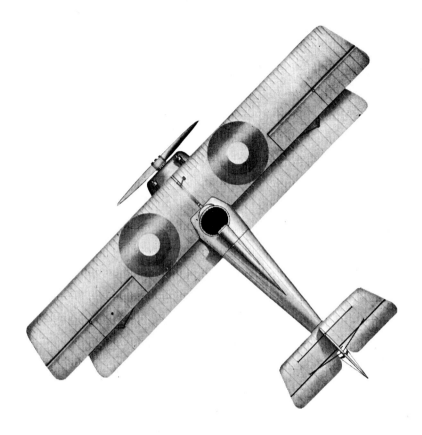

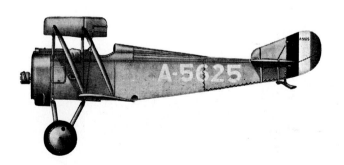

Fuselage early grey. Wings and tail surfaces silver. White number on fuselage.

THOMAS-MORSE MB-3

Overall olive-green fabric. Panelling natural metal or Navy grey. Marine insignia dark red.

VOUGHT VE-7F

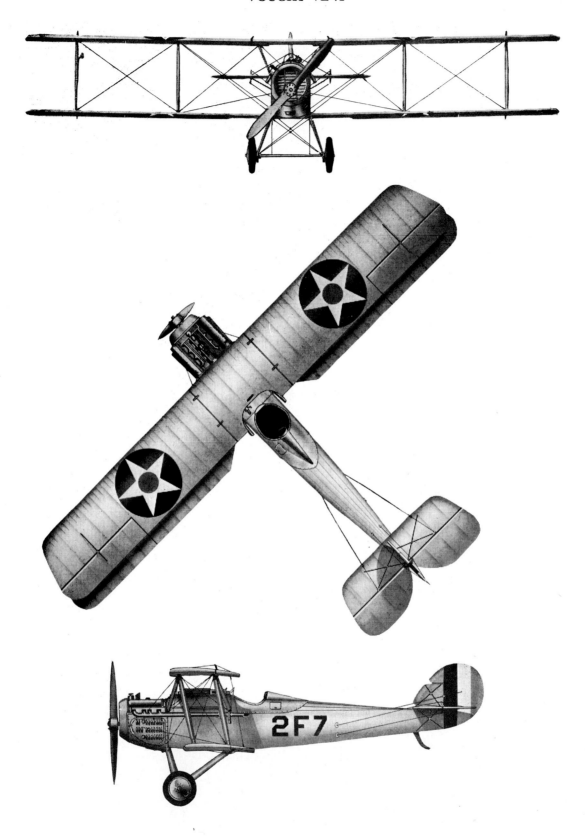

Overall silver. Cowl and coaming Navy grey. Top of upper wing and tail chrome yellow. Struts varnished spruce.

CURTISS TS-1

Squadron VF-1 from U.S.S. *Langley*. Basic scheme silver, top wing upper surface chrome, red tail.

CURTISS-HALL F4C-1

Overall silver. Top wing upper surface chrome yellow. Lettering black.

CURTISS F6C-2

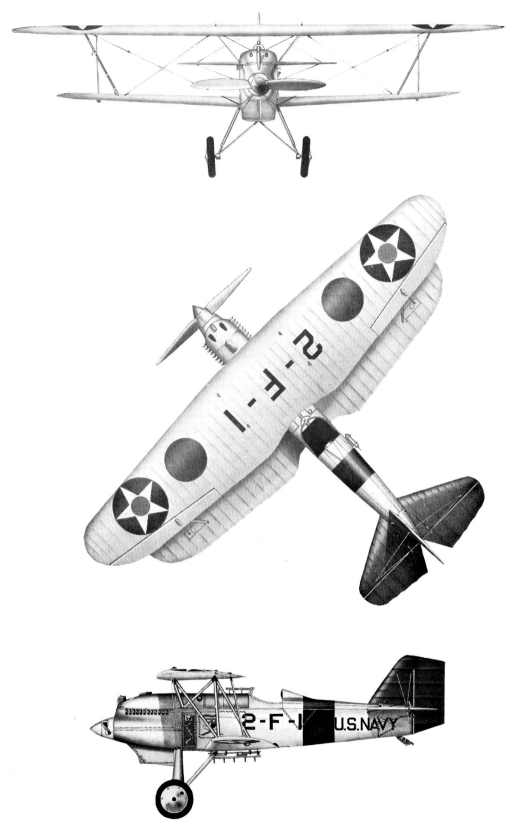

Basic scheme silver. Metal panels grey. Top wing upper surface chrome. Fuselage band, wing discs and tail unit, red.

BOEING FB-1

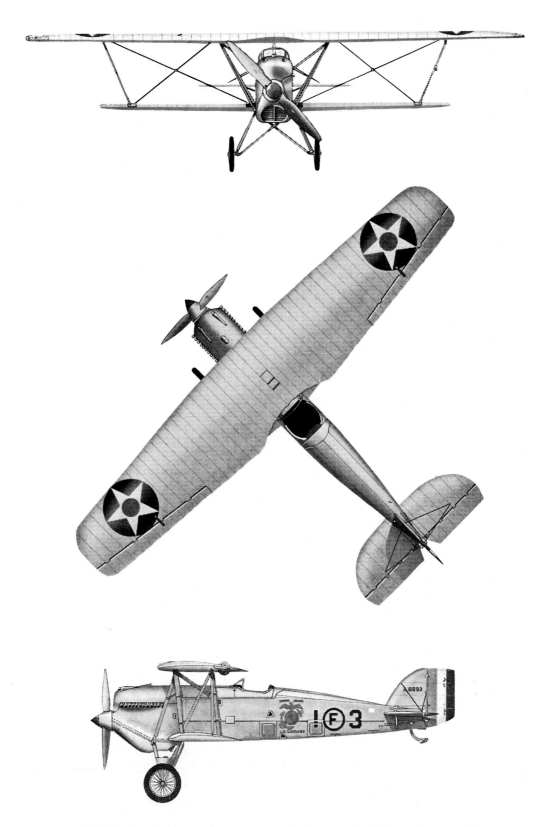

Squadron VF-1M. Basic scheme silver, top of mainplane and tailplane chrome. Spinner red.

WRIGHT F3W-1

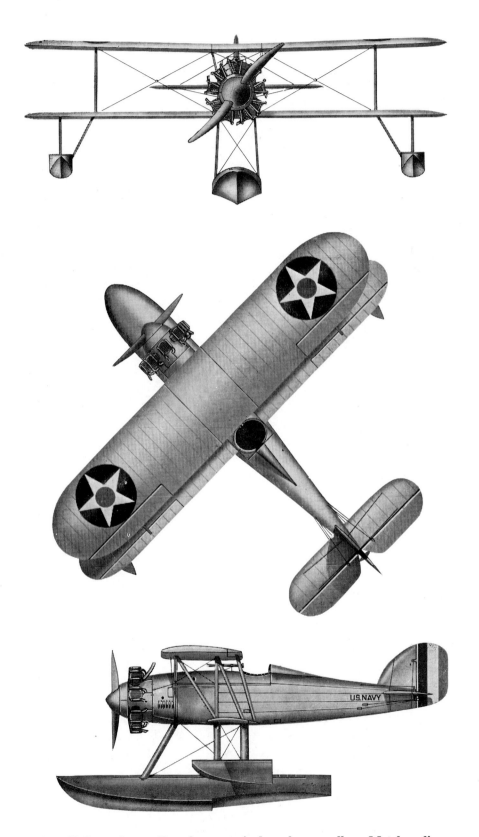

Overall silver scheme. Top of upper mainplane chrome yellow. Metal cowlings, struts and floats, early grey. Black lettering.

VOUGHT FU-1

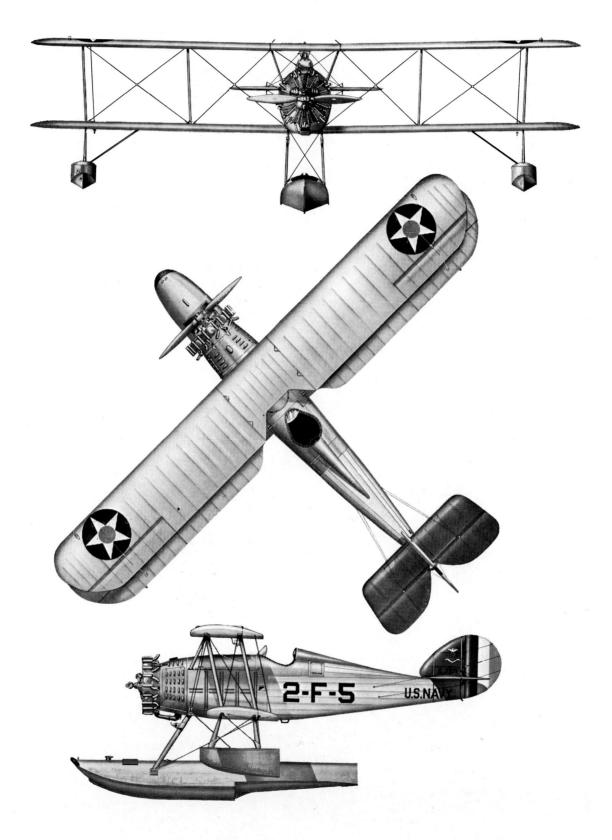

Overall silver scheme. Top wing upper surface chrome. Floats, panelling and turtle-decking Navy grey. Tail unit, insignia blue.

BOEING FB-5

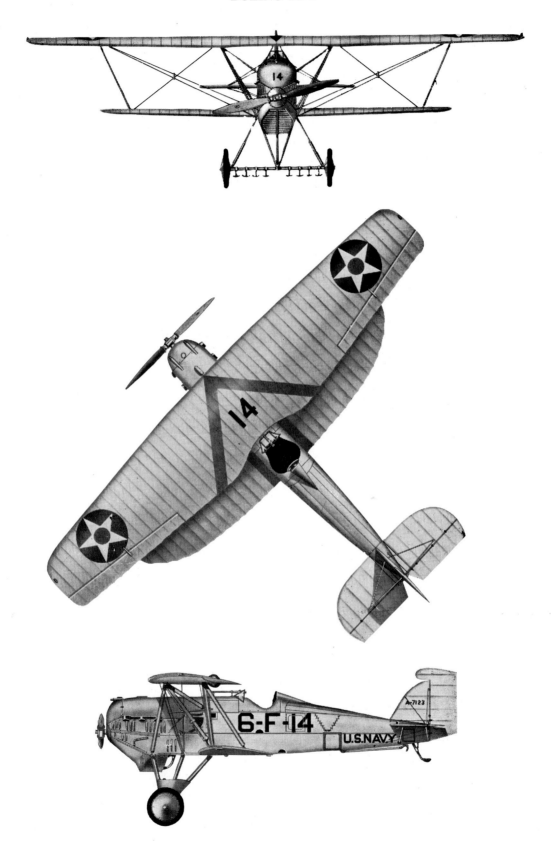

Squadron VF-6, of U.S.S. *Saratoga*. Overall silver. Top wing upper surface chrome. Chevron green. Tail unit white.

CURTISS F6C-4

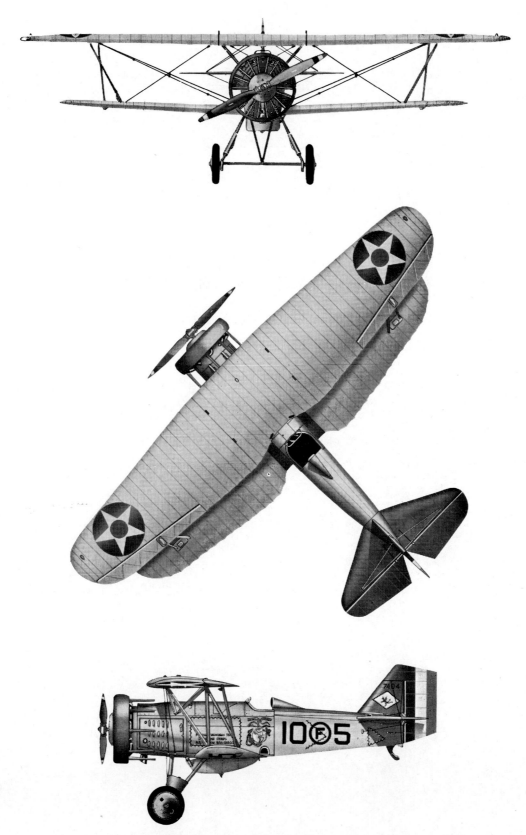

Squadron VF-10M. Overall silver. Top mainplane upper surface chrome yellow. Cowling and tail red.

BOEING F2B-1

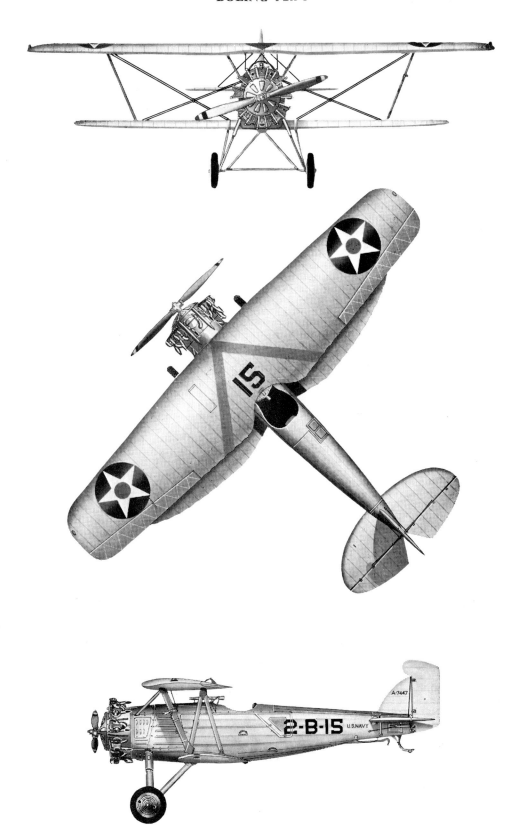

Overall silver. Top wing upper surface chrome. Chevron and lower-half of engine plate, green. Tail unit white.

CURTISS F7C-1

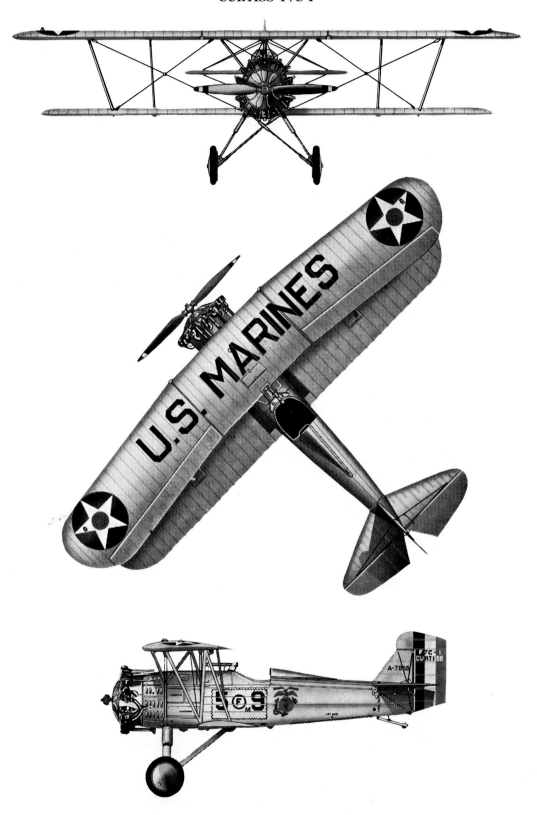

Overall silver. Top wing and tail unit upper surfaces chrome. Engine, piping and lettering black.

EBERHART XFG-1

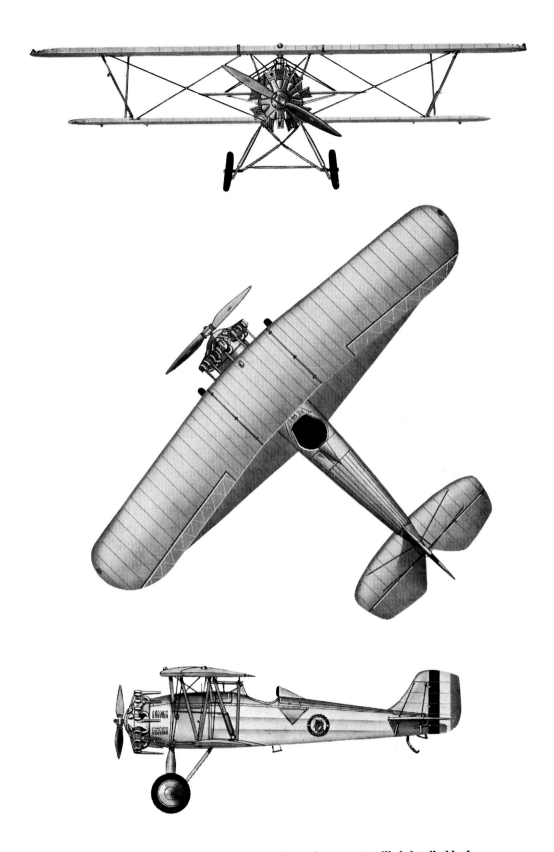

Overall silver, top wing upper surface chrome, stencilled details black.

BOEING F3B-1

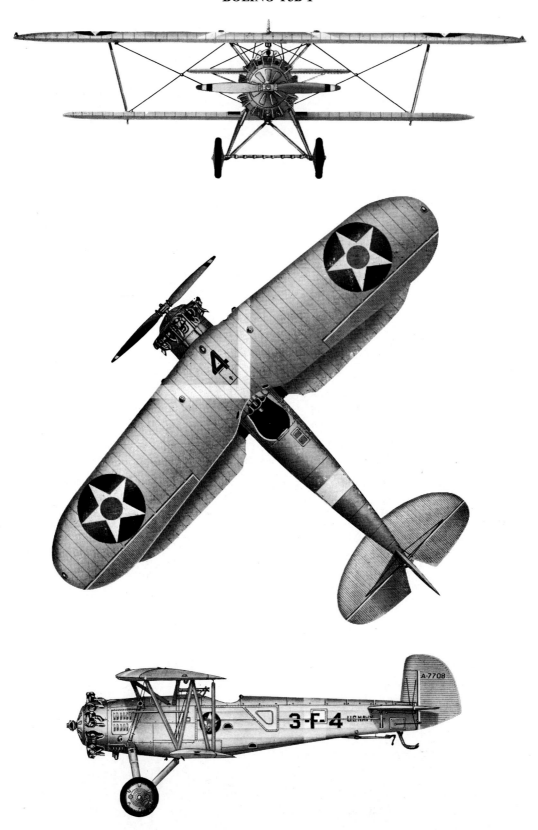

Overall silver. Top wing upper surface chrome yellow. Engine plate, fuselage band and wing chevron, white. Tail unit yellow. Forward panels and turtle-decking, Navy grey.

HALL XFH-1

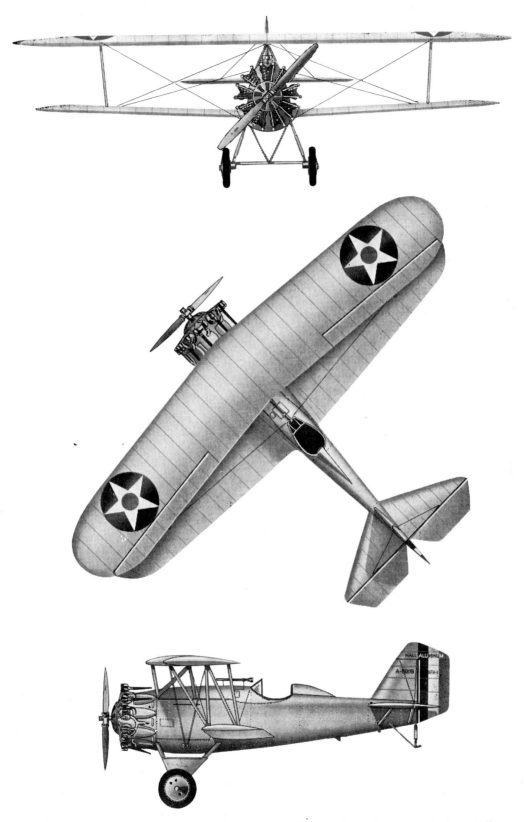

Fuselage natural aluminium. Wings silver. Top surface of upper mainplane, tailplane and elevators, chrome yellow. Lettering black or white, to contrast.

BOEING F4B-1

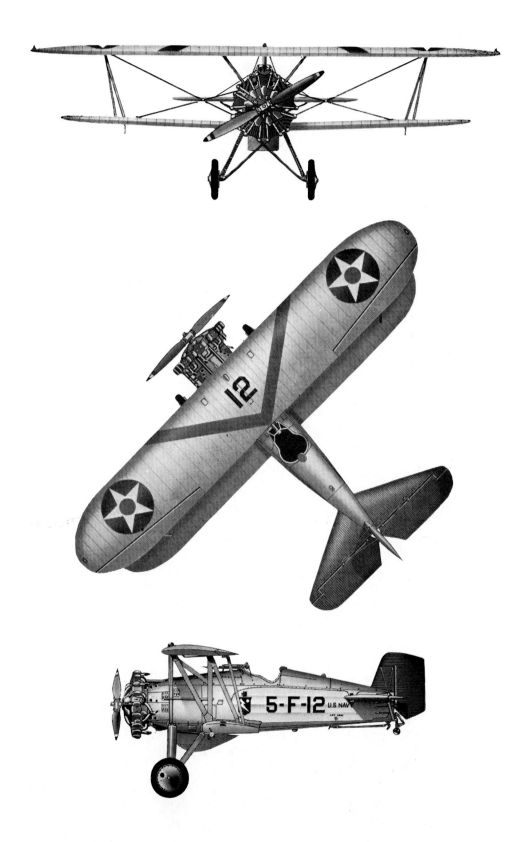

Overall silver. Top wing upper surface chrome yellow. Chevron and lower engine plate, black. Tail unit, blue.

BERLINER-JOYCE XFJ-1

Fuselage Navy grey. Wings silver. Top of upper mainplane and horizontal tail surfaces chrome yellow.

CURTISS F8C-4

Overall silver. Top wing upper surface chrome. Engine plate, chevron, fuselage band and wheel disc, green. Tail unit red.

BOEING XF5B-1

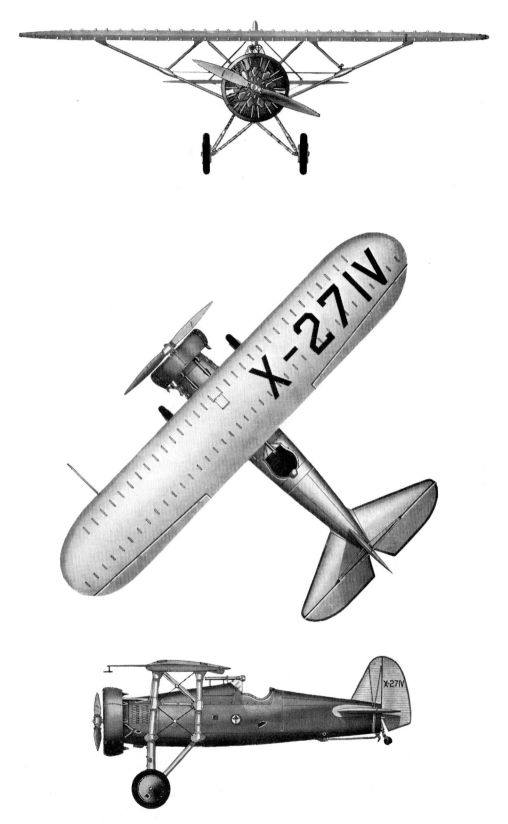

Fuselage and cowling, medium blue. Wing and all struts, chrome-yellow. Tail unit chrome, with medium blue trim lines. All lettering black.

GRUMMAN FF-1

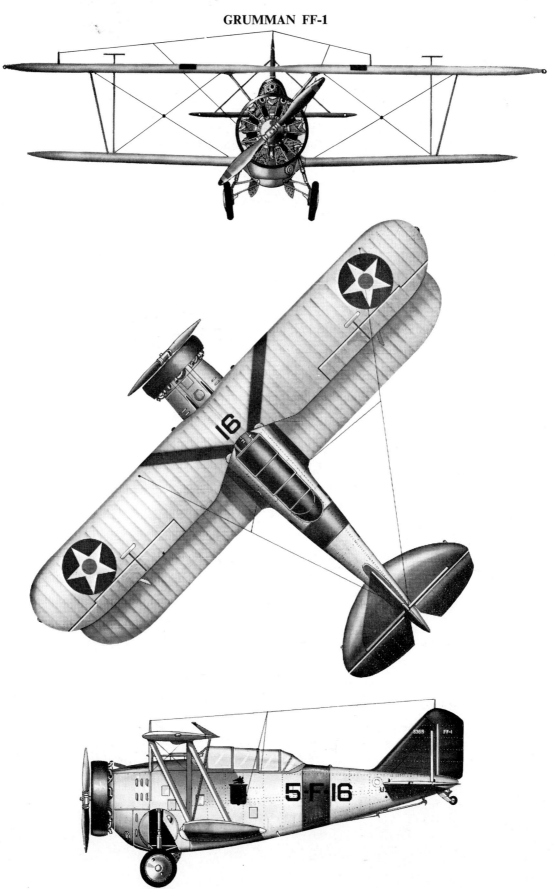

Fuselage Navy grey, wings silver with chrome top wing upper surface. Fuselage band, cowl and chevron, yellow. Tail unit, true blue.

BERLINER-JOYCE XF3J-1

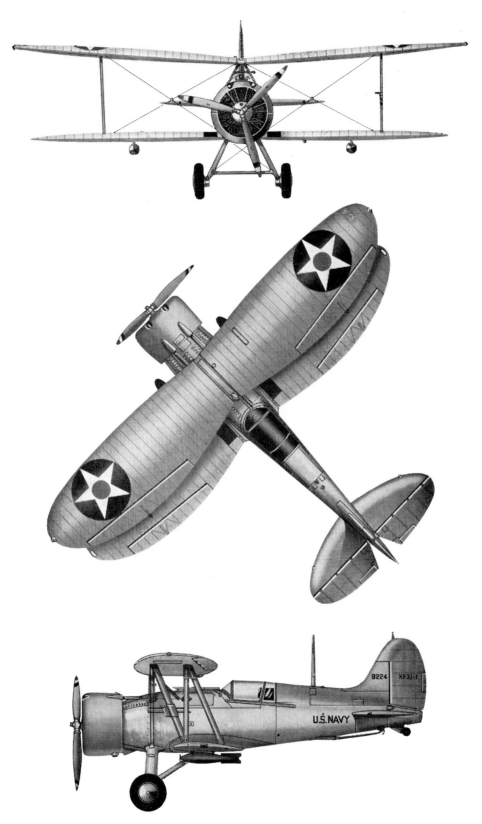

Fuselage Navy grey. Wings, elevators and rudder, silver. Top of upper mainplane chrome yellow. Lettering and wing-walks black.

CURTISS F11C-2

Overall silver. Upper surface of top wing chrome yellow. Fuselage panels, turtle-decking, tank, spats and lower cowl Navy grey. Upper cowl, chevron and tail unit red.

BOEING F4B-4

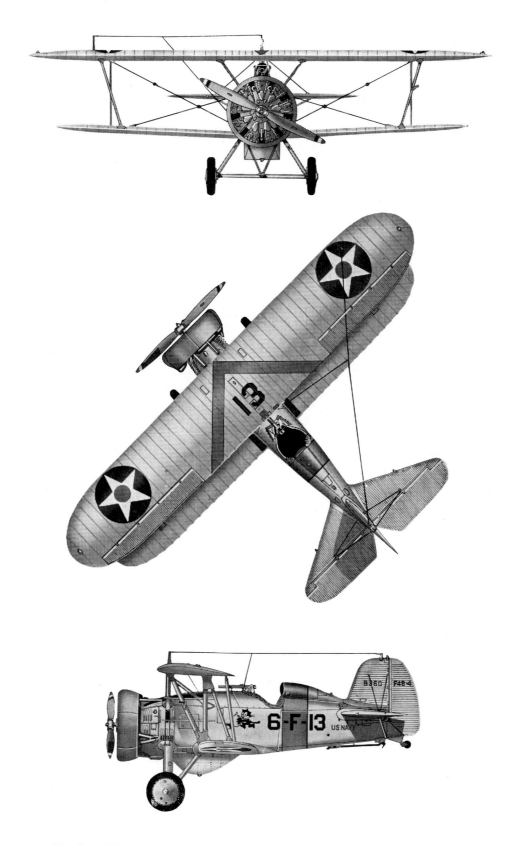

Fuselage Navy grey. Wings silver. Top wing upper surface chrome yellow. Cowling, chevron and fuselage band, green with black trim. Tail unit white.

CURTISS F9C-2

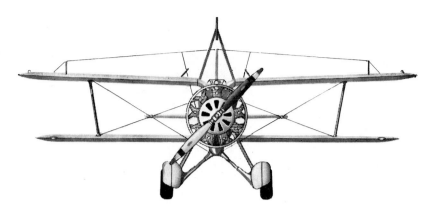

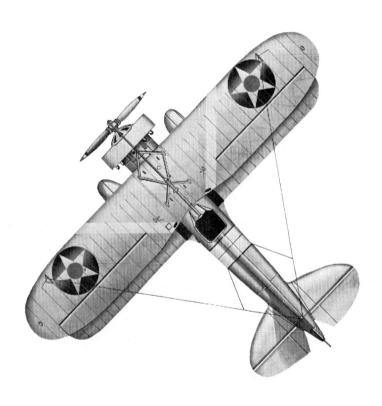

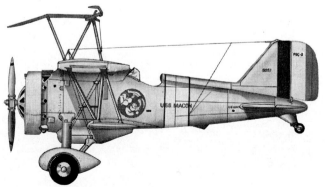

Fuselage and tail unit Navy grey. Wings silver, upper surface of top wing chrome yellow. Fuselage band, wing chevron, cowling, engine plate and spats, white.

DOUGLAS XFD-1

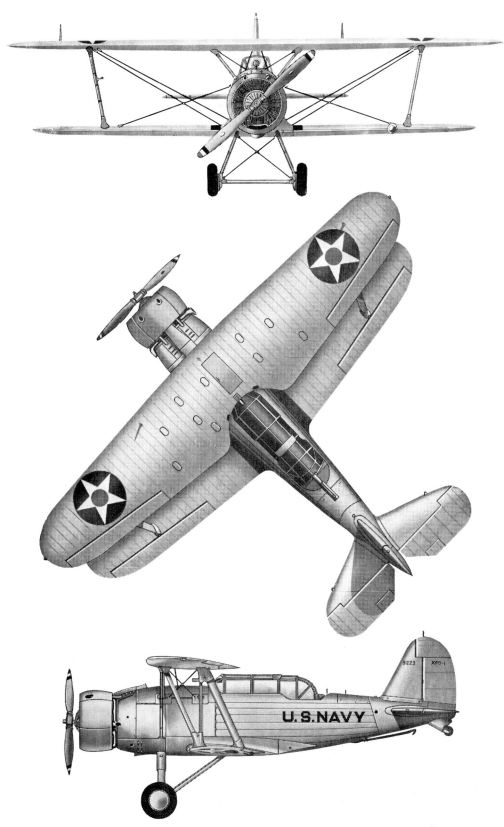

All metal areas, Navy grey. Wings, fuselage fabric area, elevators and rudder, silver. Top of upper mainplane chrome yellow. Lettering black.

CURTISS BF2C-1

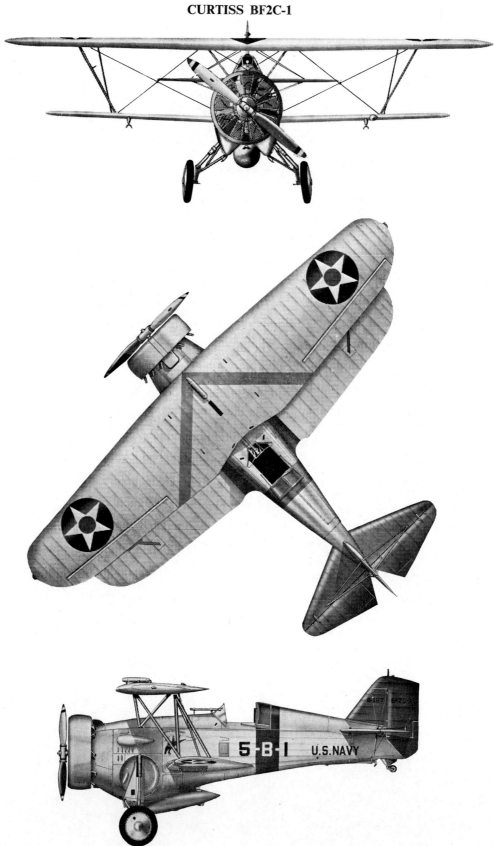

Overall silver. Top wing upper surface chrome. Fuselage band, chevron and cowling red. Fuselage forward panels, turtle decking, wheel disc and fuel tank, Navy grey. Tail unit green. All lettering black, except 'B', in white.

BOEING XF7B-1

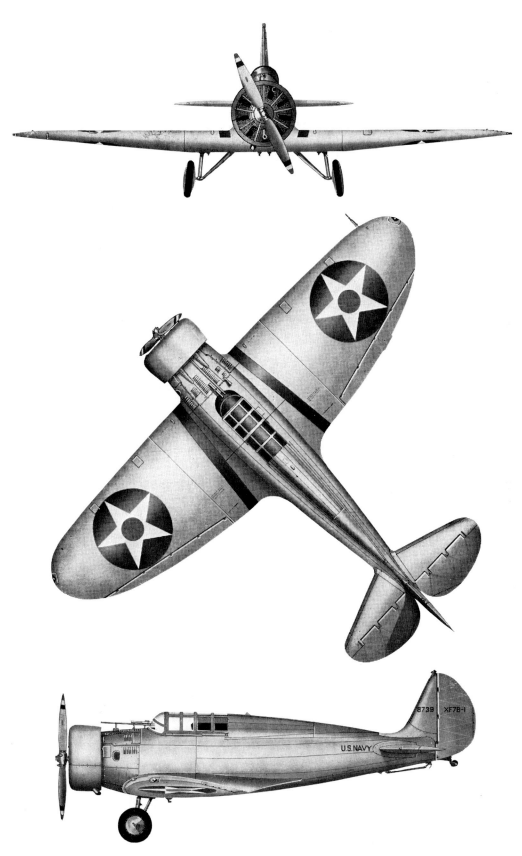

Overall Navy grey. Black lettering and wing-walks.

CURTISS XF12C-1

All metal natural finish. All control surfaces silver. Cowling highly polished. Lettering black.

GRUMMAN F2F-1

Fuselage and interplane struts Navy grey. Wings silver. Upper surface of top wing chrome yellow. Upper half of cowling, wing chevron and tail unit red.

NORTHROP XFT-1

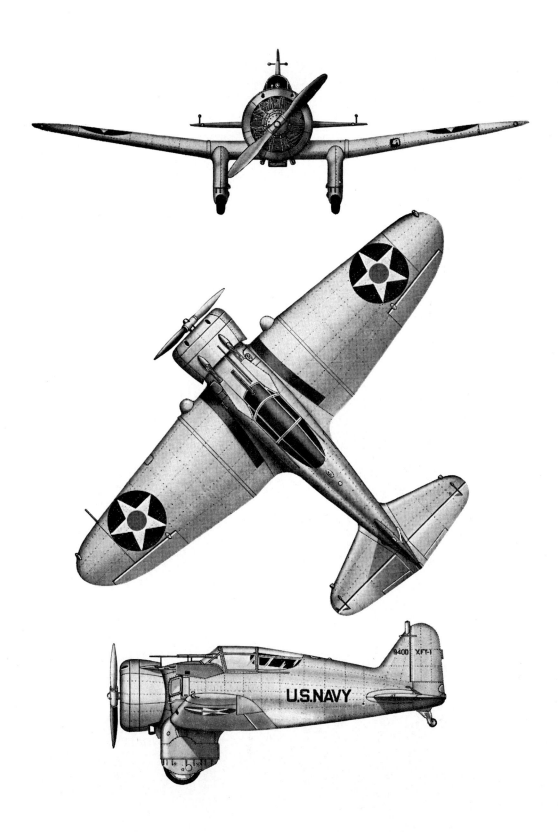

Overall Navy grey, with black lettering and wing walks.

CURTISS XF13C-1

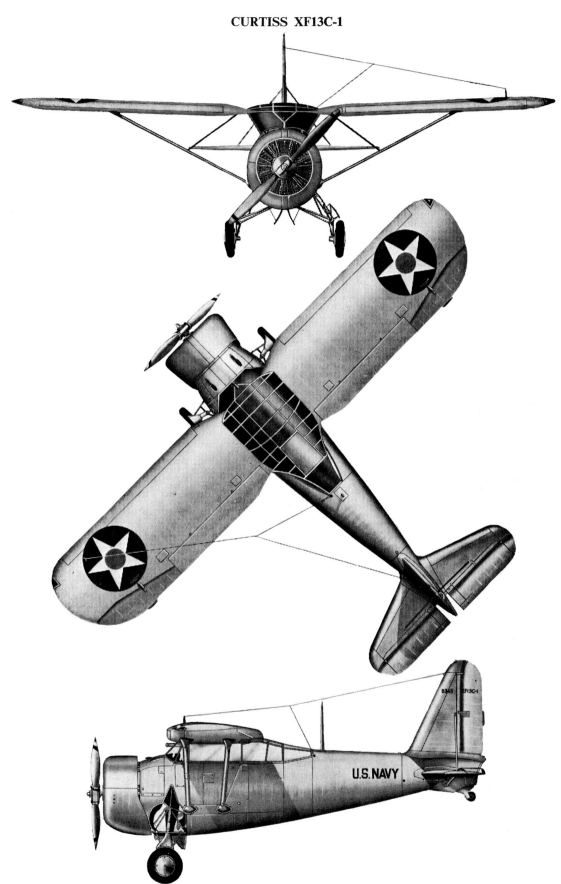

Fuselage, fin and metal parts, Navy grey. Wings and remainder of tail unit silver. Top of wing chrome.

GRUMMAN F3F-1

Fuselage Navy grey. Wings silver, top plane upper surface chrome yellow. Fuselage band, cowling and chevron true blue with white trim. Tail unit black. Codes black or white for contrast.

BREWSTER F2A-1

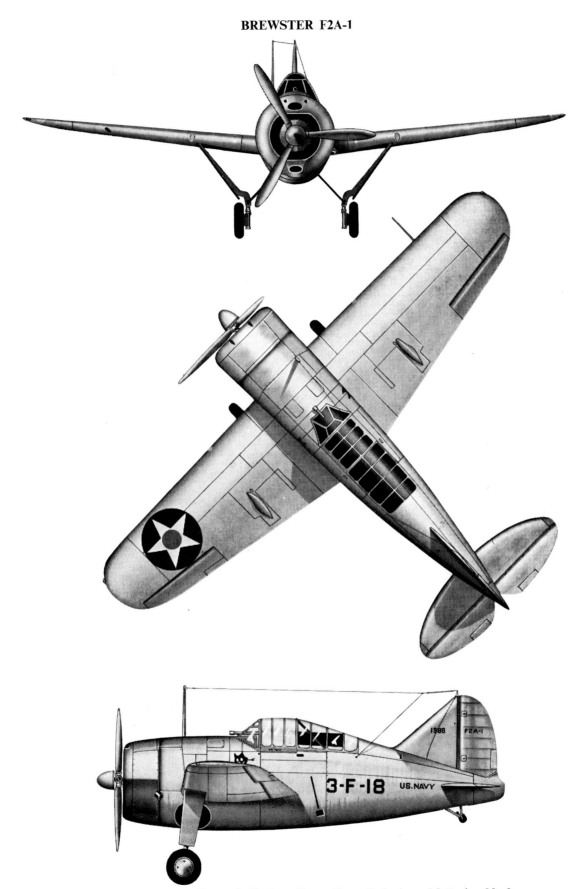

Overall early grey. Lower half of cowling yellow. Insignia and lettering black.

GRUMMAN F3F-2

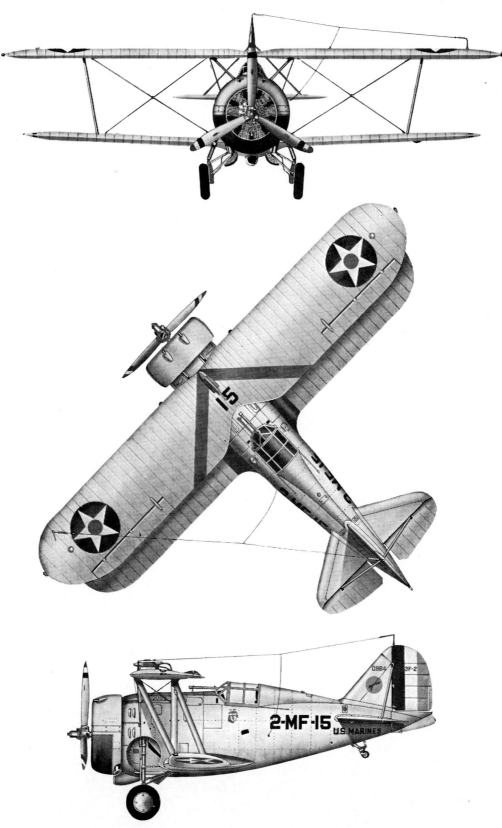

Fuselage and tail surfaces, natural aluminium. Silver wings, top surface of upper mainplane chrome yellow. Lower half of cowl, and wing chevron green. All codes black.

BELL XFL-1

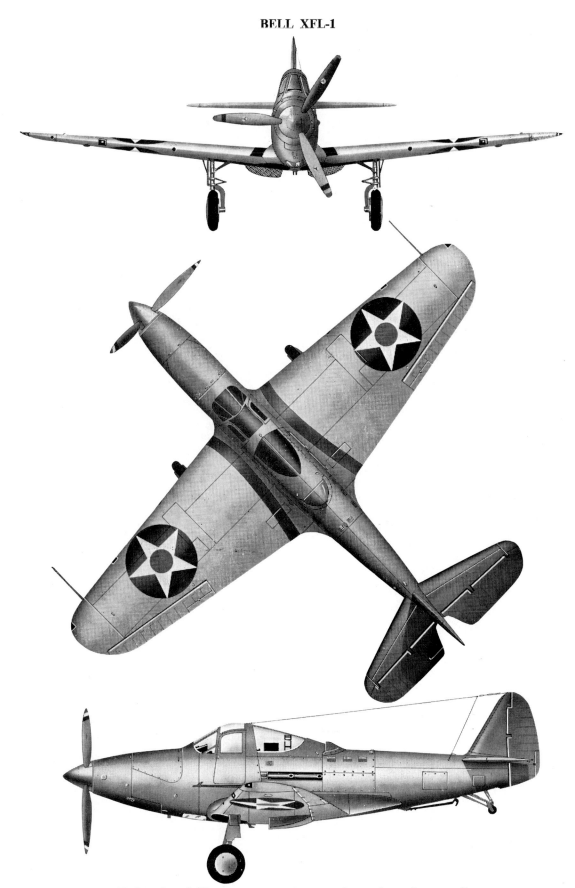

Entire aircraft Navy grey, except upper wing surface, chrome yellow.

GRUMMAN F4F-3

Lt. E. H. ('Butch') O'Hare's aircraft of VF-3. Upper surfaces dark blue, lower surfaces very pale blue, almost white.

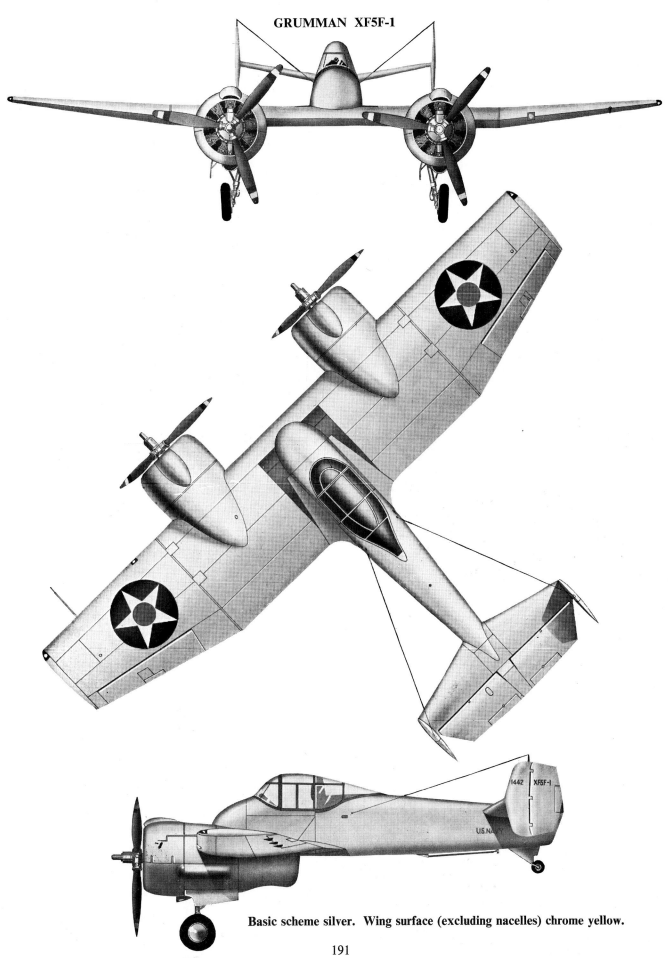

GRUMMAN XF5F-1

Basic scheme silver. Wing surface (excluding nacelles) chrome yellow.

CHANCE VOUGHT F4U-1D

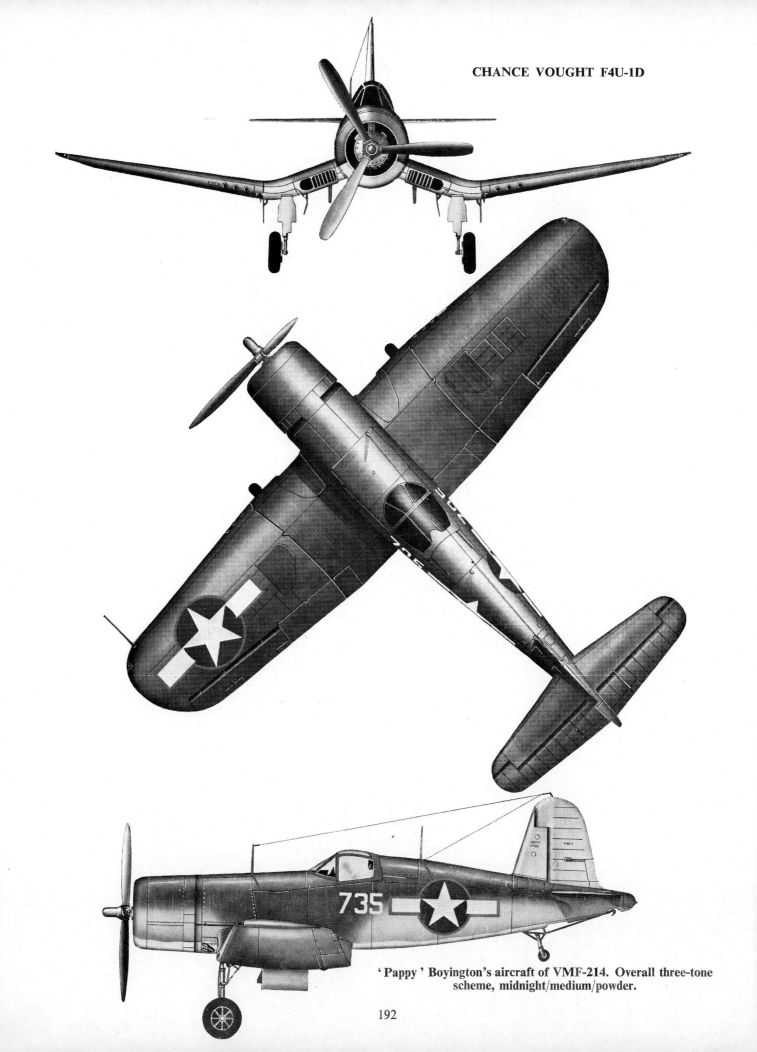

'Pappy' Boyington's aircraft of VMF-214. Overall three-tone scheme, midnight/medium/powder.

GRUMMAN F6F-3

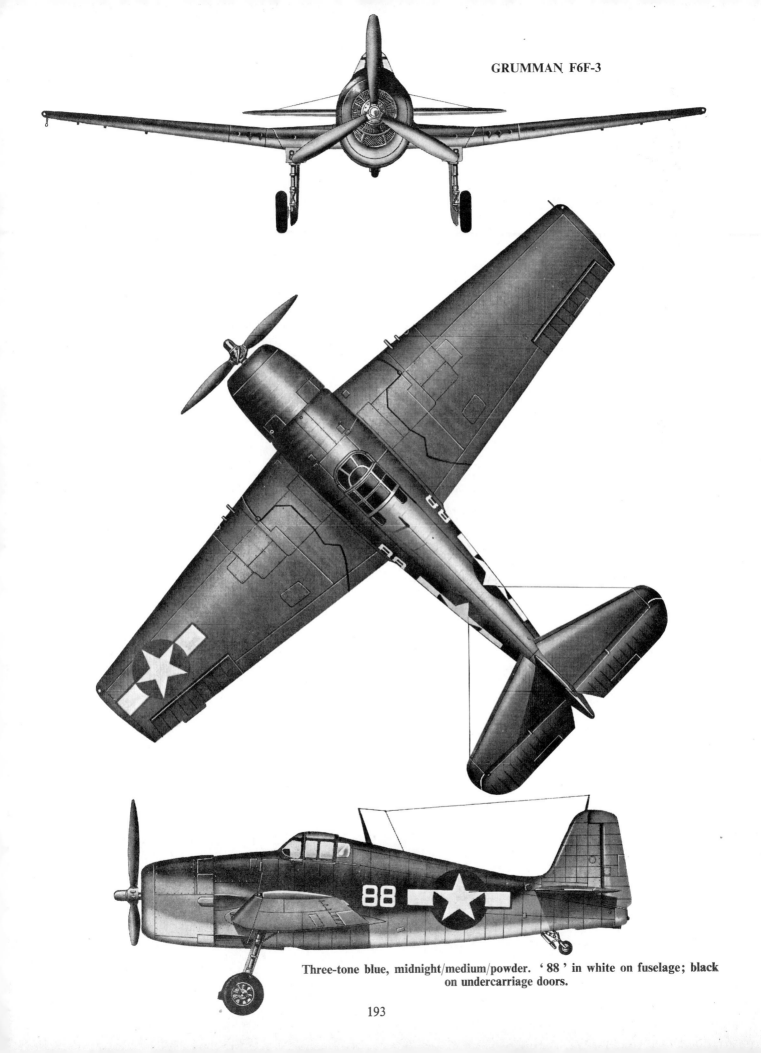

Three-tone blue, midnight/medium/powder. '88' in white on fuselage; black on undercarriage doors.

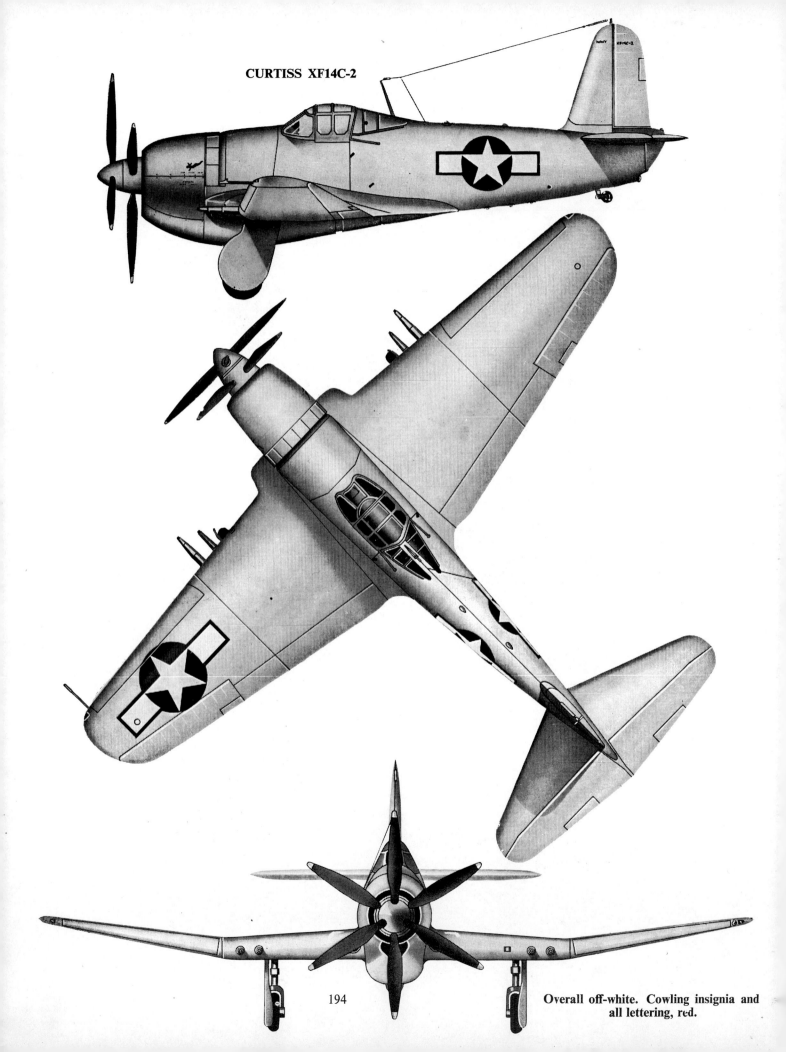

GRUMMAN F8F-1

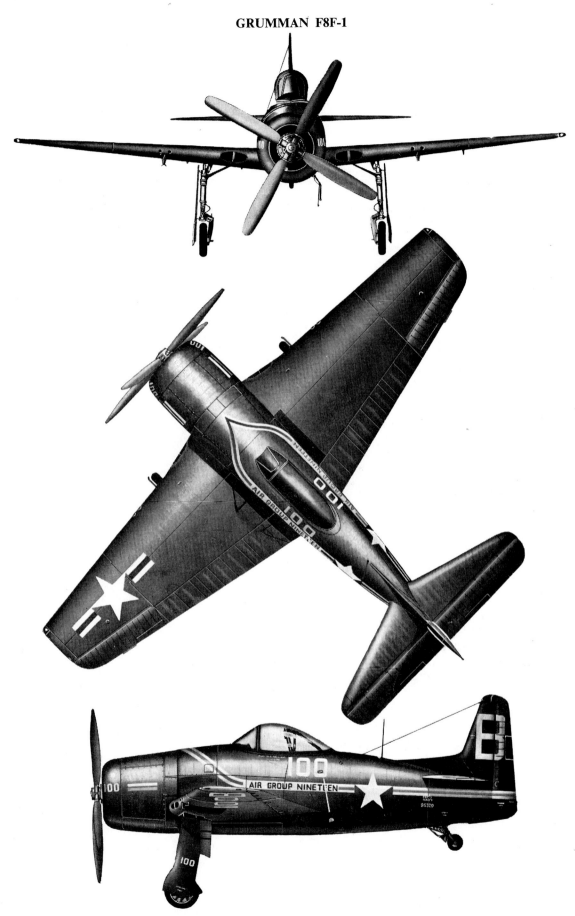

Aircraft of Cmdr. M. E. Cash Jr. Overall colour scheme midnight blue.
Fuselage stripe red and white, with blue lettering. Codes, etc. white.

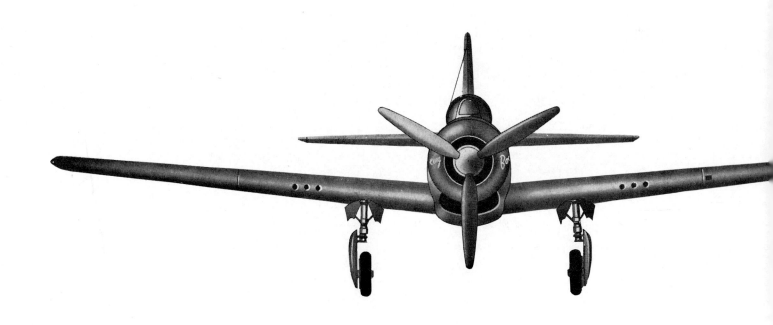

BOEING XF8B-1

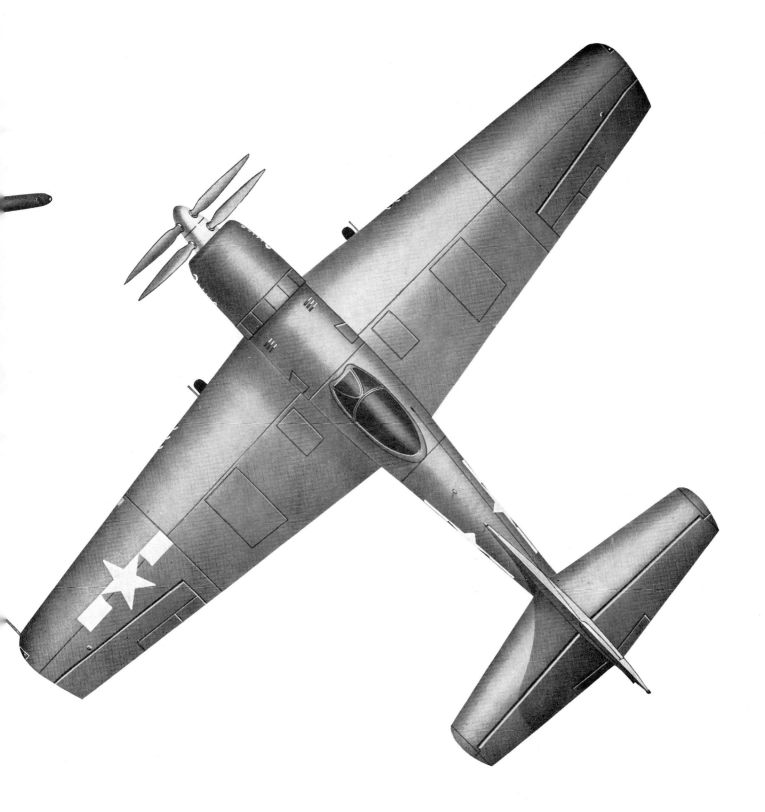

Overall glossy dark blue finish, with white lettering.

RYAN FR-1

Overall glossy dark blue. Lettering on fuselage and wings, chrome yellow. Tail letters white.

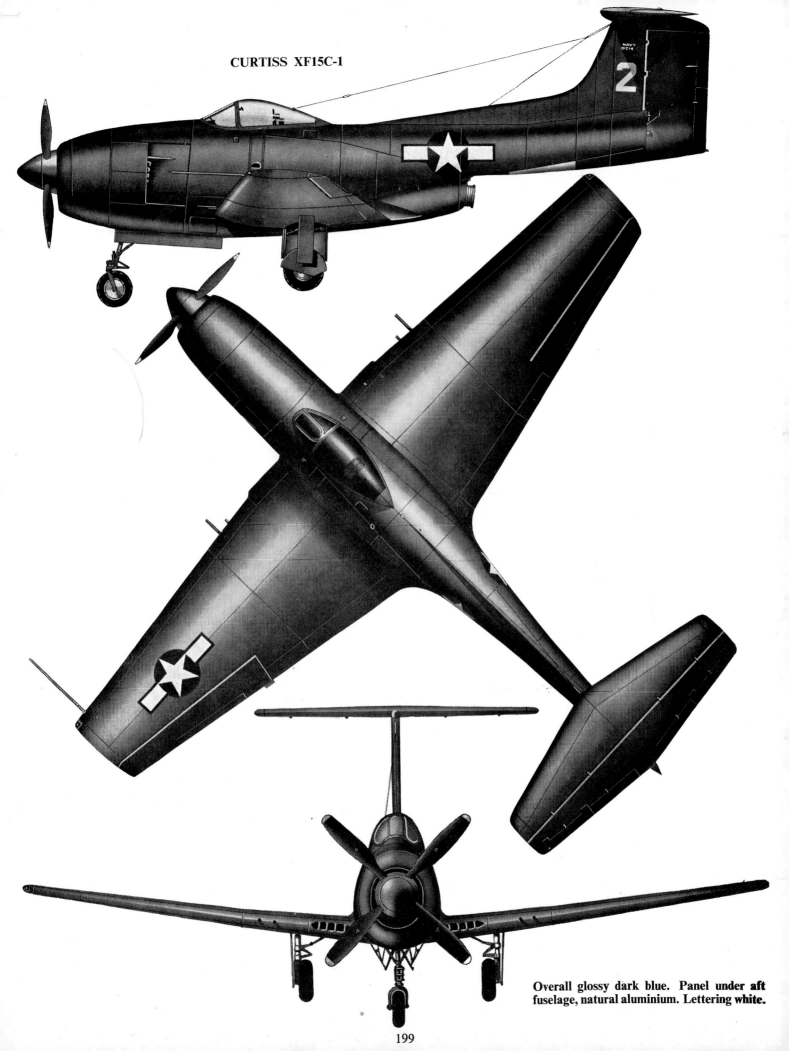

RYAN XF2R-1

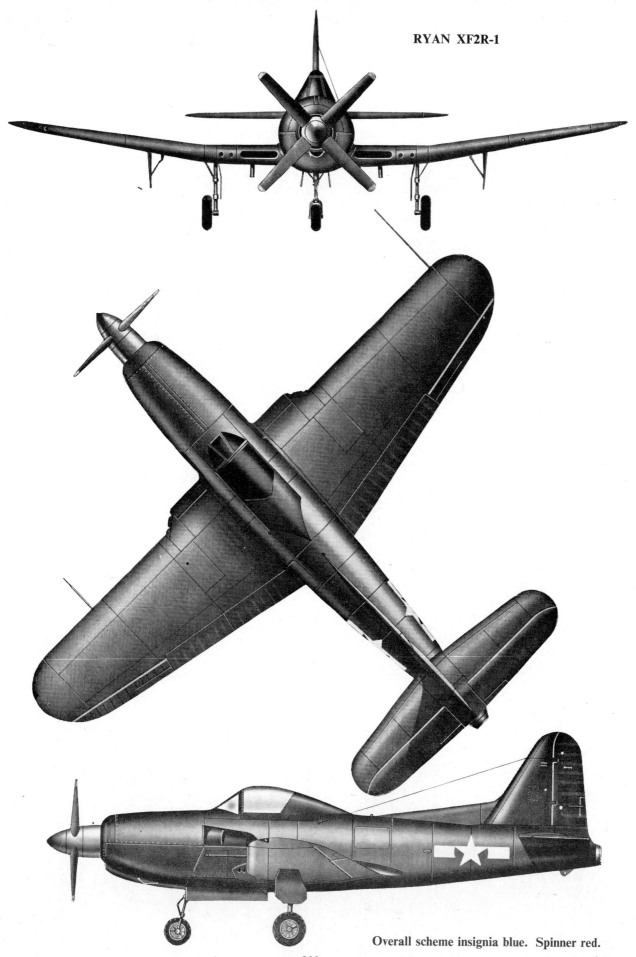

Overall scheme insignia blue. Spinner red.

McDONNELL FH-1

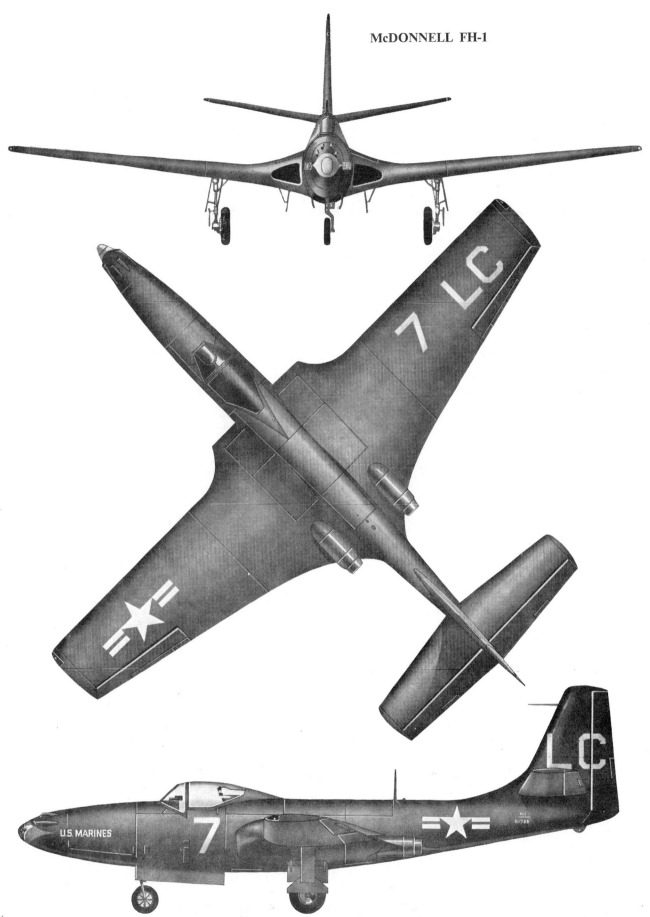

Squadron VMF-122. Overall scheme glossy dark blue. Lettering and trim in white.

VOUGHT XF5U-1

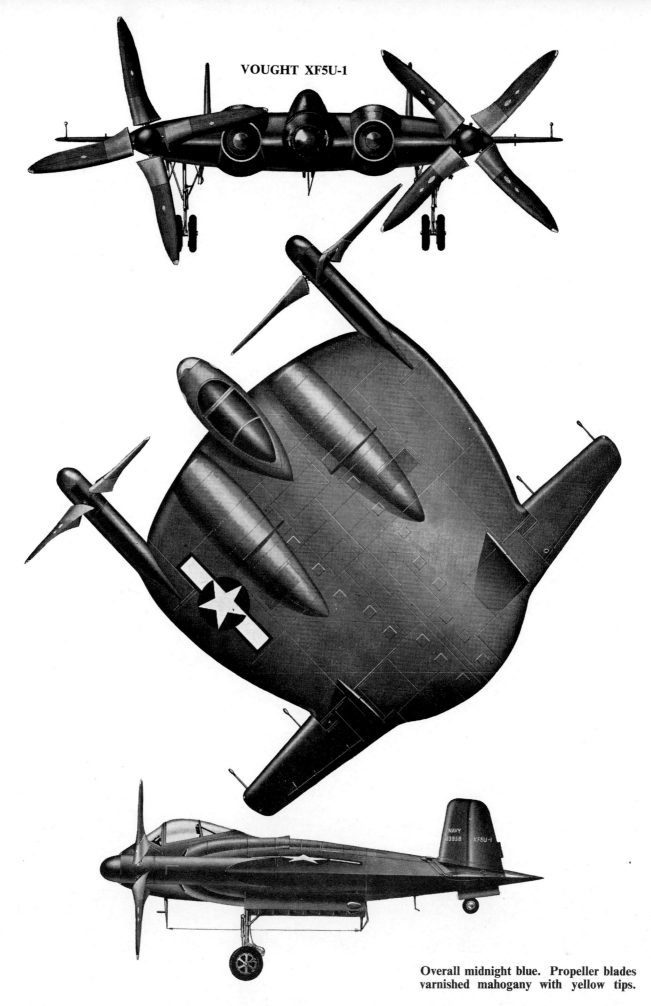

Overall midnight blue. Propeller blades varnished mahogany with yellow tips.

NORTH AMERICAN FJ-1

Cmdr. Aurand's aircraft, VF-5A. Overall insignia blue. White lettering.

VOUGHT F6U-1

Overall scheme glossy dark blue. All lettering white. Anti-glare panel olive-drab.

VOUGHT F4U-5N

Lt. Guy Bordelon's aircraft of VC-3, in Korea. Basic overall scheme midnight blue. All codes, etc. pale blue. Aircraft name and victory stars in red. Mission bombs, cameo insignia and radome nose, white.

GRUMMAN F7F-3N

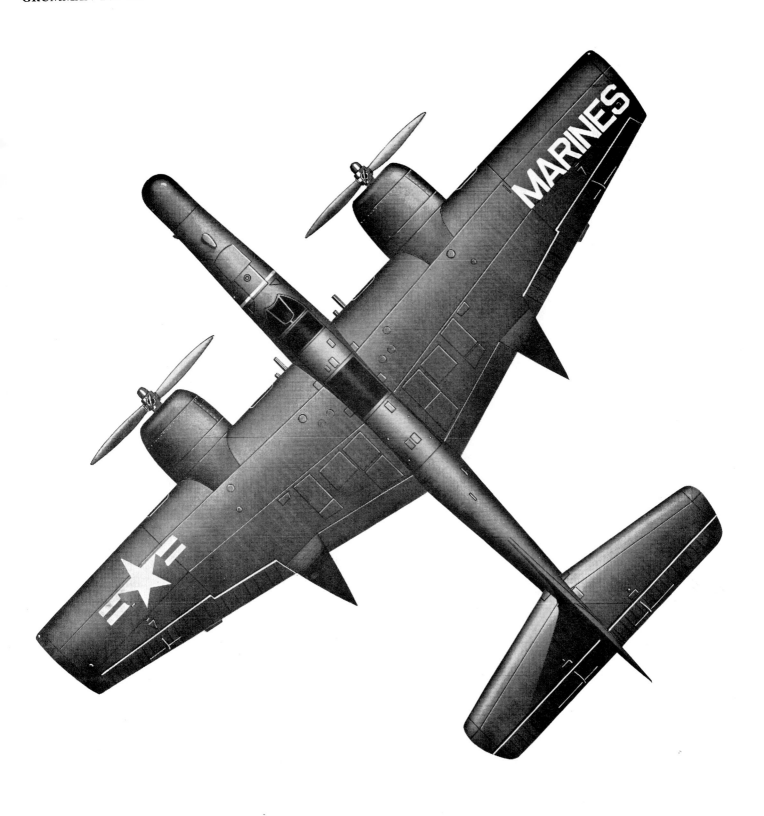

Overall glossy dark blue. All lettering white.

DOUGLAS F3D-2

Overall flat black finish, with all lettering in dull red. Standard insignia.

GRUMMAN F9F-2

Overall midnight blue. White trim and letters. Nose flash red.

McDONNELL F2H-2

Overall glossy dark blue. White lettering.
All surface tips and tank flash yellow.

CONVAIR XFY-1

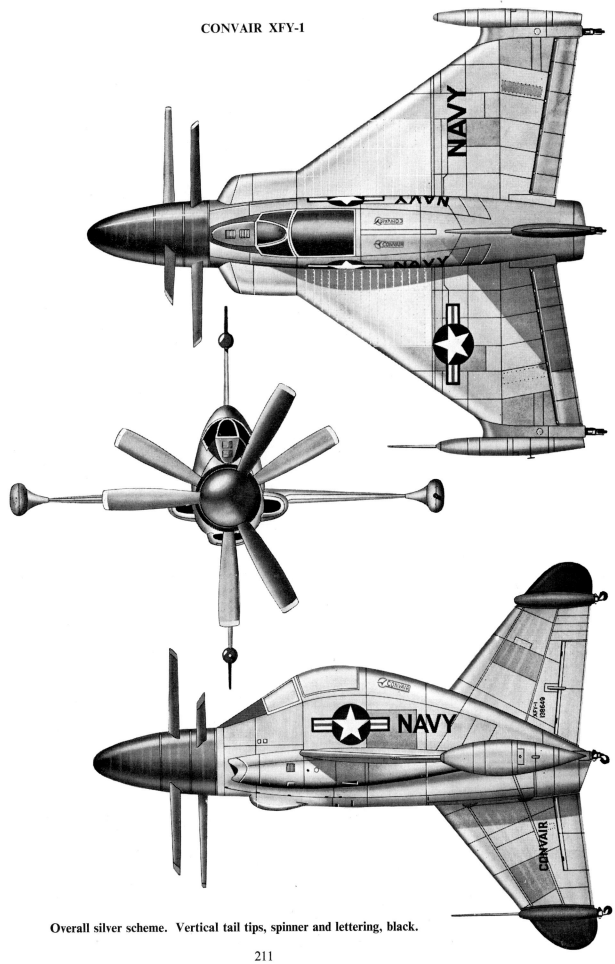

Overall silver scheme. Vertical tail tips, spinner and lettering, black.

CONVAIR XF2Y-1

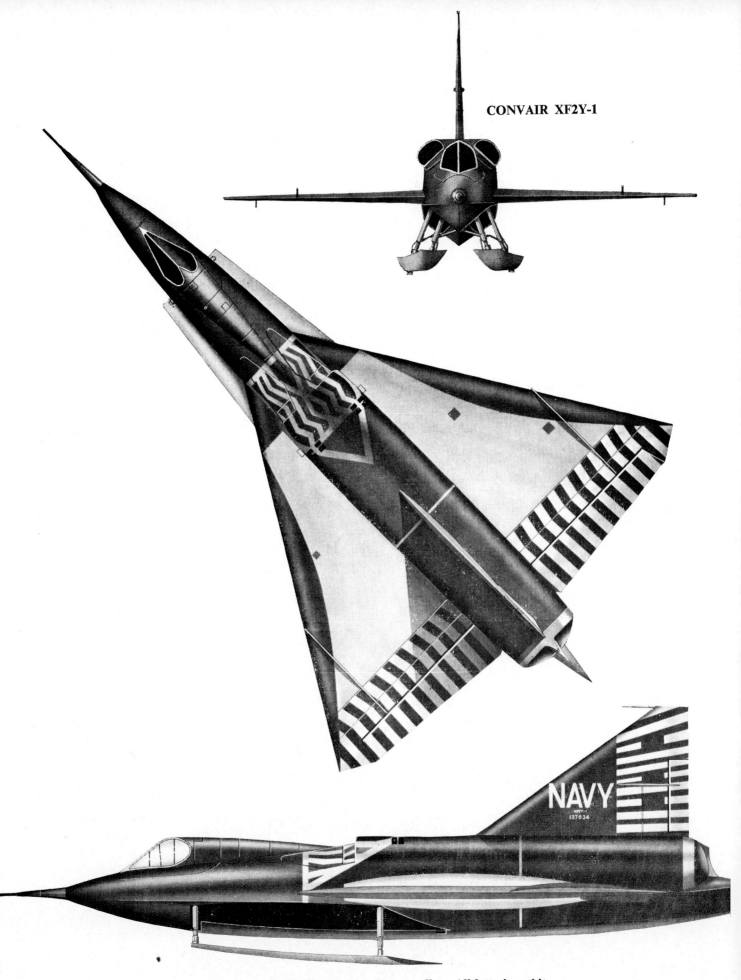

Overall scheme midnight blue, and chrome yellow. All lettering white.

GRUMMAN F9F-8

Upper surfaces grey. Lower surfaces glossy white. Black lettering.

GRUMMAN XF10F-1

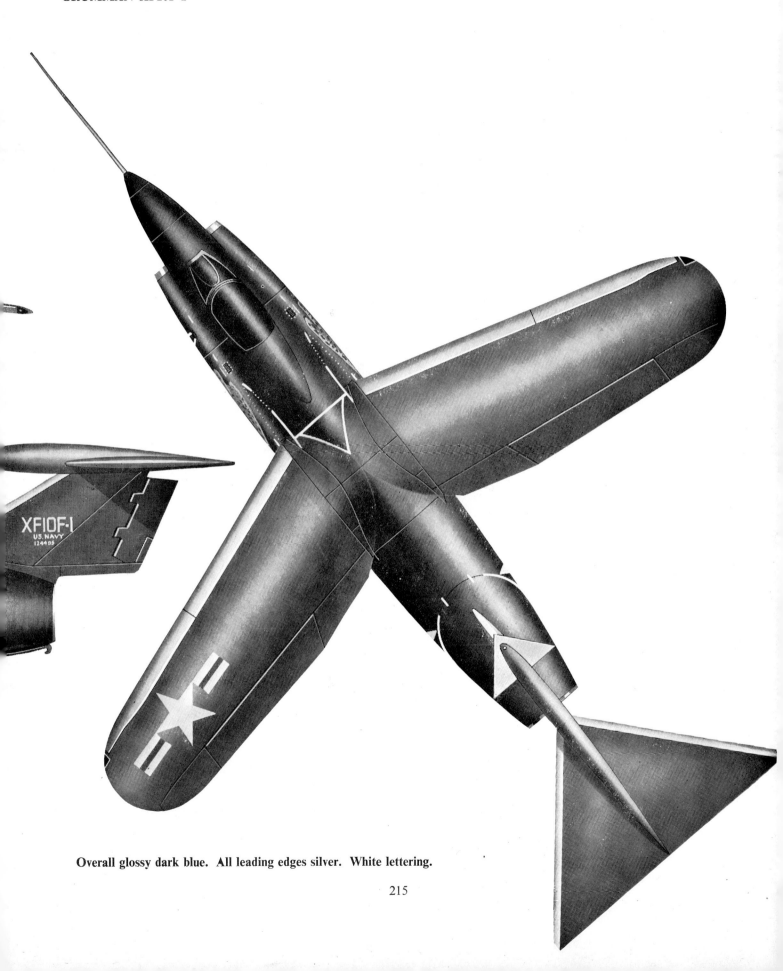

Overall glossy dark blue. All leading edges silver. White lettering.

NORTH AMERICAN FJ-3

Overall glossy dark blue. All leading edges natural metal.
All lettering white.

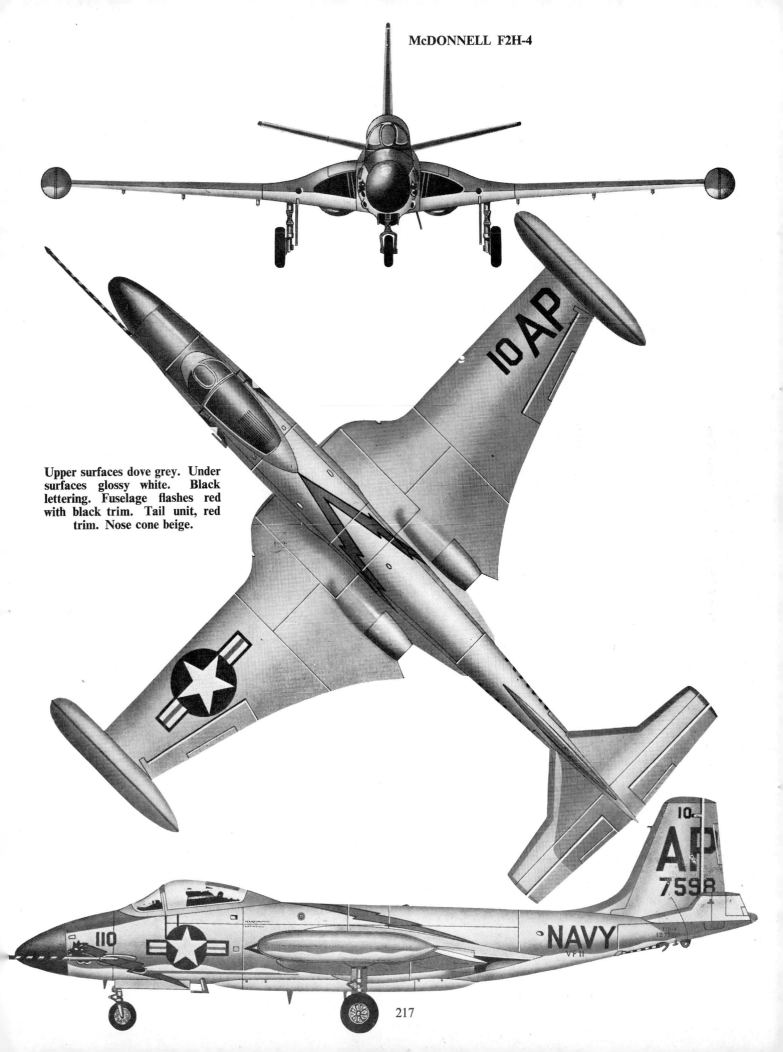

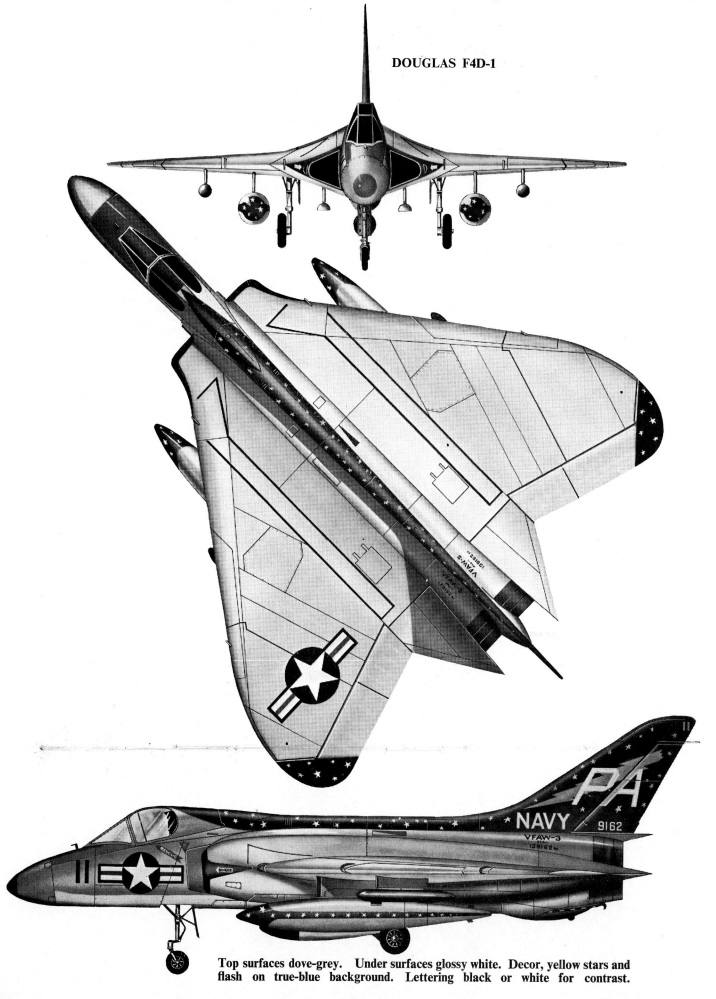

DOUGLAS F4D-1

Top surfaces dove-grey. Under surfaces glossy white. Decor, yellow stars and flash on true-blue background. Lettering black or white for contrast.

VOUGHT F7U-3

Overall natural aluminium. Lettering, wing-walks and tip of vertical tail, black. Yellow tail insignia.

NORTH AMERICAN FJ-4

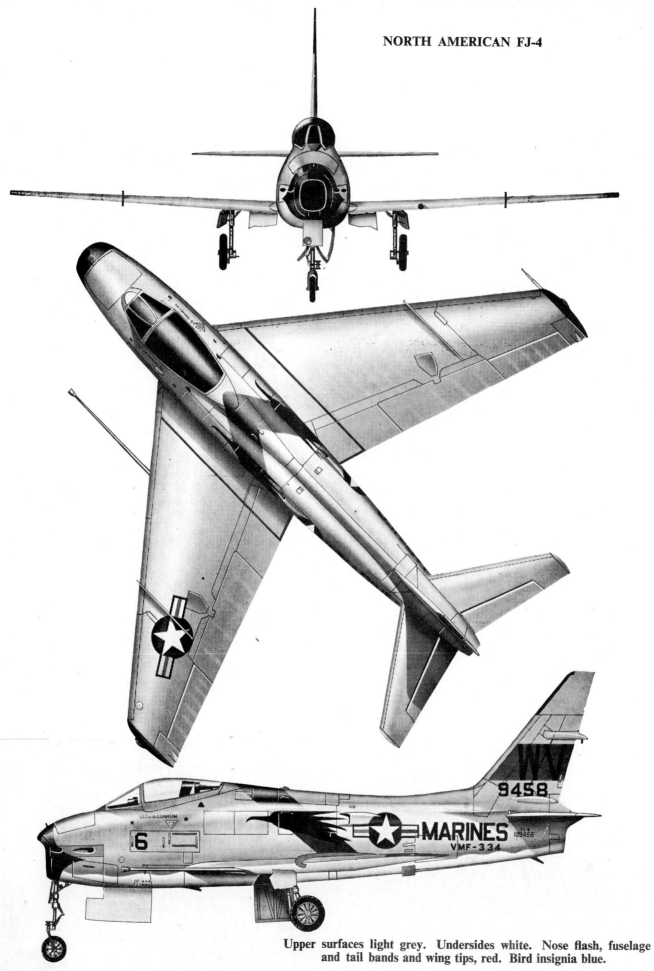

Upper surfaces light grey. Undersides white. Nose flash, fuselage and tail bands and wing tips, red. Bird insignia blue.

GRUMMAN F11F-1

Upper surfaces dove grey. Undersides white. Rudder and fin tip blue. Nose cone brown, 'lips' yellow. All other trim red. Lettering black.

McDONNELL F3H-2N

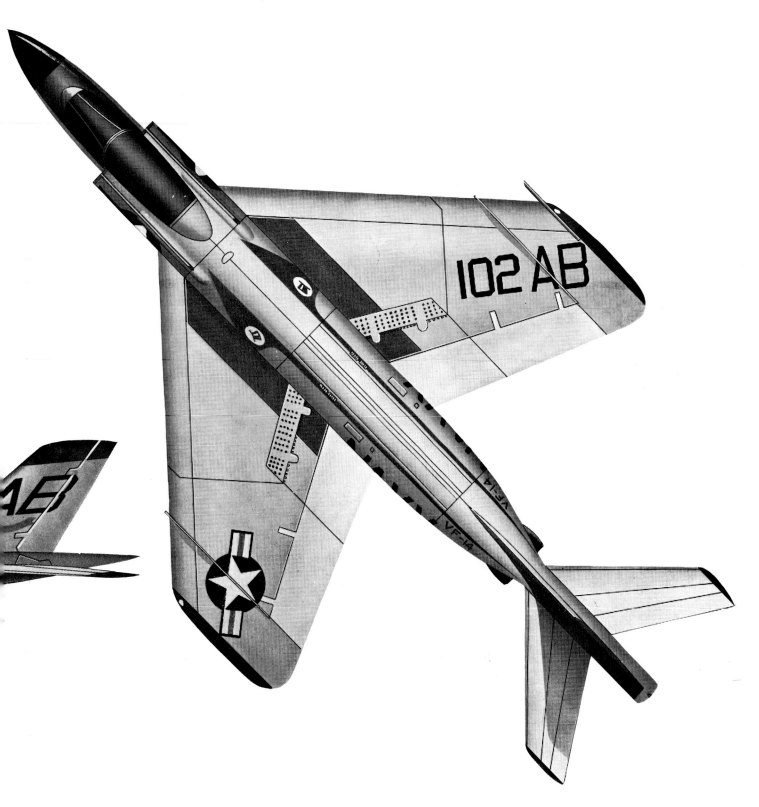

Upper surfaces dove grey. Undersides, ailerons and horizontal tail surfaces, glossy white. Rudder and wing tips and fuselage arrow, true blue.

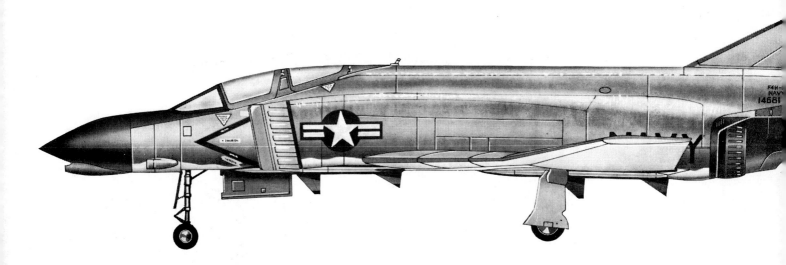

McDONNELL F4H-1

Upper surfaces dove grey. Under surfaces glossy white. Lettering black. Nose cone beige. Ailerons overall white. Intake warning ' vee ', red.

A representative selection of Navy and

228

Marine Corps fighter-unit insignia

FROM WAR TO WAR

Top to bottom, left: Curtiss TR-1 and pilot, winners of the 1922 Curtiss Marine Trophy Race; Eberhart XFG-1 in landplane [upper] and floatplane [lower] forms; Boeing F4B-4 with an F4B-3 rudder. \A Grumman F3F-2 in early wartime colour scheme. Right: Dayton-Wright DH-4, A-3280 of U.S.M.C., France 1918; Naval Aircraft Factory TF-1; Curtiss F6C-3 of VF-1M at Quantico, 1927; Boeing F2B-1, A-7440, in VF-1 markings; Curtiss BF2C-1, 9607, at Buffalo, N.Y. in 1936.

RACERS AND CHASERS

Top to bottom, left: Wright F3W-1 'Apache'; Wright F2W-1 as flown in the 1923 St. Louis Races; Curtiss R3C-2 floatplane for Schneider Trophy Races; Chance Vought UO-1 of VO-6; Chance Vought FU-2 was FU-1 modified for utility use on U.S.S. Langley. Right: Curtiss CR-3 racer for Pulitzer and Schneider Trophy Races; Curtiss R3C-1, 6778, with Lt. Al Williams in Pulitzer Race; Thomas-Morse MB-7, to U.S.M.C., ex-A.S.64373; Curtiss F6C-1, Hawk of VF-2; Douglas XFD-1, two seat fighter.

PACIFIC PURSUITS

Left, top to bottom: Early Grumman F4F Wildcats and Chance Vought Corsairs on a strike mission. Corsair using J.A.T.O. on take-off from carrier. F4U-1 Corsairs taxi out from Pacific island dispersal. Grumman Hellcat lines up for deck take-off. Right, top to bottom: Grumman Wildcats parked on Guadalcanal airfield. Lt. Ira Kepford's Corsair, decorated for sixteen victories. First Marine Corps Corsairs operating from Munda, only one week after the airstrip was captured. Hellcats on hangar-deck of U.S.S. Bennington. *Hellcat which landed in flames is sprayed with foam hoses.*

ROUNDEL REVIEW

Left, top to bottom: Corsair I, serial JT104, with early cockpit canopy. Later models had modified semi-bubble canopy as seen here on New Zealand Corsair, serial NZ5552 over field at Ardmore, in 1945. Grumman Hellcat displaying small South East Asia Command roundels is 'waved off'. As one Hellcat is successfully arrested, another prepares to land. Right, top to bottom: R.A.F. Brewster Buffalo, originally destined for Belgium. Grumman F4F supplied as Martlet I. Later F4F-3A came as Martlet III. Royal Navy Corsair gets 'chop' signal. Hellcat FN327 was one of first batch for Royal Navy.

MAINLY EXPERIMENTAL

Top to bottom, left: Curtiss F7C-1 used to evaluate leading-edge slats; Curtiss OC-1, re-designated from F8C-1 [compare sign with rudder]; Curtiss XF8C-4 in pre-production configuration; Curtiss XF11C-3 of 1933; Grumman XF4F-3S Wildcat 4038 in floatplane experiments. Right: Curtiss XF8C-2, the original 'Helldiver'. Curtiss OC-2 originally designated F8C-3; Curtiss F8C-5, Helldiver; Grumman XF3F-1 on final Navy trials; Bell L-39-2 swept wing Kingcobra.

MISCELLANEOUS MONOPLANES

Top to bottom, left: Ryan FR-1 Fireballs of VF-66; Grumman XF5F-1 Skyrocket with folding wings; Grumman F9F-6 Cougar of VMF-10; Convair YF2Y-1 Sea Dart; Chance Vought F7U-3 Cutlass—the fourteenth production aircraft under test. Right: Chance Vought XF4U-3B Corsair, produced as F4U-4; Chance Vought XF5U-1 Flapjack; Grumman F9F-5 Panther; North American P-51A and P-51H Mustangs, which were evaluated by U.S. Navy.

TABLE OF UNITED STATES

Abbreviations: Conv = Converted, Deliv = Delivered, FF = First Flight, Bi = Biplane,
N.B.—Quantities given in brackets indicate modified airframes previously recorded

Type No.	Firm	Crew and Type	Significant date	Engine h.p.*	Engine Type	Top Speed (m.p.h.)	Wing Span (ft. in.)	Length (ft. in.)	Loaded Weight (lb.)	Quan.
XFA-1	Atlantic (Fokker)	1 Bi	Deliv 5 Mar. 32	403	P. & W. R-985	170	25 6	20 6	2525	1
XFL-1	Bell	1 LWM	FF 13 May 40	1150	Allison XV-1710-6	335	35 0	29 9	6213	1
F2L-1K	Bell	1 LWM	Deliv in 46	1200	Allison V-1710-85	385	34 0	30 2	8250	2
XFJ-1	Berliner-Joyce	1 Bi	Deliv May 30	450	P. & W. R-1340-C	177	28 0	19 11	2797	1
XFJ-2	Berliner-Joyce	1 Bi	Re-deliv 22 May 31	450	P. & W. R-1340-92	193	28 0	19 11	2847	(1)
XF2J-1	Berliner-Joyce	2 Bi	Deliv May 33	625	Wright R-1510-92	193	36 0	28 10	4520	1
XF3J-1	Berliner-Joyce	1 Bi	Deliv Sep. 34	625	Wright XR-1510-26	209	29 0	22 11	4264	1
FB-1	Boeing 16	1 Bi	Deliv 1 Dec. 25	400	Curtiss D-12	167	32 0	23 5	2835	10
FB-2	Boeing	1 Bi	Deliv 8 Dec. 25	400	Curtiss D-12	164	32 0	23 5	3039	2
FB-3	Boeing 55	1 Bi	Deliv mid-26	510	Packard 1A-1500	165	32 0	23 5	3039	3
FB-4	Boeing 47	1 Bi	Conv 27	450	Wright P-1	157	32 0	22 11	2817	1
FB-5	Boeing	1 Bi	Deliv 10 Jan. 27	525	Packard 2A-1500	169	32 0	23 5	3249	27
FB-6	Boeing 54	1 Bi	Conv 27	450	P. & W. R-1340-B	159	32 0	22 10	2737	(1)
XF2B-1	Boeing 69	1 Bi	Deliv 12 Dec. 26	450	P. & W. R-1340-B	155	30 1	22 11	2670	1
F2B-1	Boeing 69B	1 Bi	Deliv Oct. 27	450	P. & W. R-1340-B	158	30 1	22 11	2830	32
XF3B-1	Boeing 74	1 Bi	Deliv 7 Mar. 27	450	P. & W. R-1340-B	157	30 1	22 9	2715	1
F3B-1	Boeing 74	1 Bi	Deliv 27 Oct. 27	450	P. & W. R-1340-80	157	33 0	24 10	2945	73
XF4B-1	Boeing 83/89	1 Bi	FF 25 June 28	500	P. & W. R-1340-C	169	30 0	20 1	2557	2
F4B-1	Boeing 99	1 Bi	FF 6 May 29	500	P. & W. R-1340-C	176	30 0	20 1	2724	27
F4B-2	Boeing 223	1 Bi	Deliv 2 Jan. 31	500	P. & W. R-1340-C	186	30 0	20 1	2800	46
F4B-3	Boeing 235	1 Bi	Deliv 24 Dec. 31	500	P. & W. R-1340-D	187	30 0	20 5	2989	21
F4B-4	Boeing 235	1 Bi	Deliv 28 July 32	550	P. & W. R-1340-16	184	30 0	20 5	3107	93
F4B-4A	Boeing 234	1 Bi	Conv 40	550	P. & W. R-1340-16	189	30 0	20 5	2300	23
XF5B-1	Boeing 205	1 HWM	Deliv 14 Feb. 30	485	P. & W. R-1340-B/C	183	30 6	21 0	2848	1
XF6B-1	Boeing 236	1 Bi	Deliv May 33	625	P. & W. R-1535-44	195	28 6	22 2	3704	1
XF7B-1	Boeing 273	1 LWM	FF 14 Sep. 33	550	P. & W. R-1340-30	233	32 0	27 7	3868	1
XF8B-1	Boeing 400	1 LWM	FF 27 Nov. 44	3250	P. & W. XR-4360-10	411	54 0	43 3	18000	3
XF2A-1	Brewster 139	1 LWM	FF Dec. 37	850	Wright XR-1820-22	277	35 0	25 6	4830	1
F2A-1	Brewster 239	1 LWM	Deliv Jun. 39	850	Wright R-1820-34	304	35 0	26 0	5040	11
XF2A-2	Brewster	1 LWM	Deliv July 39	1200	Wright R-1820-40	325	35 0	25 6	5395	(1)
F2A-2	Brewster 239	1 LWM	Deliv Sep. 40	1200	Wright R-1820-40	323	35 0	25 6	5945	43
F2A-3	Brewster 439	1 LWM	Deliv July 41	1200	Wright R-1820-40	321	35 0	26 4	6320	108
XF2A-4	Brewster	1 LWM	Reworked Sep. 41	1200	Wright R-1820-40	298	35 0	26 4	7150	(1)
F3A-1	Brewster	1 LWM	Deliv July 43	2000	P. & W. R-2800-8	417	41 0	33 4	12039	735
XFY-1	Convair	1 MWM	FF 1 Aug. 54	5800	Allison T-40A-A-14	500+	25 8	34 10	10000	2
XF2Y-1	Convair	1 LWM	FF 9 Apr. 53	3000*	2 × J34-WE-42	650+	30 6	41 2	22000	2
YF2Y-1	Convair	1 LWM	Deliv mid-54	6500*	2 × J46-WE-16	?	30 6	43 6	?	3
HA	Curtiss 16	2 Bi	FF 21 Mar. 18	380	Liberty 12	132	36 0	30 9	4012	3
18-T	Curtiss 15/15A	2 Tri	FF 5 July 18	400	Kirkham K-12	168	31 11	23 3	2864	2
TS-1	Curtiss 28/N.A.F.	1 Bi	Deliv 9 May 22	220	Lawrence J-1	124	25 0	22 1	1929	39
TS-2	Curtiss/N.A.F.	1 Bi	Deliv late 22	210	Aeromarine U-8-D	132	25 0	22 1	2136	2
TS-3	Curtiss/N.A.F.	1 Bi	Deliv early 23	180	Wright E-2	122	25 0	22 1	2105	2
F4C-1	Curtiss-Hall 34/39	1 Bi	FF 24 Sep. 24	200	Lawrence J-1	131	25 0	18 4	1656	2
F6C-1	Curtiss 34C	1 Bi	Deliv Sep. 25	400	Curtiss D-12	164	31 6	22 8	2802	9
F6C-2	Curtiss 34D	1 Bi	Conv late 25	400	Curtiss D-12	159	31 6	22 8	2871	(4)
F6C-3	Curtiss 34E	1 Bi	Deliv early 27	400	Curtiss D-12	153	31 6	22 10	2963	35
F6C-4	Curtiss 34H	1 Bi	FF Sep. 26	425	P. & W. R-1340	158	31 6	21 10	2785	31
XF6C-5	Curtiss	1 Bi	Conv 28	525	P. & W. R-1690	159	31 6	22 6	2960	(1)
XF6C-6	Curtiss	1 HWM	Conv June 30	600	Curtiss Conqueror	210	31 6	22 7	2831	(1)
XF6C-7	Curtiss	1 Bi	Conv 29	420	Ranger V-770	157	31 6	22 3	2950	(1)
XF7C-1	Curtiss 43	1 Bi	FF 28 Feb. 27	450	P. & W. R-1340-B	155	30 8	22 7	2892	1
F7C-1	Curtiss 43A	1 Bi	Deliv July 27	450	P. & W. R-1340-B	150	32 8	22 2	2782	16
XF8C-1	Curtiss 37D	2 Bi	Deliv Jan. 28	425	P. & W. R-1340	144	38 0	28 0	3918	2

* For jet engines figures given as pounds of static thrust. † Armament for Berliner-Joyce, Boeing and Curtiss designs, unless otherwise

NAVY AND MARINE CORPS FIGHTERS

Tri = Triplane, MWM, HWM, LWM = Mid, High and Low Wing Monoplane respectively
under another type or model, or quantity cancelled if so qualified in remarks

Serial Numbers	Remarks including unofficial or popular name†
8732 only	Built to same specification as Curtiss F9C, BuAer Design 96. Firm taken over by General Aviation Corp.
1588 only	'Airabonita'. Navalised version of U.S.A.A.C. P-39 'Airacobra'. Conventional undercarriage fitted.
91102-91103	'Airacobra'. Ex-U.S.A.A.F. P-39Q-5 No. 42-20807 and P-39Q-10 No. 42-19976 as target drones.
8288 only	Built to BuAer design with underslung lower wing. Modified into XFJ-2.
8288 only	XFJ-1 modified. Cowling, wheel spats and large spinner added. Utility use until late 1938.
8973 only	Two-seat fighter. Original version with open cockpits, had canopies fitted later.
9224 only	Last U.S.N. fixed u/c fighter, elliptical 'Butterfly' wings. Provision for carriage of 4 × 116 lb. G.P. bombs.
6884-6893	As Army Air Service PW-9 with navalised cockpit. 2 × ·300 m.gs. or 1 × ·300 + 1 × ·50 m.gs.
6894-6895	Modified FB-1s equipped for carriers to assess suitability. Convertible sea/land.
6897, 7089, 7090	FB-1 with engine change. L.P./F.P. 6897 crashed before delivery, 31 Dec. 1925. A7089-90 reverted to FB-1.
6896 only	Modified FB-1 with engine change. Convertible L.P./S.P. Equipped for carrier operations.
7101-7127	Production models of FB-3. First true carrier based VF type by Boeing.
6896 only	FB-4 with engine change. Early evaluation of Pratt & Whitney R-1300 series of engines.
7385 only	First fully-aerobatic VF service type. Radial engine permitted sustained inverted flight.
7424-7455	Production series based on FB design. Standard carrier VF type 1926/27. Delivery completed 24 Feb. 1928.
7674 only	Development of F2B. First type to utilise early flotation gear. Convertible sea/land.
7675-91, 7708-63	Much modified production model. Dural tail, slightly swept-back, straight chord upper wing.
8128-8129	Tested as private venture. Navy evaluation led to U.S.A.A.C. P-12. A-8129 Model 83, A-8128 Model 89.
8130-8156	Originally as fighter-bomber. No.8133 converted to executive aircraft for Assistant Secretary of Navy.
8613-39, 8791-809	As F4B-1 with cowling as standard. Frise ailerons. Later fitted with F4B-4 tails.
8891-8911	Production models of Boeing 218 private venture. First metal covered fuselage models for U.S.N. service.
See remarks	F4B-3 with enlarged vertical tail area and headrest. 72 delivered to U.S.N., 21 to U.S.M.C. BuAer Nos. 8912-20, 9009-53 and 9226-63 from production and 9719 built from spares at Marine Base, Quantico.
2489-2511	Surplus A.A.C. P-12Es converted to radio controlled target drones. Stripped of equipment.
8640 only	First fighter monoplane fully tested by U.S.N. Basically U.S.A.A.C. XP-15 with arrester hook.
8975 only	Equal span wings. Metal covered fuselage. First fighter-bomber initially designed as such. Later as XBFB-1.
9378 only	First low wing fighter type fully tested by U.S.N. Shipboard variant of U.S.A.A.C. YP-29A.
57984-57986	All-purpose fighter-bomber. Internal bomb bay. No.57985 to Army Air Force in 1946.
0451 only	'Buffalo'. First VF monoplane in service. Standard armament, provision for addition of 2 × ·50 m.gs. in wings.
1386-1396	'Buffalo'. Original order for 54. Eleven delivered as F2A-1, remainder delivered as F2A-2.
0451 only	'Buffalo'. Modified XF2A-2 with engine change.
1397-1439	'Buffalo'. Production of improved XF2A-2. Some fitted full armament of 4 × ·50 m.gs. Released to Finland.
1516-1623	'Buffalo'. Heaviest wing loading of any U.S.N. fighter to date. Modified to fighter-trainers.
1516	'Buffalo'. High altitude, pressurised cabin experiment with original F2A-3 model. Project abandoned.
See remarks	'Corsair'. As Vought F4U-1 built by Brewster. BuAer Nos. 04515-4774, 08550-8797, 11067-11293 includes 366 to U.K. as Corsair III. 369 delivered to U.S.N.
138649-138650	'Pogo'. Made first free VTO flight. N.B. T-40A unit = 2 × T-38 units. Not armed. 138650 was not flown.
137634-137635	'Sea Dart'. First fighter with retractable hydro-ski u/c. 4 × 20 mm. cannons and/or air-to-air rockets.
135763-135765	'Sea Dart'. Service evaluation of XF2Y-1 with engine change. BuAer No. 137635 changed to 135762 and aircraft redesignated YF2Y-1, but crashed on 4 Nov. 1954. Nos. 135763-5 still in storage 1962.
2278, 4110, 4111	'Dunkirk Fighter'. First U.S.N. Fighter. Floatplane. 2 fixed + 2 Scarff-mounted ·300 guns planned.
3325-3326	'Kirkham' or 'Wasp'. No.3326 held alt. record 1919. Planned armament as HA. Span increased to 40′ 7″.
6248-70, 6300-15	First fighter used in sqdn. strength. N.A.F. designed. First radial-engined fighter for U.S.N.⎫ Some
6446-6447	As TS-1 with engine change. Built under contract by Curtiss. Convertible wheels/floats. ⎬ re-designated
6448-6449	As TS-1 with engine change. Re-designated as racers, TR-3 and 3A. ⎭ TR for racing
6689-6690	Based on TS-1 design. All-metal lightweight construction for comparative tests with TS series.
6968-6976	'Hawk'. As standard U.S.A.S. P-1 series. No change from standard Army version.
6973-6976	'Hawk'. As F6C-1 modified for carriers. Heavier undercarriage. Early dive-bomber concept.
7128-7162	'Hawk'. First fully navalised 'Hawk'. No.7147 became famous as a racer.
7393-7423	'Hawk'. F6C-3 with engine change. First VF production type to use P. & W. engine. Mainly to U.S.M.C.
7403 only	'Hawk'. Converted F6C-4 with engine change.
7147 only	'Hawk'. 'Page Racer', modified F6C-3, parasol monoplane, crashed in 1930 Thompson Trophy race.
7403 only	'Hawk'. Ex-F6C-5 with engine change (inverted V-770). Abandoned in favour of radial engine.
7653 only	'Seahawk'. First Curtiss VF design from inception. Convertible, shipboard/catapult concept.
7654-7668, 7670	'Seahawk'. Modified production model. Used in several experiments. All to U.S.M.C.
7671-7672	'Falcon'. Two-seat VF, based on Army O-1. 7671 re-designated OC-1, 7672 modified to XOC-3.

† stated was standard for period of 2 × ·300 Browning m.gs. or optional 1 × ·300 m.g. + 1 × ·50 m.g. N.B. F4B-5 Boeing Model 216 not built.

TABLE OF UNITED STATES

Abbreviations: Conv = Converted, Deliv = Delivered, FF = First Flight, Bi = Biplane,
N.B.—Quantities given in brackets indicate modified airframes previously recorded

Type No.	Firm	Crew and Type	Significant date	Engine h.p.*	Engine Type	Top Speed (m.p.h.)	Wing Span (ft. in.)	Length (ft. in.)	Loaded Weight (lb.)	Quan.
F8C-1	Curtiss 37D	2 Bi	Deliv Feb. 28	435	P. & W. R-1340	137	38 0	25 11	3918	4
XF8C-2	Curtiss 49	2 Bi	Deliv Aug. 29	450	P. & W. R-1340-80	149	32 0	26 0	3338	1
XF8C-3	Curtiss 37	2 Bi	Deliv mid 28	425	P. & W. R-1340	137	38 0	28 0	4191	1
F8C-3	Curtiss 37E	2 Bi	Deliv July 28	425	P. & W. R-1340	137	38 0	28 0	4191	21
XF8C-4	Curtiss 49A	2 Bi	Deliv Jan. 30	450	P. & W. R-1340-88	148	32 0	26 0	3679	1
F8C-4	Curtiss 49D/E	2 Bi	Deliv May 30	450	P. & W. R-1340-88	137	32 0	26 0	3783	25
F8C-5	Curtiss 49B	2 Bi	Deliv Sep. 30	450	P. & W. R-1340-88	147	32 0	26 0	4020	20
XF8C-6	Curtiss	2 Bi	Re-worked Oct. 30	450	P. & W. R-1340-88	147	32 0	26 0	3886	(2)
XF8C-7	Curtiss 49C	2 Bi	Deliv 6 Nov. 30	575	Wright R-1820-E	178	32 0	26 2	4275	1
XF8C-8	Curtiss 49C	2 Bi	Deliv Nov. 31	575	Wright R-1820-E	178	32 0	25 8	4475	3
XF9C-1	Curtiss 58	1 Bi	FF 12 Feb. 31	423	Wright R-975-C	176	25 6	20 1	2502	1
XF9C-2	Curtiss 58A	1 Bi	Deliv Dec. 31	400	Wright R-975-C	176	25 6	20 1	2770	1
F9C-2	Curtiss 58B	1 Bi	Deliv 3 May 32	420	Wright R-975-E3	176	25 6	20 1	2770	6
XF10C-1	Curtiss 61	2 Bi	Re-designated 31	575	Wright R-1820-E	174	32 0	25 8	4627	(1)
XF11C-1	Curtiss 64	1 Bi	FF Mar. 32	600	Wright R-1510-98	203	31 6	23 1	4368	1
XF11C-2	Curtiss 64A	1 Bi	FF 25 Mar. 32	700	Wright R-1820-78	202	31 6	25 0	4132	1
F11C-2	Curtiss 64B	1 Bi	Deliv Nov. 32	700	Wright R-1820-78	205	31 6	25 0	4120	17
BFC-2	Curtiss 64B	1 Bi	Deliv Feb. 33	700	Wright R-1820-78	205	31 6	25 0	4120	10
XF11C-3	Curtiss 67	1 Bi	Deliv 27 May 33	700	Wright R-1820-78	216	31 6	23 0	4495	1
BF2C-1	Curtiss 67A	1 Bi	Deliv 7 Oct. 34	700	Wright R-1820-78	225	31 6	23 0	4555	27
XF12C-1	Curtiss 73	2 HWM	Deliv Oct. 33	625	Wright R-1510-92	217	41 6	29 1	5379	1
XF13C-1	Curtiss 70	1 HWM	FF Dec. 33	600	Wright R-1510-94	224	35 0	25 9	4423	1
XF13C-2	Curtiss 70	1 Bi	FF Jan. 34	600	Wright R-1510-94	205	35 0	25 9	3956	(1)
XF13C-3	Curtiss 70	1 HWM	Deliv May 35	700	Wright XR-1510-12	233	35 0	25 9	4721	(1)
XF14C-1	Curtiss 94A	1 LWM	Projected late 42	2200	Lycoming XH-2470-4	374	46 0	38 2	12691	(1)
XF14C-2	Curtiss 94B	1 LWM	Deliv July 44	2300	Wright XR-3350-16	398	46 0	37 9	13405	1
XF14C-3	Curtiss 94C	1 LWM	Projected 44	2300	—	400	46 0	37 9	—	(2)
XF15C-1	Curtiss 99	1 LWM	Deliv Mar. 45	2700* / 2100	De Havilland H-1B / P. & W. R-2800-34W	450	48 0	44 0	16650	3
XFD-1	Douglas	2 Bi	Deliv 18 June 33	700	P. & W. R-1535-64	204	31 6	25 4	4745	1
XF3D-1	Douglas	2 MWM	FF 23 Mar. 48	3000*	2 × J34-WE-22	510	50 0	45 5	22000	3
F3D-1	Douglas	2 MWM	FF 13 Feb. 50	3250*	2 × J34-WE-32	540	50 0	45 5	27362	28
F3D-2	Douglas	2 MWM	FF 14 Feb. 51	3600*	2 × J34-WE-36	565	50 0	45 5	30000	237
XF3D-3	Douglas	2 HWM	Projected 1950	4050*	2 × J46-WE-3	630	42 5	49 0	34000	—
XF4D-1	Douglas	1 MWM	FF 23 Jan. 51	5000*	1 × J35-A-17	500	33 6	45 8	20000	2
F4D-1	Douglas	1 MWM	FF 5 June 54	13500*	1 × J57-P-8	M.1	33 6	45 8	20000	420
F5D-1	Douglas	1 LWM	FF 21 Apr. 56	14500*	1 × J57-P-12	990	34 0	50 0	30000	4
XF6D-1	Douglas	1 LWM	Contracted July 60	8250*	2 × TF30-T-2 (P. & W.)	—	—	—	—	(2)
XFG-1	Eberhart	1 Bi	Deliv 26 June 27	425	P. & W. R-1340-D	154	32 0	22 6	2938	1
FM-1	Eastern	1 MWM	FF 31 Aug. 42	1200	P. & W. R-1830-86	318	38 0	28 9	7404	1060
FM-2	Eastern General Motors Corp.,	1 MWM	Deliv late 43	1350	Wright R-1820-56	320	38 0	28 11	7431	4777
XF2M-1	Eastern Eastern	1 MWM	Proposed 44	1500	Wright XR-1820-70	340	38 0	28 11	7950	(3)
F3M-1	Eastern Division	1 LWM	Proposed 45	2100	P. & W. R-2800-34W	421	35 10	28 3	9386	(1876)
FG-1	Goodyear	1 LWM	Deliv Apr. 43	2000	P. & W. R-2800-8	417	41 0	33 4	12040	4007
FG-4	Goodyear	1 LWM	Proposed 1944	2100	P. & W. R-2800-18W	446	41 0	33 4	12420	(2500)
F2G-1	Goodyear	1 LWM	Deliv June 45	3000	P. & W. R-4360-4	435	41 0	33 10	13290	5
F2G-2	Goodyear	1 LWM	Deliv Nov. 19	3000	P. & W. R-4360-4	431	41 0	33 10	13346	5
XFF-1	Grumman G-23	2 Bi	Deliv 29 Dec. 31	620	Wright R-1820-E	195	34 6	24 6	3933	1
FF-1	Grumman G-23	2 Bi	Deliv. May 33	700	Wright R-1820-78	207	34 6	24 6	4677	27
XF2F-1	Grumman	1 Bi	FF 10 Oct. 33	625	P. & W. XR-1535-44	229	28 6	21 1	3490	1
F2F-1	Grumman	1 Bi	Deliv Jan. 35	700	P. & W. R-1535-72	231	28 6	21 5	3847	55
XF3F-1	Grumman	1 Bi	FF 22 Mar. 35	700	P. & W. R-1535-72	226	32 0	23 0	4094	1
F3F-1	Grumman	1 Bi	Deliv Jan. 36	700	P. & W. R-1535-84	231	32 0	23 3	4170	54

* For jet engines figures given are pounds of static thrust.

NAVY AND MARINE CORPS FIGHTERS (continued)

Tri = Triplane, MWM, HWM, LWM = Mid, High and Low Wing Monoplane respectively
under another type or model, or quantity cancelled if so qualified in remarks

Serial Numbers	Remarks including unofficial or popular name†
7945–7948	'Falcon'. Originally VF class but changed to observation (OC-1) within months of procurement.
7673 only	'Helldiver'. 2-seat VF class. Although designated in F8C series it was a completely new design.
7669 only	'Falcon'. Re-evaluation of Falcon design. Standard armament with 2 × ·300 m.gs. in lower wing.
7949–7969	'Falcon'. Additional 2-seat VF type. All changed to OC-2. To U.S.M.C. for observation role.
8314 only	'Helldiver'. Improved XF8C-2 as production standard with more power and improved stressing.
8421–8445	'Helldiver'. Initially as VF-1 class. VF-1 only VF squadron to be equipped with type. Re-assigned as O2C-1.
8446–56, 8589–97	'Helldiver'. Improved F8C-4, less carrier gear. Initially 18 as VF. 2 as O2C-1. All O2C-1 by 1931.
8446–47	'Helldiver'. F8C-4 models, modified. Wing flap and Frise aileron tests.
8845 only	'Helldiver'. F8C-5 with engine change. Much modified as personal 'hack' aircraft.
8847–8849	'Helldiver'. Re-order of XF8C-7 model to evaluate engine change. No.8847 to 'paper' designation XF10C-1.
8731 only	'Sparrowhawk'. BuAer design 96, for compact light-weight shipboard fighter.
9264 only	'Sparrowhawk'. Modified design 96. Originally Curtiss sponsored prototype for airship 'parasite' fighter.
9056–9061	'Sparrowhawk'. Production L.T.A. fighter. Earlier 'X' models were modified to production standard.
8847 only	See XF8C-8 above. Original XF8C-8, modified and re-designated O2C-2. Later as XS3C-1.
9219 only	'Goshawk'. Modified 'Hawk' with long chord cowl and 3-blade propeller. Re-designated XBFC-1.
9213 only	'Goshawk'. XF11C-1 design with engine change and short chord cowl. 2-bladed propeller. Became XBFC-2.
9265–68, 9270–82	'Goshawk'. Production Goshawk, combining features of 'X' models. Re-designated BFC-2 in March 1934.
9331–9340	'Hawk'. Modified F11C-2 models for fighter-bomber role. New cockpit arrangement including canopy.
9269 only	'Hawk'. As an F11C-2 with retractable gear. Experimental model for BF2C-1. Later as XBF2C-1.
9586–9612	'Hawk'. Last of 'Hawk' series. Much modified production version of XF11C-3. All F11Cs to 'BF' role.
9225 only	Experimental parasol monoplane with folding wings, retractable undercarriage, wing slots and flaps.
9343 only	Convertible biplane/monoplane in monoplane configuration. Retractable undercarriage.
9343 only	Convertible biplane/monoplane in biplane configuration. Retractable undercarriage.
9343 only	BuAer No.9343 above, modified from test results. U.S.A.A.C. tested, U.S.N. carrier trials. Rejected as fighter.
03184 allotted	Not built. Estimated performance poor and in-line engine not favoured. Basic design to XF14C-2.
03183 only	XF14C-1 with engine change. 4 × 20 mm. cannons with 166 r.p.g. planned. Failed to meet specifications.
30297–98 allotted	Proposed high altitude pressurised cabin version. Cannon-armed. Cancelled 1945. None built.
01213–1215	Last Curtiss project for U.S. Navy. Composite power, high altitude heavily armed fighter. No.01213 destroyed. No.01215 U.S.A.A.F. tested. Contract terminated 10 Oct. 1946.
9223 only	2-seat fighter. BuAer design 113 as per XF3U-1. Became engine test bed for Pratt & Whitney.
121457–121459	'Skyknight'. All-weather fighter, carrier/land based. 4 × 20 mm. cannons, bombs and rockets.
123741–123768	'Skyknight'. Most to U.S.M.C. Relegated to training role.
See remarks	'Skyknight'. 124595–664, 125783–882, 127019–85. First U.S. jet to shoot down jet at night, 11 Feb. 1952.
125883–92, 130463–739	'Skyknight'. Swept-wing project terminated Feb. 1952. Designed as improved night fighter. Cancelled.
124586–124587	'Skyray'. Modified delta configuration. A.W. interceptor. Held speed and climb records. 124586 FF with J34. 124587 had XJ40-WE-8 fitted. 4 × 20 mm. cannons + 2 × 1000 lb. bombs or rockets.
See remarks	'Skyray'. Engine change. BuAer Nos. 130740–130751, 134744–134973, 139030–139207.
139208A–9, 142349A–50B	'Skylancer'. On 'paper' as F4D-2. Improved A.W. night fighter, 4 × 20 mm. cannons and homing missiles.
Cancelled	'Missileer'. Project terminated 25 Apr. 1961. Missile launching concept for 'Eagle' A.A.M.
7944 only	'Comanche'. Land/sea convertible. Length with float, 28 ft. 6 in. Tested as contractor-owned aircraft.
14992–15951, 46738–837	'Wildcat'. As Grumman F4F-4. 312 allotted to U.K. as Wildcat V. 839 to U.S.N.
See remarks	'Wildcat'. Improved F4F-4, many to U.K. as Wildcat VI, 4437 to U.S.N. BuAer Nos. (including U.K. diversions) 15952–16791, 46838–47437, 55050–55649, 56684–57083, 73499–75158, 86297–86973.
82855–82857 allotted	Improved 'Wildcat'. Cancelled V-J day. One only part-built.
109273–11148 allotted	'Bearcat'. Grumman F8F-1 to have been built under contract. Cancelled V-J day before construction.
See remarks	'Corsair'. Chance Vought F4U-1 built under contract, 2007 as FG-1D, 2 as FG-1A. BuAer Nos. incl. 475 to U.K. as Corsair IV: 12992–14991, 67055–67099, 76139–76739, 87788–88453, 92007–701.
See remarks	'Corsair'. Planned production of Chance Vought F4U-4. BuAer Nos. 67255–754 and 106876–7875 allotted.
88454–88458	'Corsair'. Modified F4U-4 with engine change and bubble canopy. 413 cancelled after V-J day.
88459–88463	'Corsair'. F2G-1 equipped for carriers as Kamikaze deterrent. Low-level fighter. Five others cancelled.
8878 only	'Fi-Fi'. 2-seat fighter. First production VF type with retractable undercarriage. To Anacostia as utility.
9350–9376	'Fi-Fi'. Successful in VF, VB and VS roles. Fitted with dual controls. As FF-2 in reserve status.
9342 only	Short, stubby fighter based on FF-1. Retractable undercarriage. Equipped for carriers.
9623–9676, 9997	First of long line Grumman Navy fighters. Mainly to Navy squadrons. Two to U.S.M.C. in 1938.
9727 only	Improved, larger F2F. Original model crashed, and a replacement was built.
0211–64	Production model of XF3F-1 with improved engine. Six to U.S.M.C., 48 to U.S.N. Standard pre-war armament.

† Armament of pre-war type unless otherwise stated was standard 2 × ·300 m.gs. or optional 1 × ·300 m.g. + 1 × ·50 m.g.

TABLE OF UNITED STATES

Abbreviations: Conv = Converted, Deliv = Delivered, FF = First Flight, Bi = Biplane,
N.B.—Quantities given in brackets indicate modified airframes previously recorded

Type No.	Firm	Crew and Type	Significant date	Engine h.p.*	Engine Type	Top Speed (m.p.h.)	Wing Span (ft. in.)	Length (ft. in.)	Loaded Weight (lb.)	Quan.
XF3F-2	Grumman	1 Bi	Deliv Jan. 37	850	Wright XR-1820-22	260	32 0	23 2	4495	1
F3F-2	Grumman	1 Bi	Deliv Dec. 1937	950	Wright R-1820-22	260	32 0	23 2	4453	81
XF3F-3	Grumman	1 Bi	Reworked early 38	950	Wright R-1820-22	264	32 0	23 2	4495	(1)
F3F-3	Grumman	1 Bi	Deliv Dec. 38	950	Wright R-1820-22	264	32 0	23 2	4543	27
XF4F-2	Grumman G-36	1 MWM	FF 2 Sep. 37	1050	P. & W. R-1830-66	290	34 0	26 5	5386	1
XF4F-3	Grumman G-36A	1 MWM	FF 12 Feb. 39	1200	P. & W. R-1830-76	334	38 0	28 10	6099	(1)
F4F-3	Grumman G-36A	1 MWM	FF Feb. 40	1200	P. & W. R-1830-76	331	38 0	28 9	7065	280
XF4F-4	Grumman G-36B	1 MWM	FF 14 Apr. 41	1200	P. & W. R-1830-76	326	38 0	28 9	7489	1
F4F-4	Grumman G-36B	1 MWM	Deliv mid 41	1200	P. & W. R-1830-86	318	38 0	28 9	7964	1169
XF4F-5	Grumman	1 MWM	FF June 40	1200	Wright R-1820-40	306	38 0	28 10	6063	2
XF4F-6	Grumman	1 MWM	FF Oct. 1940	1200	P. & W. R-1830-90	319	38 0	28 9	7056	1
F4F-7	Grumman	1 MWM	FF 30 Dec. 41	1200	P. & W. R-1830-8t	310	38 0	29 10	10328	21
XF4F-8	Grumman	1 MWM	FF 8 Nov. 42	1350	Wright XR-1820-56	321	38 0	29 10	5365	2
XF5F-1	Grumman G-34	1 LWM	FF 1 Apr. 40	1200	2 × Wright R-1820-40	358	42 0	28 11	9758	1
XF6F-1	Grumman	1 LWM	Deliv 43	2000	P. & W. R-2800-27	385	42 2	33 7	12800	1
XF6F-2	Grumman	1 LWM	Proposed as test	2100	P. & W. R-2800-21	—	42 10	33 7	—	(1)
XF6F-3	Grumman	1 LWM	FF 26 June 42	2000	P. & W. R-2800-10W	380	42 10	33 7	10558	1
F6F-3	Grumman	1 LWM	Deliv Nov. 42	2000	P. & W. R-2800-10W	375	42 10	33 7	12440	4402
XF6F-4	Grumman	1 LWM	Deliv Mar. 43	2000	P. & W. R-2800-27	385	42 10	33 7	12800	(1)
F6F-5	Grumman	1 LWM	Deliv May 44	2000	P. & W. R-2800-10W	366	42 10	33 7	12900	7870
XF6F-6	Grumman	1 LWM	Deliv July 44	2100	P. & W. R-2800-18W	417	42 10	33 7	12768	2
XF7F-1	Grumman G-51	1 MWM	FF 3 Nov. 43	1800	2 × Wright XR-2600-14	398	51 6	45 4	19500	2
F7F-1	Grumman G-51	1 MWM	Deliv Apr. 44	2100	2 × P. & W. R-2800-22W	425	51 6	45 4	21425	34
XF7F-2	Grumman	2 MWM	Deliv July 44	2100	2 × P. & W. R-2800-22W	420	51 6	45 4	21850	1
F7F-2	Grumman	2 MWM	Deliv Nov. 44	2100	2 × P. & W. R-2800-34W	421	51 6	45 4	21860	65
F7F-3	Grumman	1/2 MWM	Deliv Mar. 45	2100	2 × P. & W. R-2800-34W	435	51 6	45 4	21725	249
F7F-4	Grumman	2 MWM	Deliv June 46	2100	2 × P. & W. R-2800-34W	430	51 6	46 10	21950	13
XF8F-1	Grumman	1 LWM	FF June 44	2100	P. & W. R-2800-22W	424	35 6	28 8	8785	3
F8F-1	Grumman	1 LWM	Deliv Feb. 45	2100	P. & W. R-2800-32W	421–	35 10	28 3	9385 }	896
					-34W(F8F-B), -22W(F8F-1D)	424	35 10	28 3	9398 }	
F8F-2	Grumman	1 LWM	Deliv Nov. 46	2300	P. & W. R-2800-34W	428	35 6	27 6	10426	365
XF9F-2	Grumman G-79	1 LWM	FF 24 Nov. 47	5000*	Rolls-Royce Nene I	550	35 3	37 8	10840	2
F9F-2	Grumman	1 LWM	Deliv Nov. 48	5000*	P. & W. J-42-P-6	525	38 0	37 3	16450	520
XF9F-3	Grumman	1 LWM	FF 24 Aug. 48	4600*	Allison J33-A-8	520	38 0	37 3	16400	1
F9F-3	Grumman	1 LWM	Deliv Sep. 48	4600*	Allison J33-A-8/10	520	38 0	37 3	16400	101
F9F-4	Grumman	1 LWM	Deliv Nov. 49	5500*	Allison J33-A-16A	570	38 0	38 10	17500	109
XF9F-5	Grumman	1 LWM	FF 21 Dec. 49	6250*	P. & W. J48-P-6	575	38 0	38 10	17695	(2)
F9F-5	Grumman	1 LWM	Deliv Dec. 49	6250*	P. & W. J48-P-6	575	38 0	38 10	17765	655
XF9F-6	Grumman G-93	1 LWM	FF 20 Sep. 51	6250*	P. & W. J48-P-8A	650	34 6	41 7	17500	3
F9F-6	Grumman	1 LWM	Deliv first 52	6500*	P. & W. J48-P-6A/8	650	34 6	41 7	17618	706
F9F-7	Grumman	1 LWM	Deliv first 52	6500*	Allison J33-A-16A	650	34 6	41 7	17620	168
F9F-8	Grumman	1 LWM	FF 18 Dec. 53	7250*	P. & W. J48-P-8A	714	34 6	41 7	19500	1111
F9F-9	Grumman	1 LWM	FF 30 July 54	11200†	Wright J65-W-4	850	31 8	40 10	13850	3
XF10F-1	Grumman	1 HWM	FF 19 May 53	11600†	West. J40-WE-8	720	50 7	55 7	32000	2
F11F-1	Grumman G-98	1 LWM	Deliv 56	12000†	1 × J65-W-18	1000+	31 8	42 6	17200	198
XFH-1	Hall-Aluminum	1 Bi	Deliv 18 June 29	450	P. & W. R¹1340-B	152	32 0	22 6	2517	1
FO-1	Lockheed	1 MWM	Procured 43	1425	2 × Allison V-1710-89/91	414	52 0	37 10	15500	4
XFV-1	Lockheed	1 MWM	Deliv late 54	5800	Allison T-40-A	500	30 11	36 10	10000	2

* In the case of jet engines figure given is for static thrust in pounds.   † Static thrust with afterburner in operation.

NAVY AND MARINE CORPS FIGHTERS (continued)

Tri = Triplane, MWM, HWM, LWM = Mid, High and Low Wing Monoplane respectively
under another type or model, or quantity cancelled if so qualified in remarks

Serial Numbers	Remarks including unofficial or popular name
0452 only	Larger, improved F3F-1 with engine change. Carrier equipment. Special built model as 'Gulfhawk II'.
0967–1047	Production models. Six to VF-6 U.S.N., all others to U.S.M.C.
1031 only	Modified F3F-2, with redesigned cowling. N.A.C.A. tests.
1444–1470	VF, VMF replacement equipment. Re-designated F3F-2 later. Last U.S. naval service VF biplane.
0383 only	XF4F-1 (biplane) not built and contract changed to XF4F-2. All flying surfaces rounded in planform.
0383 only	'Wildcat'. Enlarged, modified F4F-2, engine change. Squared off-flying surface tips. Improved performance.
1844–45, 1848–96, 2512–38, 3856–4057	'Wildcat'. Modified production version. Some to U.K. as Martlets, only naval combat VF type 1940–43. First service aircraft in world, with two-speed supercharged engine. Quantity includes 95 as F4F-3A, see XF4F-6.
1897 only	'Wildcat'. As F4F-3 with hydraulically folding wings which was abandoned on production aircraft.
See remarks	'Wildcat'. Improved production model, 220 to U.K. as 'Wildcat IV' not included in BuAer Nos. as follows: 4058–4098, 5030–5262, 01991–2152, 03385–3544, 11655–12227.
1846–1847	'Wildcat'. 3rd and 4th F4F-3 diverted for engine change evaluation. Single stage/two-speed supercharger.
7031 only	'Wildcat'. Original designation for F4F-3A with single-stage supercharger. Production as F4F-3A.
5263–5283	'Wildcat'. Long-range P.R. version, 685 galls. of fuel. Armament deleted. Converted to F4F-4s.
12228–12229	'Wildcat'. As F4F-4 with engine change and slotted wing flaps. Produced later by Eastern as FM-2.
1442 only	'Skyrocket'. Twin engined, high powered, heavily armed interceptor.
2981 only	'Hellcat'. Re-designated XF6F-4. F6F-1 model on 'paper' only ordered for Navy 30 June 1941.
66244 only	'Hellcat'. Finally built as production F6F-3.
02982 only	'Hellcat'. Prototype for 'Wildcat' replacement series. Wheels retracted into wings. Delivered Nov. 41.
See remarks	'Hellcat'. Includes, 125 F6F-3N, 18 F6F-3E. Few 'Field modified' as '3P'. Lowest wing loading of service VF type. Serials 04775–958, 08798–9047, 25721–6195, 65890–6244, 39999–43137 do not include 252 to U.K.
02981 only	'Hellcat'. Ex-XF6F-1, engine change over production versions. Later refitted as an F6F-3.
See remarks	Improved version, modified cowl, windshield and increased armour. BuAer Nos. 58000–999, 69992–70462, 70463–2991, 77259–80258, 93652–94521 including 5529 as F6F-5N, excluding 930 to U.K. as Hellcat II.
70188 and 70913	'Hellcat'. F6F-5 with engine change and four-bladed propeller.
03549–3350	'Tigercat'. Based on basic XF5F-1 design. Tricycle gear, heavily armed. No.03549 crashed 1 May 1944.
80259–60, 80262–93	'Tigercat'. Most to F7F-1N as concept changed to single-seat night fighter. Two to U.K. as TT348–9.
80261 only	'Tigercat'. Two-seat night fighter version with added nose radar and nose armament deleted.
80294–80358	'Tigercat'. Improved version. All designated F7F-2N and delivered to U.S.M.C.
80359–547, 80549–608	'Tigercat'. Includes 60 as two-seat F7F-3N. All to U.S.M.C., carrier gear not fitted.
80548, 80609–80620	'Tigercat'. All designated F7F-4N. To U.S.N., U.S.M.C. Added and improved radar. Carrier equipment.
90460–90462	'Bearcat'. Successor to 'Hellcat'. Light-weight interceptor. 90460 crashed, replaced by 90462.
See remarks	'Bearcat'. BuAer Nos. 90437–59, 94752–95498, 122087–152, 121463–522 include 126 F8F-1B, one to F8F-1C (No.90440), 12 as F8F-1N. Four-bladed propeller. 7253 cancelled. 4 × ·50 m.gs + bombs/rockets.
121523–792, 122614–708	'Bearcat'. Improved model with taller tail. 4 × 20 mm. cannons. 12 as 2N. 60 as 2P with 2 × 20 mm. cannons.
122475 and 122477	'Panther'. XF9F-1 night fighter not built. XF9F-2 was first Grumman jet aircraft.
123397–713, 125083–155, 127086–127215	'Panther'. Day fighter with permanent tip-tanks. Includes 'B' versions and eight as 'P' versions. 4 × 20 mm. cannons, 2 × 500 lb. bomb or rockets. Late models used J42-P-8 engine.
122476 only	'Panther'. F9F-2 with engine change.
122560–89, 123016–86	'Panther'. Converted later to F9F-2s and re-designated as standard F9F-2 models.
125081, 125156–227, 125913–48	'Panther'. Merged into F9F-5 series. Improved engine and modifications.
123084–123085	'Panther'. Improved version with engine change (J48 as Rolls-Royce Tay). Taller tail. F9F-3s modified.
125080, 125082 and as remarks	'Panther'. Production day fighter. BuAer Nos. 125228–321, 125447–76, 125489–99, 125533–648, 125893–912, 125949–6256, 126627–72, 126265–90, 127471–2 includes 'B' and 36 'P' versions.
126670–126672	'Cougar'. As F9F-5 model with sweptwings and horizontal tail. Boundary layer control wings.
126257–64, 127216–470, 127473–92 and as remarks	'Cougar'. First sweptwing in carrier service. 59 as 'P' version. 4 × 20 mm., six H.V.A.R. rockets or four Sidewinders. Some to 'K' and 'PD' Drones. Nos. 128055–310, 131252–5, 130920–1062, 134446–65.
130752–130919	'Cougar'. F9F-6 model with engine change. Re-designated F9F-6.
131063–251, 134234–44, 138823–98 and as remarks	'Cougar'. Modified, 'Saw-tooth'. 4 × 20 mm. cannon + 4000 lb. bomb/rocket load. 110 'P', 400 'T', 410 'B'. Nos. 141030–229, 141648–727, 142437–532, 142954–3012, 144271–426, 146342–425, 147270–429.
138604–138606	'Tiger'. Re-designated F11F-1. First to use area rule concept. 'Shot' itself down 21 Sep. 1956.
124435–124436	'Jaguar'. Variable sweptwing. 20°–40°, 50ft. 7in. to 36ft. 8in. No.124435 only flew. 30 'X' models cancelled.
138607–47, 141728–884	'Tiger'. 138646–47 as F11F-1F version with J-79 engine, one used J-64. Later to VA status. Evaluated for N.A.T.O. use. All to Navy units. Diverted to training role. Was F9F-9.
8009 only	BuAer design, jettisonable undercarriage. Water-tight all-metal fuselage. Production considered impractical.
01209–01212	'Lightning'. P-38s procured from U.S.A.A.F. for photo work in N. Africa. Others 'borrowed' subsequently.
138657–138658	'Salmon'. World's second true successful V.T.O. aircraft. Straight wing, 'X' tail. 138658 not completed.

TABLE OF UNITED STATES

Abbreviations: Conv = Converted, Deliv = Delivered, FF = First Flight, Bi = Biplane,
N.B.—Quantities given in brackets indicate modified airframes previously recorded

Type No.	Firm	Crew and Type	Significant date	Engine h.p.*	Engine Type	Top Speed (m.p.h.)	Wing Span (ft. in.)	Length (ft. in.)	Loaded Weight (lb.)	Quan.
XFL-1	Loening	1 Bi	Proposed 33	625	Wright XR-1510-26	200	28 0	20 9	4250	(1)
XFD-1	McDonnell	1 LWM	FF Oct. 44	1365*	2 × WE19-XB-2B	479	40 9	37 2	8825	2
FH-1	McDonnell	1 LWM	Deliv Jan. 47	1600*	2 × J30-WE-20	487	40 9	38 9	10035	60
XF2D-1	McDonnell	1 LWM	FF 11 Jan. 47	3000*	2 × J34-WE-22	585	41 6	39 0	13000	2
F2H-1	McDonnell	1 LWM	Deliv Aug. 48	3000*	2 × J34-WE-30/34	585	41 6	39 0	14300	56
F2H-2	McDonnell	1 LWM	Deliv Aug. 49	3150*	2 × J34-WE-34	535	44 10	40 2	20615	436
F2H-3	McDonnell	1 LWM	Deliv Apr. 52	3600*	2 × J34-WE-36	590	44 10	47 6	21300	250
F2H-4	McDonnell	1 LWM	Deliv early 53	3600*	2 × J34-WE-38	610	44 11	48 2	19200	150
XF3H-1	McDonnell	1 LWM	FF 7 Aug. 51	7200*	1 × XJ40-WE-6	700	34 2	58 4	33000	2
F3H-1	McDonnell	1 LWM	Deliv 8 Jan. 54	8000*	1 × J40-WE-22	800	35 4	58 4	33500	56
F3H-2	McDonnell	1 LWM	Deliv June 55	14250†	1 × J71-A-2	M1+	35 4	58 11	33900	460
F4H-1	McDonnell	2 LWM	FF 27 May 58	17000†	2 × J79-GE-2A(40) 2 × J79-GE-8	M2+	38 5 38 5	58 3	40000+	In prod.
TF-1	Naval A/c Factory	3 Bi	Deliv 22	300	2 × Wright-Hispano	107	60 0	44 5	8846	3
XFN-1	Naval A/c Factory	1 Bi	Ordered 6 June 31		Not determined	—	—	—	—	(1)
XFT-1	Northrop/Douglas	1 LWM	Deliv Mar. 34	650	Wright XR-1510-8	235	32 0	21 4	4749	1
XFT-2	Northrop/Douglas	1 LWM	Re-deliv Apr. 36	650	P. & W. R-1535-72	234	32 0	21 1	4770	(1)
F2T-1	Northrop	2 HWM	Deliv late 45	2000	2 × P. & W. R-2800-65	365	66 9	49 7	35000	12
XFJ-1	North American	1 LWM	FF Sep. 1946	3800*	1 × J35-A5	542	38 1	33 5	12135	3
FJ-1	North American	1 LWM	Deliv mid 47	4000*	1 × J35-A-2	547	38 2	34 5	15115	30
XFJ-2	North American	1 LWM	FF 27 Dec. 51	6000*	1 × J47-GE-2	650	37 1	37 6	15000	2
FJ-2	North American	1 LWM	Deliv mid 52	6000*	1 × J47-GE-2	650	37 1	37 6	15300	201
FJ-3	North American	1 LWM	FF 3 July 53	7200*	1 × J65-W-2/16	650	37 1	37 7	18000	538
FJ-4	North American	1 LWM	FF 28 Oct. 54	7700*	1 × J65-W-16A	695	39 1	36 5	19003	366
XFR-1	Ryan 28	1 LWM	FF 25 June 44	1600* 1350	G.E. I-16 Wright R-1820-56	426	40 0	32 4	9862	3
FR-1	Ryan 28	1 LWM	Deliv Feb. 45	1600* 1425	J31-GE-3 Wright R-1820-72W	400	40 0	32 4	9958	66
XFR-2	Ryan	1 LWM	Proposed 45	1600* 1500	GE I-16 Wright R-1820-74W	—	40 0	32 4	—	—
XFR-3	Ryan	1 LWM	Proposed 45	2000* 1500	GE I-20 Wright R-2820-74W	—	40 0	32 4	—	—
XFR-4	Ryan	1 LWM	Deliv late 44	3000* 1450	J34-WE-22 Wright R-1820-74W	425	40 0	32 4	10100	(1)
XF2R-1	Ryan 29-30	1 LWM	Deliv Nov. 46	2300* 1600	GE XT31-GE-2 (TG-100) J31-GE-3	500	40 0	36 0	11000	(1)
XNF-1	Seversky	1 LWM	Deliv 24 Sep. 37	950	Wright R-1820-22	250	36 0	25 2	5230	1
VE-7F	Vought/N.A.F.	1/2 Bi	Deliv 20	180	Wright-Hispano E-2	121	34 3	24 2	2100	39
VE-7GF	Vought/N.A.F.	2 Bi	Deliv 20	180	Wright-Hispano E-2	112	34 3	24 2	2098	1
VE-7SF	Vought/N.A.F.	2 Bi	Deliv 21	180	Wright-Hispano E-2	117	34 3	24 2	2100	50
UF-1	Chance Vought	1 Bi	Deliv 22	220	Lawrence J-1	132	26 0	20 1	1992	18
FU-1	Chance Vought	1 Bi	Deliv Jan. 27	220	Wright J-5	125	34 4	24 5	2409	20
XF2U-1	Chance Vought	2 Bi	FF 21 June 29	450	P. & W. R-1340-C	146	36 0	27 0	3886	1
XF3U-1	Chance Vought	2 Bi	FF May 33	700	P. & W. R-1535-64	214	31 6	26 6	4643	1
XF4U-1	Vought-Sikorsky	1 LWM	FF 29 May 40	1850	P. & W. XR-2800-2	405	41 11	31 11	9146	1
F4U-1	Vought-Sikorsky	1 LWM	FF 25 June 42	2000	P. & W. R-2800-8	417	41 0	33 4	12050	4700
F4U-2	Vought-Sikorsky	1 LWM	Modified early 43	2000	P. & W. R-2800-8W	380	41 0	33 4	11445	(12)
XF4U-3	Vought-Sikorsky	1 LWM	Modified Mar. 42	2000	P. & W. XR-2800-16	412	41 0	33 4	11650	(3)
XF4U-4	Chance Vought	1 LWM	Deliv Oct. 44	2100	P. & W. R-2800-18W	452	42 0	33 8	12250	5

* In the case of jet engines figure given is for static thrust in pounds.   † Static thrust with afterburner in operation.

NAVY AND MARINE CORPS FIGHTERS (continued)

Tri = Triplane, MWM, HWM, LWM = Mid, High and Low Wing Monoplane respectively
under another type or model, or quantity cancelled if so qualified in remarks

Serial Numbers	Remarks including unofficial or popular name
9346 only	Not built, contract study of BuAer design 120. Cancelled by joint Navy/Loening agreement.
48235–48236	'Phantom'. First contracted Navy jet. Carrier equipped. Re-designated XFH-1.
11749–11808	'Phantom'. First carrier jet squadron type (VF-17). Most to U.S.M.C. 4 × ·50 m.gs. 130 cancelled V-J Day.
99858 and 99860	'Banshee'. Improved, larger FH-1. Carrier equipped day fighter. Also as XF2H-1.
122528–58, 122991–3015	'Banshee'. Production. As 'X' model with stabilizer dihedral eliminated. No tip-tanks. Carrier equipment.
See remarks	'Banshee'. Improved F2H model, additional integral tip-tanks. Serials include 123204–382, 124940–5079, 125649–706, 126673–95, 128857–86. 14 'N', 89 'P' and two 'B' versions.
See remarks	'Banshee'. Improved, engine change, increased internal fuel capacity, 39 sold to Royal Canadian Navy. BuAer Nos. 126291–350, 126354–489, 127493–546. No.126319 modified to F2H-4 model prototype.
126351–126353, 127547–127693	'Banshee.' All weather version, additional and improved radar and armament.
125444–125445	'Demon'. New fighter design, 4 × 20 mm. cannons.
133489–133544	'Demon'. Delivered as F3H-1N version. 29 to F3H-2N.
See remarks	'Demon'. Standard carrier based all weather fighter aircraft. BuAer Nos. 133545–639, 136966–7095, 143403–92, 145202–306, 146328–39, 146709–40, including 146 as 'N' versions, and 79 'M' versions.
142259A–60, 143388–92, 145307–17, 146817–21, 148363–410, etc.	'Phantom II'. All-purpose attack fighter based on weapons delivery system. Missile armament of six Sparrow III or Sidewinders. Evaluation by U.S.A.F. over F-105/108 concept. Project started as XAH-1 attack aircraft. First two as prototypes. No 'X' assignment in VF role. 11th became F4H-1F for ground attack role.
5576–5578	Based on NC flying boat. Tandem engines. No.5578 had 2 × 510 h.p. Packard 1A-1500.
8978 only	Not built, assignment of BuAer number and designation, August 1931. Cancelled September 1931.
9400 only	Based on Company's commercial designs. Large spatted wheels. Standard pre-war armament.
9400 only	XFT-1 with engine change, modified cowling and landing gear. Crashed. U.S.N. equity refunded.
52750–52761	'Black Widow'. Ex-U.S.A.A.F. Procured by U.S.M.C. as radar equipped night fighter trainer.
39053–39055	'Fury'. Based on F-86 Sabre. Modified with straight wings. Navalised with carrier gear.
120342–120371	'Fury'. Production. Straight jet ducting. First pure jet VF to operate from carrier. 4 × 50 m.gs.
133754–133755	'Fury'. Navalised version of F-86E. Swept folding wings, carrier equipment.
131927–2126, 133756	'Fury'. Most to U.S.M.C. No.133756 to XFJ-2B and prototype for FJ-3 models. 4 × 20 mm. cannons.
See remarks	'Fury'. Improved version with engine change. To Navy and U.S.M.C. BuAer Nos. 135446, 135774–6162, 139210–78, 141364–443 include 45 'M' versions.
139279–555, 141444–89, 143493–643	'Fury'. Re-designed model. Improved performance for naval use. First aircraft to deploy with 'Bullpups'. First two as prototypes. 147 as FJ-4, 217 as FJ-4B. Two converted to FJ-4F.
48232–48234	'Fireball'. Composite power, all-altitude fighter. 2 × ·50 m.gs. and 1000 lb. bombs or rockets.
39647–39712	'Fireball'. Production. Served VF-66 Squadron only. One to XFR-4, one to XF2R-1. 1234 FR-1/2 cancelled.
104572–105175	'Fireball'. Improved version. 600 cancelled V-J Day. None built.
Not assigned	'Fireball'. Proposed more powerful version. Not built.
39665 only	'Fireball'. Modified FR-1 with engine change. Flush fuselage type. Intake ducting and modified tail.
39661 only	'Darkshark'. Modified FR-1 with engine change, taller tail and 4-bladed propeller.
Not assigned	Navalised commercial design with carrier gear. Civil registered as X-1254. Company sponsored project.
5692–5700, 5942–71	Convertible, one/two seat (single-seat as VF type) Vought design. N.A.F. built.
5691 only	Two-seat ground fighter as experimental status to U.S.M.C. Built at the N.A.F.
5912–5941, 6011–6030	Convertible scout-fighters, 10 built by N.A.F. Landplane/floatplane convertible.
6482–6499	Convertible interim equipment based on UO-1 observation type. Catapult stressed. All re-designated UO-1.
7361–7380	Interim fighter. Convertible S.P./L.P. Originally UO-3, became FU-2 with second cockpit in training role.
7692 only	Fighter project in competition with Curtiss XF8C-2. First employment of long chord N.A.C.A. cowl.
9222 only	Project in competition with Douglas XFD-1. Finally modified as SBU-1.
01443 only (V–166 B Model)	'Corsair'. Successful VF design around 2000 h.p. engine. Inverted gull wings. First VF to top 400 m.p.h.
02153–736, 03802–841, 17392–8191, 49660–50659 + ‡	'Corsair'. Late F4U-1 models had modified 'clear-view' semi-bubble canopy. Serial numbers include 605 for Royal Navy and 370 New Zealand, and incorporate 197 F4U-1C and 1777 F4U-1D versions.
From F4U-1 numbers	'Corsair'. Converted to experimental night fighters from F4U-1. Initial trials with airborne radar.
02157, 17516, 49664	'Corsair'. Converted F4U-1s, turbo-supercharged engine for high altitude experiments.
80759–80763	'Corsair'. Modified F4U-1 with supercharged engine and 4-bladed propeller.

‡ 55784–6483, 57084–983, 82178–852. N.B.—XF3R-1 projected V.T.O. fighter with J-40 engine cancelled.

TABLE OF UNITED STATES

Abbreviations: Conv = Converted, Deliv = Delivered, FF = First Flight, Bi = Biplane,
N.B.—Quantities given in brackets indicate modified airframes previously recorded

Type No.	Firm	Crew and Type	Significant date	Engine h.p.*	Engine Type	Top Speed (m.p.h.)	Wing Span (ft. in.)	Length (ft. in.)	Loaded Weight (lb.)	Quan.
F4U-4	Chance Vought	1 LWM	Deliv Dec. 44	2100	P. & W. R-2800-18W	446	42 0	33 8	12400	2351
XF4U-5	Chance Vought	1 LWM	Modified July 46	2300	P. & W. R-2800-32W	460	41 0	34 6	12500	(3)
F4U-5	Chance Vought	1 LWM	Deliv mid 47	2300	P. & W. R-2800-32W	465	41 0	34 6	13000	568
XF4U-6	Chance Vought	1 LWM	Deliv mid 52	2300	P. & W. R-2800-83W	438	41 0	34 1	18979	111
XF5U-1	Chance Vought	1 All wing	Built 1944	1200	2 × P. & W. XR-2000-7	388	32 6	28 7	16500	1
XF6U-1	Chance Vought	1 LWM	FF 2 Oct. 46	2700*	J34-WE-22/30A	525	32 10	33 9	9305	3
F6U-1	Chance Vought	1 LWM	FF July 49	4000†	J34-WE-22	595	35 8	37 7	9200	30
XF7U-1	Chance Vought	1 LWM	FF 29 Sep. 48	4200†	2 × J34-WE-32	625	38 8	39 7	14500	3
F7U-1	Chance Vought	1 LWM	FF 1 Mar. 50	4200†	2 × J34-WE-32	650	38 8	40 11	14700	14
F7U-3	Chance Vought	1 LWM	FF 20 Dec. 51	6100†	2 × J46-WE-8A	700	38 8	44 4	27350	290
XF8U-1	Chance Vought	1 HWM	FF 25 Mar. 55	13200†	P. & W. J57-P-4	1000+	35 9	54 3	22000	2
F8U-1	Chance Vought	1 HWM	FF 20 Sep. 55	15000†	P. & W. J57-P-12	1000+	35 9	54 3	25000	592
F8U-2	Chance Vought	1 HWM	FF Dec. 57	17500†	P. & W. J57-P-16/20	M2	35 9	55 2	26000	339
F8U-3	Chance Vought	1 HWM	FF 2 June 58	23500†	P. & W. J75	M2-3	39 11	58 9	29000	4
WP-1	Wright Aeronautical	1 HWM	Deliv 1923	300	Wright H-3	162	32 10	24 5	2674	1
F2W-1/2	Wright Aeronautical	1 Bi	Deliv early 24	710	Wright T-3	230	22 6	27 7	1935	2
F3W-1	Wright Aeronautical	1 Bi	Deliv 5 May 26	450	P. & W. R-1340-B	165	27 4	25 4	2128	1
NW-1/2	Wright Aeronautical	1 Bi	Deliv 1922	600	Wright T-2	220	28 0	21 0	3000	2

U.S. Navy and Marine Corps Aircraft procured as VF types and used in temporary fighter role.

Type No.	Firm	Crew and Type	Significant date	Engine h.p.*	Engine Type	Top Speed	Wing Span	Length	Loaded Weight	Quan.
Bulldog	Bristol 105	1 Bi	FF 10 Oct. 29	490	Bristol Jupiter VIIF	174	33 11	25 0	3503	2
D-1	Dornier	1 Bi	Deliv 1919	185	B.M.W.	125	25 8	21 0	1914	1
C-1	Fokker	2 Bi	Deliv 1921	243	B.M.W.	112	34 10	23 8	2576	3
D-7	Fokker-Netherlands	1 Bi	Deliv 1921	350	Packard 1-A-1237	125	29 4	22 9	2462	6
SE-5	Curtiss	1 Bi	Deliv 1922	150	Hispano-Suiza	122	28 0	21 4	1930	2
HD-1	Hanriot/N.A.F.	1 Bi	Built N.A.F. 18	130	Le Rhône Rotary	108	28 6	19 3	1325	10
'Pup'	Sopwith	1 Bi	Deliv 1919	100	Gnôme Monosoupape	110	26 6	19 4	1297	2
'Camel'	Sopwith 2F.1	1 Bi	Deliv 18/19	150	Gnôme Monosoupape	120	28 0	18 1	1500	6
E-1	Standard	1 Bi	Deliv 1919	80	Le Rhône Rotary	100	24 0	18 10	1144	10
N-28	Nieuport	1 Bi	Deliv mid-1919	165	Gnôme Monosoupape	122	26 3	20 4	1625	12
MB-3	Thomas-Morse	1 Bi	Deliv 1922	340	Wright-Hispano H	152	26 0	19 11	1818	11
MB-7	Thomas-Morse	1 HWM	Deliv 1922	400	Wright-Hispano H-3	180	24 0	18 6	2000	1
HPS-1	Handley Page 21	1 LWM	Built Sep. 1923	230	Gwynne BR-2	145	29 2	21 5	1920	1
P-80A	Lockheed	1 LWM	Deliv 45/46	5200†	J33-GE-9	558	39 0	34 6	13000	3
P-80C	Lockheed	1 LWM	Deliv 48	6000†	J33-A-23	578	39 0	34 6	15336	50
F-104A	Lockheed	1 LWM	Assigned 1959	17000†	1 × J79-GE-3	1400	25 0	58 4	19200	3
YP-59A	Bell 27	1 MWM	Deliv 1943	2500†	2 × J31-GE-	413	45 6	38 2	12700	2
YP-59B	Bell	1 MWM	Deliv (1) 45, (2) 46	2500†	2 × J31-GE-	415	45 6	38 10	13000	3
P-63A	Bell	1 LWM	Deliv 1946	1325	Allison V-1710-93	422	38 4	32 8	10000	2
P-51A	N. American NA-73	1 LWM	Deliv 17 May 43	1200	Allison V-1710-81	390	37 0	32 3	7900	1
P-51D	N. American	1 LWM	Loaned 1944	1490	Packard V-1650-7	437	37 0	32 3	11800	1
P-51H-5	N. American	1 LWM	Deliv 31 Aug. 48	1380	Packard V-1650-9	487	37 0	33 4	11000	1
F-84	Republic	1 MWM	Transferred 58	5600†	1 × J35-A-15	550	36 5	37 5	14230	4

U.S. Navy and Marine Corps Aircraft procured on VF funds but not designated or assigned in VF class.

Type No.	Firm	Crew and Type	Significant date	Engine h.p.*	Engine Type	Top Speed	Wing Span	Length	Loaded Weight	Quan.
BR-1	Booth	1 LWM	Deliv 1922	390	Wright H-3	188	28 1	21 1	2020	1
BR-2	Booth	1 LWM	Deliv 1922	390	Wright H-3	177	30 1	21 1	2020	1
CR-1	Curtiss 23 (L-17-1)	1 Bi	Ordered 16 Jun. 21	400	Curtiss D-12	185	22 8	21 0	2095	1
CR-2/3	Curtiss 27 (L-17-3)	1 Bi	Deliv 1921	400	Curtiss D-12	194	22 8	25 0	2593	1
R2C-1	Curtiss 32	1 Bi	Deliv 1923	488	Curtiss D-12A	267	22 0	19 8	2150	2
R2C-2	Curtiss 32A	1 Bi	Deliv 1923	500	Curtiss D-12A	227	23 0	22 7	2640	(1)
R3C-1	Curtiss 42	1 Bi	Deliv 1925	565	Curtiss V-1440	265	22 0	20 0	2176	3
R3C-2	Curtiss 42A	1 Bi	Deliv 1925	565	Curtiss V-1440	238	22 0	22 7	2733	(3)
R3C-3	Curtiss 42A	1 Bi	Modified 1925	685	Packard 2A-1500	242	22 0	22 7	—	(1)
R3C-4	Curtiss 42A	1 Bi	Modified 1926	685	Curtiss V-1500	—	22 0	22 7	—	(1)

* In the case of jet engines figure given is for static thrust in pounds.   †Static thrust with afterburner in operation.

NAVY AND MARINE CORPS FIGHTERS (continued)
Tri = Triplane, MWM, HWM, LWM = Mid, High and Low Wing Monoplane respectively
under another type or model, or quantity cancelled if so qualified in remarks

Serial Numbers	Remarks including unofficial or popular name
96752–7531	'Corsair'. F4U-4C, -4D, -4E, -4N and -4P cancelled V-J Day. Nos. 62915–3071, 80764–2177 also built.
97296, 97415, 97364	'Corsair'. Diverted from F4U-4 series. No.97296 crashed 8 July 1946, replaced by No.97364.
121793–2066, 122153–206	'Corsair'. Includes 315 F4U-5N and 30 F4U-5P. 123144–203, 124441–560, 124665–724.
129320+	'Corsair'. Low altitude, close support concept. Re-designated as AU-1 in attack category.
33958 (33959 not built)	'Flapjack'. Design based on V-173, circular wing planform. 'Metalite' construction. Not flown.
33532–33534	'Pirate'. First installation of afterburner on production. 4 × 20 mm. cannon. 'Metalite' construction.
122483–122512	'Pirate'. Limited to (VC) composite squadrons. Training and evaluation of use of afterburner.
122472–122475	'Cutlass'. All wing, high performance interceptor. First fighter with afterburner from original concept.
124415–124428	'Cutlass'. Pre-production limited to composite squadrons. 4 × 20 mm. cannon and Mighty Mouse rockets.
128451–78, 129545–756, 139868–917	'Cutlass'. First 16 had J35-A-29 engines. VF later to VA status. Instability led to abandonment. Includes 100-3M and 12–3P versions.
138899–900	'Crusader'. First 1000 m.p.h. carrier-borne aircraft. Variable incidence wing, day fighter concept.
140444–48, 141336–62, 142408–15, 143677–821 and see remarks	'Crusader'. Production front-line VF type, No. 140447 to prototype F8U-2, No.141363 prototype of 144 F8U-1P version, No.145318 first of F8U-1E block. Nos. 144427–61, 144607–25, 145318–545, 145604–47, 146822–901, include F8U-1P, F8U-1T and 448 F8U-1 and F8U-1E versions.
145546+ 147035+ & subsequent	'Crusader'. Improved all weather version, includes 187 F8U-2, 152 F8U-2N, 190 F8U-2NE on order.
146340–41, 147085–86	'Crusader III'. All missile concept, provision for 'Rocketdyne' rocket engine installation. Only three flew.
6748 only	'Falke'. Swiss built, all-metal experiment of Dornier design. Designation changed to FW-1.
6743–6744	As landplanes in 1923 St. Louis races. Converted F2W-2 S.P. with T-2 engine for 1923 Pulitzer races.
7223 only	'Apache'. Held altitude record. Was XF3W-1 under re-designation in 1928. Initial use of P. & W. 'Wasp'.
6543–6544	Racers of BuAer design, Chance-Vought built. NW-1 F.P., NW-2 F.P. with Lamblin radiators.
8485, 8607	Procured to study high tensile steel construction. No.8485 as Mk. II, No.8607 as Mk. IIA, No.8485 crashed.
6058 only	One of two brought to U.S. after Armistice to evaluate all-metal structure.
5887–5889	To U.S.M.C. as trainers for proficiency in VF type. Armament not fitted.
5843–5848	Ex-Army A.S. to U.S.M.C. for fighter training. Unarmed. Used 1921–1924. First 2 only assembled.
5588–5589	Built from parts supplied by R.A.F. Evaluated for immediate VF needs.
5620–5629	Built at N.A.F. from components supplied by Hanriot. Fighter status. Used in catapult tests.
5655–5656	Supplied in crated form by Sopwith but never uncrated.
5658–9, 5721–2, 5729–30	Standard 2F.1 models used in early battleship turret catapult experiments.
4218–4227	Ex-Army Air Service. Turned over to U.S.M.C. for fighter and proficiency training.
5794–805	As new equipment, constituted entire Navy VF category until July 1920. Later used as racers.
6060–6070	All to U.S.M.C. Ex-Army aircraft. Used to form first Marine fighter squadron. Armed.
6071 (Ex-A.S.64373)	Ex-Army racer as surplus. To U.S.M.C. in VF category.
6402 (6403–04 cancelled)	'Handley Page Shipboard-One'. Evaluation of H.P. slotted wing. Delivered less armament.
29667–68, 29689	'Shooting Star'. U.S.A.F. aircraft evaluated as jet VF type. No. 29667 underwent carrier trials.
33821–70	'Shooting Star'. To U.S.M.C. as TO-1 (TV-1) trainers.
None assigned‡	'Starfighter'. Used for high-speed Sidewinder tests at N.O.T.S., China Lake, 1960–61. All withdrawn.
63960–1 (Ex-42-108778–9)	'Airacomet'. Used for jet aircraft evaluation, including carrier trials.
64100, 64108–09	'Airacomet'. Further evaluation of jet VF type. Ex-44-22651, 44-22658, 44-22657 respectively.
90060–90061	'Kingcobra'. Sweptwing VF. Dimensions given as prior to modification. To N.A.C.A. under Navy contract.
57987 (Ex-U.S.A.A.F.)	'Mustang'. Evaluation of design as naval fighter. Modified for carrier trials. 4 × ·50 m.gs. Deleted July 1947.
None assigned	Ex-44-14017. Carrier equipment. Tested on U.S.S. *Shangri-La* (CV-38) 1944.
109064 (Ex-44-64192A)	'Mustang'. Used by BuAer Research Unit. Flew 659 hours on U.S.N. service. Deleted 11 Aug. 1953.
142269–142272	'Thunderjet'. Ex-U.S.A.F. F-84B-26RE, used as experimental target drones. Re-designated F-84KX.
6429 only	'Beeline Racer'. Retractable undercarriage, Lamblin radiators, cantilever wings. 1922 Pulitzer races.
6430 only	'Beeline Racer'. Wing-skin radiators, retractable gear. 1922 Pulitzer races. As BR-1 basic design.
6080 only	'Curtiss Racer'. Classified as VF on official account sheets.
6080–6081	'Curtiss Racer'. Both CR-1/2 originally as L.P., converted to F.P., CR-3 for Pulitzer and Schneider Trophy.
6691–6692	F2C-1 'paper' designation. Winner 1923 Pulitzer race as L.P. Evolved from CR series.
6692 only	Landplane converted to twin-float seaplane for 1924 Schneider races. Winner 1923 Pulitzer as landplane.
6978–6979, 7054	Built for Pulitzer and Schneider Cup races. F3C-1 'paper' designation.
6978–6979, 7054	Ex-R3C-1s modified as floatplane. Reconditioned for Schneider Cup Trophy.
7054 only	Ex-R3C-2 modified as floatplane. All R3C series built for Pulitzer or Schneider Trophy races.
7054 only	Ex-R3C-3 modified as floatplane. Surveyed as XR3C-4, November 1928.

‡ Ex-U.S.A.F. 55–956/56–740/60–757. N.B.—F7U-2 proposed strengthened version of F7U-1 not built.

GENERAL INDEX AND GLOSSARY

PART ONE *Alphabetical Index by Manufacturer*

ATLANTIC (FOKKER)		F7C-1	41, 45, 166, 234, 236	XF4F-3	73, 77, 240	**NAVAL AIRCRAFT FACTORY**	
XFA-1	54–7, 236	XF8C-1	44, 236	F4F-3/3A	77–82, 140, 190, 232–4, 240	TF-1	21, 230, 242
BELL		F8C-1	44–6, 234, 238	XF4F-4	79, 240	XFN-1	51, 242
XFL-1	75–6, 189, 234, 236	XF8C-2	44–5, 234, 238	F4F-4	79–80, 82, 84–5, 240	**NORTHROP/DOUGLAS**	
F2L-1K	112, 236	XF8C-3	44, 238	XF4F-5	78, 240	XFT-1	65, 67, 184, 242
BERLINER-JOYCE		XF8C-4	45, 52, 234, 238	XF4F-6	240	XFT-2	65, 67, 242
XFJ-1	49, 51, 171, 236	F8C-4	45–6, 172, 238	F4F-7	82, 240	**NORTHROP**	
XFJ-2	51, 236	XF8C-5	45–6, 234, 238	XF4F-8	82, 85, 240	F2T-1	112, 242
XF2J-1	59, 60, 236	XF8C-6	46, 238	XF5F-1	74–6, 82, 92, 191, 235, 240	**NORTH AMERICAN**	
XF3J-1	64, 66, 175, 236	XF8C-7	46, 238	XF6F-1	83, 240	XFJ-1	103, 242
BOEING		XF8C-8	46, 59, 238	XF6F-2	83, 88, 240	FJ-1	103, 104, 107, 111, 203, 242
FB-1	31–5, 137, 141, 160, 236	XF9C-1	54–7, 238	XF6F-3	81, 83, 240	XFJ-2	115–6, 242
FB-2	33, 236	XF9C-2	55–8, 238	F6F-3	81, 83, 86–90, 141–2, 193, 232–3, 240	FJ-2	116, 124–5, 242
FB-3	32–3, 236	F9C-2	55–8, 178, 238	XF6F-4	83, 240	FJ-3	116, 124–5, 216, 242
FB-4	32–3, 38–9, 236	XF10C-1	59, 238	F6F-5	86–9, 95, 113, 122, 143, 232–3, 240	FJ-4	5, 125–8, 136, 145, 220, 242
FB-5	33, 41–2, 141, 163, 236	XF11C-1/XBFC-1	59–60, 238	XF6F-6	88, 240	**RYAN**	
FB-6	38–9, 236	XF11C-2/XBFC-2	60–1, 238	XF7F-1	92, 240	XFR-1	99–100, 242
XF2B-1	38–9, 41, 236	F11C-2	60–2, 176, 238	F7F-1	92–4, 240	FR-1	100, 146, 198, 235, 242
F2B-1	39, 40, 43, 165, 230, 236	BFC-2	60–2, 238	XF7F-2	92, 240	XFR-2	242
XF3B-1	40, 43, 236	XF11C-3	61, 63, 234, 238	F7F-2	92–3, 240	XFR-3	242
F3B-1	40, 41, 44, 47, 168, 236	BF2C-1	61–2, 66, 180, 230, 238	F7F-3	93, 96, 112, 206–7, 240	XFR-4	100, 242
XF4B-1	47, 49, 236	XF12C-1	62–4, 66, 182, 238	F7F-4	96, 240	XF2R-1	100–101, 200, 242
F4B-1	48–9, 170, 236	XF13C-1	64, 66, 185, 238	XF8F-1	93–4, 240	**SEVERSKY**	
F4B-2	5, 49, 50, 236	XF13C-2	238	F8F-1	94–6, 144, 146, 195, 240	XNF-1	71–3, 242
F4B-3	2, 50–2, 236	XF13C-3	67, 238	F8F-2	96, 144, 240	**VOUGHT/N.A.F.**	
F4B-4	49, 52–3, 67, 139, 177, 230, 236	XF14C-1	84, 86–7, 238	XF9F-1	106	VE-7	23–4, 30, 32, 242
F4B-4A	53, 236	XF14C-2	84, 87, 194, 238	XF9F-2	106–7, 240	VE-7F	23–4, 156, 242
XF5B-1	48–50, 52, 173, 236	XF14C-3	87, 238	F9F-2	106–7, 109–12, 115, 209, 240	VE-7GF	23–4, 242
XF6B-1	61, 63, 236	XF15C-1	101–2, 199, 238	XF9F-3	109, 240	VE-7SF	23–4, 242
XF7B-1	64–6, 181, 236	**DOUGLAS**		F9F-3	109–10, 115, 240	**CHANCE VOUGHT**	
XF8B-1	90–1, 97–8, 196–7, 236	XFD-1	61–2, 64, 179, 231, 238	XF9F-4	109–10, 115, 240	UF-1	30, 242
BREWSTER		XF3D-1	105, 238	XF9F-5	240	UO-1	37, 41, 54, 231
XF2A-1	71–4, 236	F3D-1	104–5, 111, 238	F9F-5	110, 112, 115, 235, 240	FU-1/2	30, 37, 162, 231, 242
F2A-1	72–5, 80–1, 140, 187, 236	F3D-2	105–7, 111–3, 208, 238	XF9F-6	115, 125, 240	XF2U-1	44, 46, 242
XF2A-2	74–5, 236	XF3D-3	112, 238	F9F-6	115, 125, 235, 240	XF3U-1	61–2, 64, 242
F2A-2	72, 74, 77, 236	XF4D-1	108–9, 114, 238	F9F-7	116–7, 124–5, 240	**VOUGHT-SIKORSKY**	
F2A-3	75–77, 79, 81, 233, 236	F4D-1	121–2, 126, 218, 238	F9F-8	124–5, 127, 136, 213, 240	XF4U-1	78–9, 82, 242
XF2A-4	77, 236	XF5D-1	121–2, 238	F9F-9	125, 127, 240	F4U-1	4, 80, 82–6, 89, 90, 96, 142, 192, 232–3, 242
F3A-1	86, 236	XF6D-1	136, 238	XF10F-1	108, 114–5, 123, 214–5, 240	F4U-2	86, 242
CONVAIR		**EBERHART**		F11F-1	127–9, 221, 240	XF4U-3B	86, 96, 235, 242
XFY-1	118–20, 211, 236	XFG-1	42, 167, 230, 238	XF12F-1	129	**CHANCE-VOUGHT**	
XF2Y-1	118–20, 212, 236	**GENERAL MOTORS (Eastern)**		**HALL**		XF4U-4	242
YF2Y-1	119–20, 235, 236	FM-1	82, 85, 238	XFH-1	42–3, 45, 169, 240	F4U-4	86, 88–90, 97, 99, 244
CURTISS		FM-2	85, 238	**LOCKHEED**		XF4U-5	97, 244
HA	13–17, 19–20, 137, 138, 152, 236	XF2M-1	89, 238	FO-1	240	F4U-5	86, 95–7, 109–10, 113, 205, 244
18-T	13, 15–17, 19–21, 24–6, 137–8, 153, 236	F3M-1	94, 238	XFV-1	118–20, 240	XF4U-6/AU-1	110–11, 113–4, 244
TS-1	24–5, 28, 30, 32, 137, 157, 236	**GOODYEAR**		**LOENING**		XF5U-1	91–2, 107, 202, 235, 244
TS-2	24, 236	FG-1	83, 85, 90, 238	XFL-1	64, 242	XF6U-1	101–2, 244
TS-3	24, 236	F2G-1	89–90, 97, 238	**McDONNELL**		F6U-1	102–3, 107, 204, 244
F4C-1	4, 28, 30, 42, 158, 236	F2G-2	90, 238	XFD-1/XFH-1	98–9, 101, 242	XF7U-1	107–8, 244
F6C-1	34–6, 39, 231, 236	**GRUMMAN**		FD-1/FH-1	96, 99, 104, 111, 201, 242	F7U-1	107–8, 113, 117, 244
F6C-2	35, 159, 236	XFF-1	59, 63, 65, 238	XF2D-1	104, 242	F7U-3	113, 117, 219, 235, 244
F6C-3	35–6, 41, 57–8, 230, 236	FF-1	59, 63, 65, 67–8, 70, 174, 238	F2H-1	104–5, 111–12, 242	XF8U-1	129–30, 244
F6C-4	38–41, 62, 141, 164, 236	FF-2	63	F2H-2	111–13, 210, 242	F8U-1	130–1, 134, 224–5, 244
XF6C-5	39, 42, 236	XF2F-1	68, 70, 238	F2H-3	111–12, 242	F8U-2	131–3, 244
XF6C-6	36, 236	F2F-1	52–3, 68–9, 71, 183, 238	F2H-4	111, 145, 217, 242	F8U-3	132–3, 135, 244
XF6C-7	39, 236	XF3F-1	69–70, 234, 238	XF3H-1	122, 242	**WRIGHT AERONAUTICAL**	
XF7C-1	41, 236	F3F-1	69–72, 74, 79, 140, 186, 238	F3H-1	122–3, 242	WP-1	29, 244
		XF3F-2	70, 240	F3H-2	122–4, 222–3, 242	F2W-1/2	27, 29, 231, 244
		F3F-2	70–2, 139, 188, 230, 240	F4H-1	8, 132–5, 226–7, 242	F3W-1	37–8, 161, 231, 244
		XF3F-3	71, 240			NW-1/2	26, 28, 244
		F3F-3	71, 240				
		XF4F-2	72–3, 76, 240				

PART TWO — Miscellaneous Navy and Marine Corps Aircraft

BELL		R2C-1	244	**FOKKER**		**NORTH AMERICAN**	
YP-59A	97–9, 104, 244	R2C-2	244	C-I	22, 24, 244	P-51A/P-51H	87, 235, 244
YP-59B	99, 244	R3C-1	27, 231, 244	D-VII	22, 23, 24, 244	**SOPWITH**	
P-63A	112, 244	R3C-2	231, 244	**HANDLEY PAGE**		Pup	18, 244
BOOTH		R3C-3	244	HPS-1	27, 244	Camel	16–18, 21–2, 55, 137, 244
BR-1	26, 244	R3C-4	244	**HANRIOT/N.A.F.**			
BR-2	26, 244	S.E.5A	19, 21–2, 244	HD-1	18–19, 137, 139, 154, 244	**STANDARD**	
BRISTOL		TR-1	25, 230	**LOCKHEED**		E-1	19, 244
Bulldog	50, 51, 244	TR-2	25	P-80A/P80C	99, 244	**THOMAS-MORSE**	
CURTISS		TR-3/3A	25	F-104A	244	MB-3/3A	22–3, 139–40, 155, 244
CR-1	25, 244	**DORNIER**		**NIEUPORT**			
CR-2/3	25–6, 30, 231, 244	D-I	19, 244	N-28	19, 22, 139, 244	MB-7	23, 231, 244

PART THREE — Personalities

Acosta, Bert	25	Dewey, Adml.	10	Lampert, Florian	31	Rodgers, Lt. John	10
Alrich, Capt. D.	90	Doolittle, Jimmy	40	Lippisch, Dr. Alexander	108	Rogers, Gen.	23
Amen, Lt. Cdr. W. T.	111			Liqued, Lt. C. N.	15	Rohlfs, Roland	13–16, 19–21
Attinello, John	110	Elder, Lt. Cdr. Bob	103–4	Loening, Grover	59	Roosevelt, Franklin D.	21, 74
Attridge, Tom	128	Ellyson, Lt. Theodore	10	Lutz, Maj. C. A.	36	Rounds, Lt. E. W.	42
Aurand, Cdr. E.	103–4	Ely, Eugene	9, 18			Rowell, Maj. Ross R.	36
		Everton, Maj.	99	MacCampbell, Cdr. D.	88		
Barrow, Lt. J. C.	124			MacCast, R. D.	72	Salman, H. R.	119
Baumeister, Charles	135	Faivre, Maj. E. N.	121	McClure, Lt. Cdr. W. H.	85	Sanderson, Lt. L. H.	26–7
Becker, R.	97	Fallon, Ens.	12	MacComsey, Lt.	49	Scmued, Ed.	103
Beisel, Rex B.	78	Fechet, Gen. J. E.	47	McDonnell, Lt. Cdr. E. O.	18	Seligman, Lt. M. T.	47
Bellinger, Rear Adml.	80	Fischer, Cdr. C. F.	81	McIlvain, Lt. William M.	10	Shryock, W. A.	135
Bolt, Lt. Col. J. F.	116	Flint, Cdr. L. E.	134–5	MacKnight, Lt. Cdr. H. C.	146	Sims, Adml.	11
Bordelon, Lt. Guy P.	113	Fokker, Anthony	29	Manby, Lt. Cdr. W. J.	126	Smith, Lt. Bernard L.	9–10, 13
Boyington, Col. G.	90	Foulois, Maj. Gen. B. D.	11	Mead, George	37	Soucek, Lt. A.	38
Bradley, Lt.	20	Fruin, Lt. Jack	104	Meade, Joe	13, 14	Spencer, Lt. Cdr. P.	135
Brady, Lt. F. X.	124			Meyer, Corwin	106, 128	Stengie, Lt. A.	19
Brice, Lt. W. O.	35	Gilmore, W. L.	15	Millar, W. H. B.	107	Storrs, Lt. A. P.	39
Brickner, Lt. J. S.	135	Glenn, Maj. J. H.	130–31	Miller, Lt. Col. T. H.	134		
Brown, C. V.	126	Gordon, Lt. R.	135	Mitchell, C. C.	146	Thaw, Russell	105
Brown, Dayton T.	72	Gorton, Lt. A. W.	25	Mitchell, R. J.	26	Thomas, Maj. F. C.	116
Brown, Ens. E. W.	109	Gray, Wireless Operator	12	Mitchell, Gen. ' Billy '	9	Tomlinson, Lt. D. W.	39, 43
Bullard, Lyman A.	78	Griffin, Lt. V. C.	24	Mitscher, Cdr. M. A.	41	Tomlinson, Lt. W. C.	36
Burroughs, Richard	91	Grumman, LeRoy	59	Moffett, Rear Adml. W. A.	32, 57	Towers, Lt. John H.	10, 11, 13, 17
		Gurney, Lt.	73			Trapnell, Lt.	49
Callaway, Lt. S. W.	27	Guyton, Boone	91	Northrop, John K.	107	Truman, Harry S.	109
Carl, Capt. Marion	80, 82						
Champion, Lt. C. C.	38	Hall, Charles W.	22, 28, 30, 42	Odom, Bill	97	Udet, Ernst	36
Chevalier, Lt. Cdr. G. de	22	Hall, Robert L.	73	Ofstie, Lt. R. A.	28, 32, 38, 49		
Cleland, Cook	96–7	Hamilton, W. Off. H. B.	82	O'Hare, Lt. E. H.	82, 88	Valencia, Lt. E. A.	88
Coburn, Lt. Cdr. F. G.	12	Hamilton, Sgt. W. J.	80	O'Keefe, Lt. J. J.	90	Vian, Vice Adml. Sir Philip	94
Coleman, J. F.	118–9	Hanson, Lt. R. M.	90	Olsen, M/Sgt. Barney	112	Vought, Chance	23
Collins, Jimmy	69	Hardison, Lt. Cdr. O. B.	36	Owen, Cdr. E.	101		
Compo, Lt.	39	Hardisty, Lt. Hunt.	134			Wagner, Lt. Cdr. F. D.	36
Conover, Al.	103	Harris, Lt. C. E.	88	Page, Capt. Arthur H.	36	Walsh, Capt. K.	90
Converse, Seldon	73	Heinemann, E. H.	105, 108	Pershing, Gen. J.	11	Watkins, Lt. Cdr. G. C.	128
Coontz, Adml. R. E.	31	Hughes, Flt. Lt. S.	136	Pihl, Lt.	49	Wehrle, Maj. Howard	23
Croft, Prof. H. O.	98			Plog, Lt. L. H.	109	Whitcomb, R. T.	127
Cruise, Lt. E. A.	47	Ingalls, Lt. D. S.	17	Potter, Ens. Stephen	12	Whiting, Lt. K.	11
Cuddihy, Lt.	28	Irvine, Lt. Rutledge	26	Pride, Lt. A. M.	41, 49	Whittle, Sir Frank	98
Cunningham, Lt. Alfred A.	9, 11, 13, 16–17					Widhelm, Cdr. W. J.	86
Curtiss, Glenn H.	10, 25	Jackson, Cdr. H. J.	146	Rahn, R. O.	108, 114, 121, 126	Willgoos, Andrew Van Dean	37
		Janazzo, Tony	97	Read, Cdr. A. C.	13	Williams, Lt. Alford	27, 40, 96
Davenport, Lt. Cdr. M. W.	96	Jeter, Lt. T. P.	47, 49	Reeves, Rear Adml. J. M.	38	Wilson, Charles E.	123
Davidson, Lt. Cdr. J.	98–9			Rentschler, F. B.	37	Windsor, Cdr. R. W.	130
Davis, Cdr. J. F.	134	Kelly, H. L. ' Herb '	52–3	Rich, Lt. W.	124		
Davis, S. V.	146	Kennedy, C. S.	108	Richbourg, C. E.	120	Young, Lt. B. R.	135
Davis, Lt. W. V.	39	Kirkham, Charles B.	15	Rittenhouse, Lt. David	26, 30		
DeEsch, Earl H.	134	Kliewer, Lt. D. D.	80	Robinson, Lt. Col.	135	Zimmerman, C. H.	91
Deubo, Lt. Cdr. R. W.	85	Konrad, John	129				
		Kurt, F. T.	81				

PART FOUR

Abbreviations and Glossary

Since this book, on an American subject, is published in Britain, the English form of notation is used throughout, except for proper nouns where the subject is purely American. Some terms are explained in the text and to avoid repetition, a reference is given to the appropriate page.

A (aircraft designation prefix): Attack role
A- (aircraft designation): Brewster Aeronautical Corporation
-A- (engine designation): Allison Division, General Motors Corporation
Aerofoil (Airfoil): The shape of the wing section
A.F.B.: Air Force Base of the U.S.A.F.
Afterburner (Re-heat): A device to increase jet thrust
Angle of Attack: The angle between aerofoil (wing) and airstream
APS: Type of airborne search radar
Area Rule: See page 128, left-hand column
A.V.G.: American Volunteer Group

B (model suffix): Fitted for bombing role
B- (aircraft designation): Boeing Airplane Corporation
Blower: Compressor or supercharger
BuAer: Bureau of Aeronautics

C (model suffix): Cannon-armed
C- (aircraft designation): Curtiss Corporation/Curtiss-Wright
Calibre: The bore diameter of a gun conditioning size of ammunition
CV: Fleet Aircraft Carrier
CVA: Attack Aircraft Carrier
CVA(N): Aircraft Carrier, nuclear powered
CVB: Large Fleet Aircraft Carrier
CVE: Escort Aircraft Carrier
CVL: Light Aircraft Carrier
CVS: Super Class, Aircraft Carrier
CVU: Utility Aircraft Carrier

D- (aircraft designation): McDonnell later Douglas
Delta Wing: Triangular wing plan-form after the Greek letter delta
Dorsal: The upper-surfaces of the fuselage

E (model suffix): Special electronic equipment

F (aircraft designation prefix): Fighter
F- (aircraft designation): Grumman Aircraft Engineering Corporation
FF: First flight

G (aircraft designation): Goodyear Aircraft Corporation

H- (aircraft designation): McDonnell Aircraft Corporation
H.M.S.: His/Her Majesty's Ship of the Royal Navy
h.p.: horse-power

J (aircraft designation prefix): Utility role
J- (aircraft designation): Berliner-Joyce (pre-war)
J- (aircraft designation): North American Aviation (post-war)
J- (engine designation prefix): Jet engine
j.g.: Junior grade

K (model designation prefix): Drone aircraft conversion

L- (aircraft designation): Loening

M- (aircraft designation): General Motors Corporation
Mach: A measure of airspeed relative to the speed of sound—Mach 1
mm.: millimetre
m.p.h.: miles per hour

N (model suffix): Night or all-weather role
N.A.C.A.: National Advisory Council for Aeronautics
N.A.F.: Naval Aircraft Factory
N.A.S.: Naval Air Station
N.A.T.C.: Naval Air Testing Centre
N.A.T.O. North Atlantic Treaty Organisation
NE (model suffix): all-weather role with special electronics
N.O.T.S.: Naval Ordnance Test Station

O (aircraft designation): Lockheed Aircraft Corporation

P (model suffix): Photographic reconnaissance role
P. & W.: Pratt & Whitney

R (aircraft designation): Ryan Aeronautical Corporation
R (engine designation prefix): Radial
R.A.A.F.: Royal Australian Air Force
R.A.F.: Royal Air Force
R.C.A.F.: Royal Canadian Air Force
Re-heat (afterburning): A device to increase jet thrust
R.F.C.: Royal Flying Corps (1912–1918)
R.N.A.S.: Royal Naval Air Service (1914–1918)
R.p.g.: Rounds per gun

Service ceiling: Height at which rate of climb drops to 100 feet per minute
S.h.p.: Shaft horse-power
Spoiler: Device on wing to spoil airflow and reduce lift
s.t.: static thrust

T (model suffix): Fitted for training role
T (engine designation prefix): Turboprop
T- (aircraft designation): Northrop Aircraft
Track: The distance between the undercarriage wheels

U- (aircraft designation): United Aircraft. (Vought Division in most cases covered by this book)
U.N.: United Nations (Organisation)
U.S.M.C.: United States Marine Corps
U.S.N.: United States Navy
U.S.S.: United States Ship

V: In general U.S. Navy use for heavier than air craft.
V (engine designation prefix): 'Vee' cylinder arrangement.
V- (aircraft designation): Lockheed Aircraft
VA: U.S. Navy attack squadron prefix
VF: U.S. Navy fighter squadron prefix
VF(N): As above for night fighter squadrons
VMF: U.S. Marine Corps fighter squadron prefix
VMF(N): As above for night fighter squadrons
VP: U.S. Navy patrol squadron prefix

W (engine model suffix): Water injection
-W- (engine designation): Westinghouse

X (aircraft/engine designation prefix): Experimental model

Y (aircraft designation prefix): Service trials model
Y- (aircraft designation): Convair

Z: In general U.S. Navy use for lighter than air craft
ZR: Lighter than air rigid aircraft

M4671

Ref. No: ....
Title: .... **Retail Price:** ....
Title: .... RP
Title: .... RP
Title: .... RP
Title: .... RP
Title: .... RP
Title: .... RP
Title: .... RP
Title: .... RP

Dear Customer,
We have pleasure in enclosing your book(s) which we hope you will enjoy. The book(s) listed above cannot be supplied at the moment for these reasons:

NYP – Not Yet Published   RP – Reprinting
NE – New Edition   OO – On Order
OP – Out of Print   NK – Not Known
ND – Not or no longer available from us.

They will be sent automatically when available, unless you instruct us otherwise.

Always quote the above Ref No. and **Title(s)** of **Book(s)** ordered when contacting us.